THE PHYSICS OF MULTIPLY AND HIGHLY CHARGED IONS

The Physics of Multiply and Highly Charged Ions

Volume 1

Sources, Applications and Fundamental Processes

Edited by

FRED J. CURRELL

Department of Physics,
The Queen's University,
Belfast, Northern Ireland

KLUWER ACADEMIC PUBLISHERS
DORDRECHT / BOSTON / LONDON

A C.I.P. Catalogue record for this book is available from the Library of Congress.

ISBN 1-4020-1565-8 (Vol. 1)
ISBN 1-4020-1582-8 (Vol. 2)
ISBN 1-4020-1566-6 (Set)

Published by Kluwer Academic Publishers,
P.O. Box 17, 3300 AA Dordrecht, The Netherlands.

Sold and distributed in North, Central and South America
by Kluwer Academic Publishers,
101 Philip Drive, Norwell, MA 02061, U.S.A.

In all other countries, sold and distributed
by Kluwer Academic Publishers,
P.O. Box 322, 3300 AH Dordrecht, The Netherlands.

Printed on acid-free paper

Printed in the Netherlands.

Contents

List of Figures

List of Tables

Preface

It is arguable that most of chemistry and a large portion of atomic physics is concerned with the behaviour of the 92 naturally occurring elements in each of 3 charge states (+1, 0, -1); 276 distinct species. The world of multiply and highly charged ions provides a further 4186 species for us to study. Over 15 times as many!

It is the nature of human beings to explore the unknown. This nature is particularly strong in physicists although this may not be readily apparent because theses explorations are undertaken in somewhat abstract 'spaces'. It is, then, no surprise that we have begun to explore the realm of multiply and highly charged ions. Over the past few decades, a consistently high quality body of work has emerged as the fruits of this exploration. This internationally based subject, pursued in universities and research laboratories worldwide, has expanded beyond its roots in atomic physics. We now see it embracing elements of surface science, nuclear physics and plasma physics as well as drawing on a wide range of technologies. This speciality offers new tests of some of our most fundamental ideas in physics and simultaneously new medical cures, new ways of fabricating electronic gadgets, a major hope for clean sustainable energy and explanations for astrophysical phenomena. It is both a deeply fundamental and a widely applicable area of investigation.

The aim of this textbook and of its companion volume is to present both an introduction and an up to date review of this emerging body of work. To this end, some thirty scientists working at the forefront of this subject have taken time to compose a set of self-contained but interlinked chapters.

The resultant volumes are intended for a wide range of interested readers with a consistent effort having been made to be instructive. To this end the volumes are suitable for anyone who has completed an undergraduate degree in physics or is in the final stage thereof, with supporting reading of the cited texts. To become a fully equipped researcher one has to make two transitions. Firstly, one must be able to understand the forefront of knowledge in a particular subject area. Secondly, one must become able to push that forefront back, revealing the unknown. The material in these volumes seeks to take the reader to that cutting edge and provide a glimpse into what might lay ahead. Much of the material is at this edge to help stimulate the first and hopefully the second transition.

So, how should a new postgraduate working in this subject area approach the material presented herein? Two approaches recommend themselves, depending on the reader's personal preference. A 'top-down' approach would be to skim the chapters in the sources and applications sections of *volume 1* before proceeding on to material of more specific interest. A 'bottom-up' approach would be to start with the chapter most closely associated with the reader's particular field of study and to work 'outwards' from there. A point to be noted is that the textbook (like the subject matter itself) is not linear. Comprehensive lists of contents, figures and tables have been provided in addition to an index to help the reader to explore the volumes as best suits. The introduction also draws out some of the common threads running through this volume.

The more experienced practitioner, already established in the field, should also find this material valuable. For this reader no advice is necessary as to how to approach the material, as a conceptual map will already be in place. This material will augment and reshape that conceptual map.

It only remains to thank the authors of the chapters for taking the time away from the cliff-face of discovery to write the following chapters. Special thanks also to Wendy Rutherford for helping typeset the volume and for doing numerous other 'little jobs' that all add up to a great deal of help and finally to Liz, the brightest star in my cosmos and a truly special person.

This book is dedicated to the reader, especially if she will use the material presented in the following pages as stimulus to take our subject further.

Introduction

Like all subjects, the physics of multiply and highly charged ions abounds with nomenclature. Even the title of the text begs the question 'What is the difference between multiply charged ions and highly charged ions?' There is no definitive answer; it is mostly a matter of emphasis.

The definition of a multiply charged ion is clear: An ion possessing a charge of +2 or more. In the context of this text it is interpreted as an atomic ion, although molecular multiply charged ions do exist. One major feature of these atomic multiply charged ions is that they can interact with other forms of matter to remove more than one electron in a single interaction. This feature of their behaviour is predominantly dealt with in the accompanying volume, "Interactions with matter", although it is also a recurring theme in the volume that you now hold. This electron scavenging nature or enhanced 'chemical reactivity' becomes more extreme as the ion becomes more highly charged.

The term 'highly charged ions' permits no such rigorous definition. Generally, it is used to emphasize atomic systems where the physics is markedly different from that of neutral atomic systems, beyond the ability to receive more than one electron during an interaction. The remainder of this introduction primarily focuses on this new physics in various manifestations.

The language of the subject reflects this comparison with neutral atomic systems through use of terms of the form *element1- like element2* where *element1* refers to the number of electrons a species possesses and *element2* refers to the nuclear charge. Hence, the term '*hydrogen-like uranium'* refers to the atomic system where one electron orbits around a uranium nucleus: the system U^{91+}. When all the electrons are removed, the system is referred to as '*bare*', denoting the fact that the nucleus is now on its own. Hence U^{92+} is referred to as *'bare uranium'*.

This nomenclature is closely related to a two dimensional parameterisation of the underlying physics as illustrated in the figure below, the 'world of atomic physics'. The parameterisation is in terms of the nuclear charge (atomic

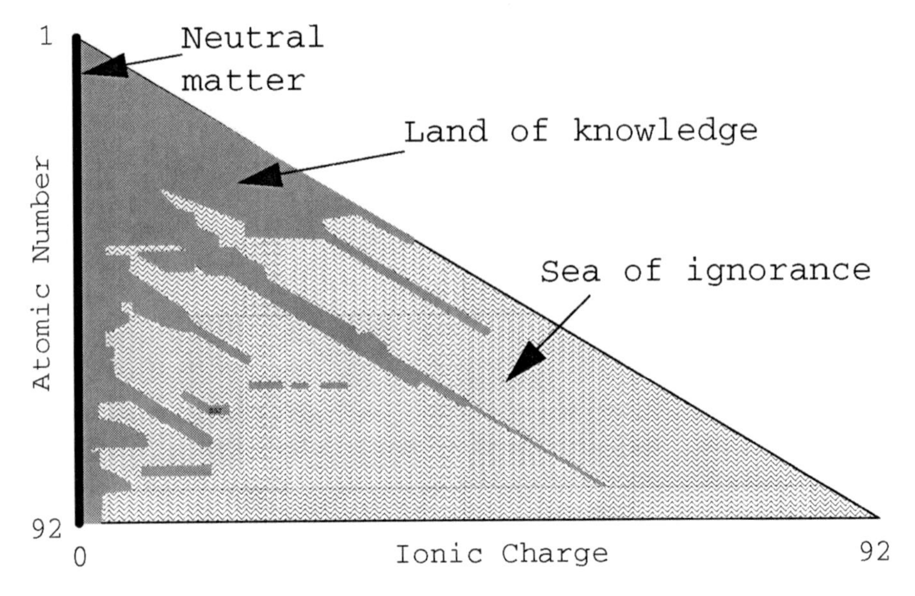

Figure I.1. Parametersation of the world of atomic physics

number) and ionic charge (the difference between the number of electrons and protons). The claim at the beginning of the preface is that most of chemistry and a large portion of atomic physics is concerned with 276 distinct species. These 276 species are contained in the thin sliver to the left hand side of this figure labelled 'neutral matter' and two equally thin slivers either side corresponding to singly charged positive and negative ions. The world of multiply and highly charged ions is the remainder of this figure. When considering the overlap between atomic and nuclear physics (for example, hyperfine interactions) one has to introduce an additional dimension to account for isotope effects related to variation in the nuclear structure. We will not consider this dimension here.

The figure above is based on a somewhat artistic reproduction of a figure which has appeared elsewhere in several guises and a very loose interpretation of what constitutes sufficient knowledge (i.e. the array of transitions electrons bound in the system can undergo has been identified). A less artistic version of this figure can be found in "On The Structure and Spectra of Multiply Charged Ions" by I. Martinson and I. Kink, in *Trapping Highly Charged Ions: Fundamentals and Applications*, ed. J. Gillaspy, Nova Scientifc (2001). Notwithstanding the looseness of the interpretation some trends are readily apparent. Most of our knowledge lies to the left hand side. Not surprisingly, these multiply charged ions are relatively easy to produce. Consequently they are the most amenable to experimental study. For the more highly charged ions, we see diagonal peninsulas of knowledge penetrating into the sea of ignorance. These correspond to isoelectronic systems, i.e. systems with the same number of electrons. In other words, *element1* is the same along one of these diagonals in our *element1-like element2* nomenclature. Again, this can largely be traced to the fact that these species are easy to create, being closed shell systems (helium-like, neon-like, argon-like), other relatively stable systems (nickel-like, hydrogen-like) or systems close to these (give or take an electron). A large part of the scientific endeavor described in this text involves creating new areas in the land of knowledge. Of course, the definitions of ignorance and knowledge are rather arbitrary. Other definitions, for example, in terms of knowledge of various processes the ions can undergo would lead to different maps but the trends described above would be the same.

So what then are the kind of processes these ions can be involved in? To begin to discuss this we need to introduce more nomenclature. For example, ***excitation*** refers to absorption of energy by an ion with a corresponding rearrangement of the electron configuration, ***ionisation*** is the removal of one or more electrons, and ***recombination*** is the addition of an electron. These terms can be augmented to express how the process is initiated. ***Electron impact excitation*** refers to excitation caused by interaction with a continuum electron whilst ***electron impact ionisation*** refers to the removal of an electron due to the interaction with a continuum electron. Chapter 9, "Electron Ion Scattering using an R-matrix Approach" and chapter 10, "Electron Impact Ionisation of Hydrogen-like Ions" contain discussions of these processes.

Analogously, ***photoexcitation*** and ***photoionisation*** refer to excitation and removal of electrons caused by interaction with photons (see chapter 6, "Photoionisation of Atomic Ions"). ***Radiative recombination*** refers to addition of an electron to an ion with the system subsequently being stabilised by emission of one or more photons. Radiative recombination is discussed along with other forms of recombination in chapter 7, "Recombination of Cooled Highly Charged Ions with Low-energy Electrons" and chapter 8, "Dielectronic Recombination in External Electromagnetic Fields". Notice the nomenclature can mask some of the underlying physics; there is nothing to suggest that photoionisation and radiative recombination are each other's time conjugate process.

A range of other terms exists to describe other interactions involving these ions. Necessarily, description of how ions sources work and of applications of multiply and highly charged ions must refer to a number of these processes so the nomenclature is introduced in a general fashion in these sections. In particular, chapter 4 of this volume, "Highly Charged Ion Collision Processes in High Temperature Fusion Plasmas" contains a catalog of these processes.

Highly charged ions (HCIs) might reasonably be dubbed "tightly bound, relativistic, quantum electrodynamical, hyper-reactive electrostatic black holes" in an attempt to summarize a myriad of interesting properties into a single catchphrase. At the heart of all of these properties is the large electrostatic potential and high internal energy of the unscreened or poorly screened nucleus. This can be seen by considering hydrogen-like uranium. The binding energy of the single electron orbiting a uranium nucleus is about 137,000eV, compared with 13.6eV for a the same electron in a hydrogen atom. The internal energy for bare uranium (i.e. the sum of successive ionisation energies for removal of all 92 electrons) comes to a value of about 0.75MeV; the removal of the final pair of 1s electrons accounts for over 1/3rd of this figure. Clearly, highly charged ion physics is high-energy atomic physics.

The single electron orbiting a uranium nucleus, has a classical velocity of about 2/3rds of the speed of light, indicating mandatory relativistic treatment. The field encountered by this electron orbiting around the bare uranium nucleus is over 10^{16} V/m, a field several orders of magnitude greater that that caused by any STM tip or for that matter found elsewhere in the laboratory, except on very short timescales. At the surface of the nucleus, this rises to over 10^{19}V/m. Highly charged ion physics is high-field physics.

Such a system exhibits both relativistic and quantum electrodynamics (QED) effects. This single bound electron in Ur^{91+} can readily emit and recapture virtual photons, giving rise to a considerable *self-energy*. Creation and annihilation of virtual electron-positron pairs is also possible in such a strong nuclear field, giving rise to a *vacuum polarization* term. Essentially, the electron is wearing a cloak of virtual particles whose screening affects the energies of transitions and the magnitude of cross-sections. Although the characteristic radius is always bigger than the nuclear radius by several orders of magnitude, the inner electron orbitals have sufficient overlap with the nucleus to exhibit Atomic-Nuclear Physics effects. Experimentally, these effects tend to be tangled up with QED effects. The coupling of ingenuity and exacting experimental techniques allows us to reveal these fascinating properties as is outlined in chapter 11 "Test of strong-field QED". chapter 7, "Recombination of Cooled Highly Charged Ions with Low-energy Electrons" also discusses test of QED through linking the energies at which recombination is enhanced to the QED-affected energy shifts.

The hyper-reactive nature of HCIs is evidenced by the large internal energy and electrostatic field mentioned above. They are the most chemically reactive species known, having a phenomenal electron scavenging capability. Accordingly, special devices are needed to create and study them; you don't buy them in bottles! Special sources are required. Common features of these sources are creation by ionising collision and through a confinement scheme. Two types of sources commonly used are Electron Cyclotron Resonance Ion Sources (described in chapter 1) and Electron Beam Ion Traps (described in chapter 2).

The traps used in the sources described above share a linear geometry with most of the action occurring close to the central axis and with both axial and radial confinement schemes being used. A different topology (roughly torroidal) is manifest in ion storage rings. These rings are injected with ions from a primary source. In some cases, passage through a thin metal foil facilitates further stripping. Again, this can be viewed as electron impact being used as the HCI creation mechanism. However, the electrons are now bound in the foil with the centre of mass transformation making higher laboratory frame interaction energies mandatory since the ion has to approach the foil with a very high velocity for an almost stationary electron bound to the foil to appear as a high-energy electron in the ion's frame.

The inverse transformation restores the balance of fortune facilitating extremely high-resolution electron-ion collision. The ions cycling round such rings have a velocity of typically tens or even hundreds of MeV/amu. Once injected, the ions can be cooled using an electron beam merged with the ion beam along one straight section of the ring. The colder electron beam is velocity matched to the co-linear ion beam so that the electron-ion coupling leads to ion cooling. The electron beam can also be used to probe electron ion interactions to great effect. Here, the inverse transformation mentioned above comes into play since any difference in relative velocity in the laboratory frame is much smaller in the laboratory frame. Fundamental processes can be studied with such devices as is outlined in chapter 7, "Recombination of Cooled Highly Charged Ions with Low-energy Electrons", chapter 8, "Dielectronic Recombination in External Electromagnetic Fields", and chapter 11 "Test of strong-field QED".

High velocity ions extracted from such rings can travel considerable distances through human tissue to deposit their high reactivity locally. The resultant localised damage can be used as a particularly effective form of therapy for inoperable tumours. This emerging medical technique and the underlying physics are discussed in chapter 5, "Radiobiological Effects of Highly Charged Ions". Nature also produces streams of fast ions in the form of solar and stellar winds. The potentially adverse effects these ions have for space exploration is also discussed in this chapter.

Indeed, the ions comprising the solar and stellar winds can be diagnosed through an understanding of the collision processes involved. Knowledge of the interactions of these ions with various neutral species can be coupled to astrophysical observations to deduce information about the composition of the solar wind. This is achieved through x-ray spectroscopy of the characteristic emissions of solar wind ions after interaction with cometary gasses as is described in chapter 3, "Collision Phenomena Involving Highly-charged Ions in Astronomical Objects". Being important to astrophysics is the understanding of various interactions involving multiply charged ions, considerable theoretical effort has gone into calculating cross-sections for relevant systems, particularly through the R-matrix formalism. Some of this work is described in chapter 6, "Photoionisation of Atomic Ions" and in chapter 9 "Electron Ion Scattering using an R-matrix Approach".

It is not surprising that the plasmas encountered in fusion reactors show many similarities to those of the Sun and other astrophysical objects. After all, the Sun is a giant fusion reactor, although the confinement mechanism differs. Many of the processes involving highly charged ions that are of astrophysical interest also occur in Tokamaks and other fusion reactors. An introduction to these devices and the relevance of highly charged ions to fusion physics is given in chapter 4, "Collision Processes in High Temperature Fusion Plasmas".

We can see from this brief overview of the subject that it is a richly interlinked tapestry, with many cross threads combining various parts of the subject. These various cross threads become apparent throughout the following chapters.

I

ION SOURCES

Chapter 1

ECR ION SOURCES

R. Trassl

Institut für Kernphysik, Universität Giessen, 35392 Giessen, Germany

Roland.H.Trassl@strz.uni-giessen.de

Abstract At the beginning of every experiment involving multiply charged ions there is a suitable ion source that is able to deliver a beam of the desired element in a specific charge state with sufficient intensity. For experiments that require intense beams of multiply charged ions in the c/w mode, Electron Cyclotron Resonance (ECR) ion sources are very commonly used. These ion sources are able to produce ions from virtually any element, they are easy to operate and have an excellent long–term stability as well as a very good reproducibility.

In this chapter, the basic mechanisms of the plasma confinement by magnetic fields and the electron heating with microwave radiation using ECR will be discussed. Furthermore, methods of improving the performance of an ECR ion source (e.g. gas mixing, afterglow effect, etc.) will be described.

There are currently two main directions in the development of ECR ion sources. On one hand, higher charges states with higher intensities are required for some experiments. In order to produce these ions, the applied microwave frequency and power as well as the necessary magnetic fields have to be increased. On the other hand, compact all–permanent magnet ECR ion sources are constructed, which have a low electrical power consumption and are therefore well suited for the use on high–voltage platforms or Van–de–Graaff accelerators. The realisation of both types of ion sources will be described.

Keywords: Ion Source, Electron Cyclotron Resonance, Gas–Mixing Effect, Afterglow Effect

1. Introduction

In order to perform experiments with highly charged ions, powerful ion sources are required that are able to supply ions in high charge states with sufficient intensities. Until 20 years ago almost solely Penning ions sources were used for the ion production. These ion sources can deliver high intensities of singly charged ions but only few ions in higher charge states. For the production of very high charge states **E**lectron **B**eam **I**on **S**ources (**EBIS**) are

F.J. Currell (ed.), The Physics of Multiply and Highly Charged Ions, Vol. 1, 3-37.

available nowadays. Some EBIS are able to deliver bare U^{92+} ions but only with a low intensity and in pulsed operation. **E**lectron **C**yclotron **R**esonance **I**on **S**ources (**ECRIS**) are a compromise between these two types: they can produce moderately charged ions with high intensities.

Since the first ECRIS was reported in 1975 [1] there has been a rapid development concerning the performance of the ion sources as well as their efficiency. Whereas the first ECRIS had an electrical power consumption of 3 MW, nowadays compact ECR ion sources have a better performance with a power consumption of only several 100 W. ECR ion sources are widely used as injectors into linear accelerators or cyclotrons, for atomic physics experiments or for the investigation and the treatment of surfaces.

The advantages of this type of ion source are:

- It can deliver continuous beams of highly charged ions
- There are practically no wearing parts that would have to be replaced from time to time and therefore it is easy to handle and can operate for very long times
- It has an excellent long–term stability and reproducibility
- It operates at pressures of about 10^{-4} $mbar$ and therefore the material consumption is very low. This is important for the production of expensive or radioactive elements

In the following article the principles of an ECRIS will be discussed. An ECR plasma is a very complex system which is not completely understood yet. Therefore, only a few basic aspects of electron heating and plasma confinement will be discussed here. A more detailed description of ECR ion sources can, for example, be found in [2].

2. Principles of ECR Operation

Fig. 1.1 shows the operation principle of an ECRIS. A plasma is confined by a magnetic field in a so–called *minimum–B–configuration*. This means that the absolute value of the magnetic field has a minimum in the centre of the ion source and from there increases in all directions. Such a configuration can be achieved by superposition of an axial mirror field and a radial multipole field. Electrons in the magnetic field B gyrate around the magnetic field lines with the cyclotron frequency. If microwave radiation of the same frequency is fed into the plasma, the electrons are accelerated resonantly. These electrons can now ionize the atoms, molecules and ions in the plasma by electron–impact ionization and a beam of these ions can be extracted through an extraction aperture.

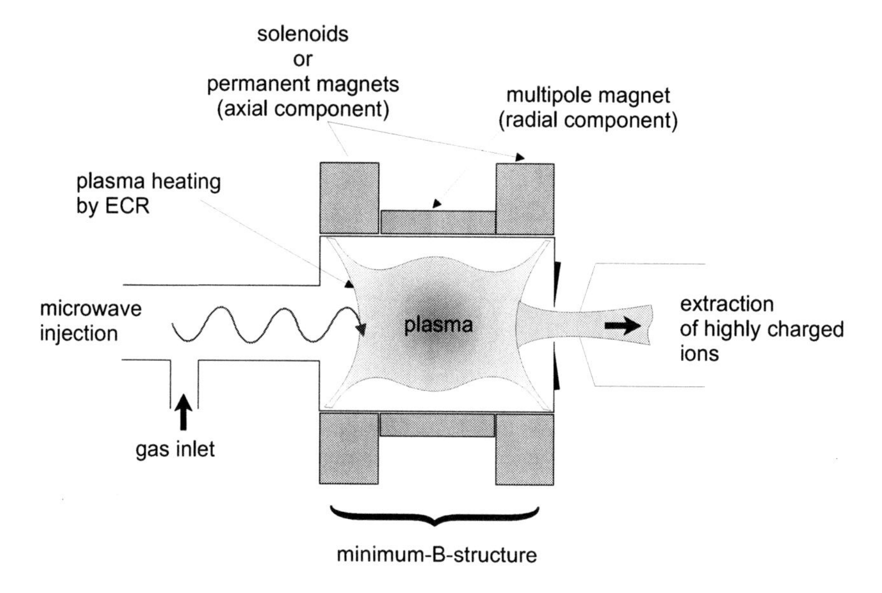

Figure 1.1. Principle of an ECR ion source.

2.1 Ionization by Electron Impact

The atoms, molecules and ions in an ECR plasma are ionized through electron–impact ionization:

$$A^{q+} + e^- \longrightarrow A^{(q+1)+} + 2e^- \tag{1.1}$$

The general condition for this process is given by the fact that the energy of the incoming electron E_e has to be greater than the ionization potential Φ_1 of the outermost electron in the atom or ion:

$$E_e > \Phi_1 \tag{1.2}$$

The energy of free electrons in a plasma is not mono–energetic but is given by a distribution function with an average energy of $3kT_e/2$. Therefore, the term *electron temperature* is widely used. Fig. 1.2 shows the cross sections for the electron–impact single–ionization of the ground states of oxygen ions in the charge states q=0 [3](left side) and q=7 [4](right side).

From the respective thresholds (eq. 1.2) the cross sections increase until they reach a maximum at an electron energy of about 3 times the ionization potential. At higher energies they decrease again exponentially. From these data it is obvious that high electron energies are needed to produce high charge states. Furthermore, the cross section for the electron–impact ionization of O^{7+}–ions

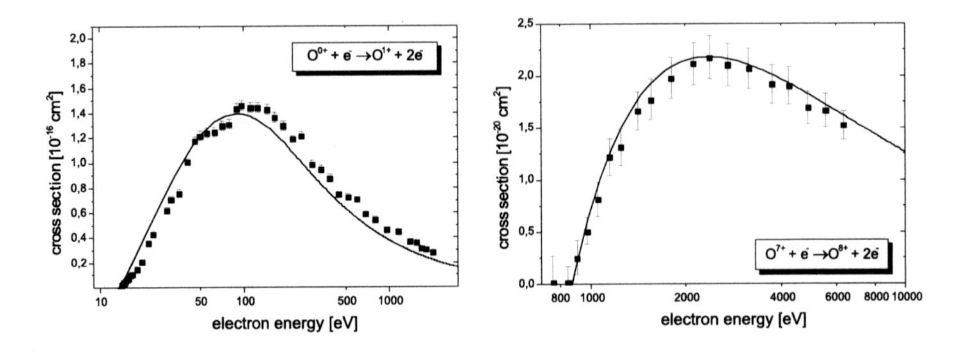

Figure 1.2. Cross sections for the electron–impact ionization of oxygen ions (left side: O^{0+} [3], right side: O^{7+} [4]), Curves: semi–empirical Lotz formula [5].

is more than 3 orders of magnitude lower than that for O^{0+}. So, not only high electron energies are necessary for the production of highly charged ions but also high electron densities are required. Also shown in the figure is a semi–empirical calculation using the Lotz formula [5] which is able to describe direct single–ionization. Due to the small data base available, especially for heavy elements, this formula is widely used among plasma physicists to estimate the cross sections. Multiple ionization by electron impact plays only a minor role since for these processes the cross sections are usually at least one order of magnitude lower than those for single ionization.

The highly charged ions in an ECR ion source are produced by successive single–ionization processes. Under this assumption an algebraic condition for the production of ions in high charge states can be derived. Defining the electron current density j_e by

$$j_e = e n_e v_e \tag{1.3}$$

where e is the electron charge, n_e the electron density and v_e the electron velocity, the most important quantitative parameter for the description of the ionization can be calculated. This parameter is called the *ionization factor I* and is defined as follows:

$$I = j_e \tau_i \tag{1.4}$$

Here τ_i denotes the confinement time of the ions in the collision region (in an ECRIS this is the plasma itself). When the ion confinement time τ_i is longer than the mean time for an ionization process by electron impact, then, on an average, all ions of the plasma in charge state q are ionized into charge state q+1, if the condition

$$j_e \tau_i \geq \left[\sigma_{q,q+1}\right]^{-1} \tag{1.5}$$

is fulfilled [6]. This condition determines the necessary plasma parameters n_e, v_e and T_e, τ_i, respectively and is very similar to the well–known Lawson–criterion in fusion research. Starting from an atom, an ionization factor of at least

$$I = i_e \tau_i \geq \sum_{q=0}^{n} \left[\sigma_{q,q+1}\right]^{-1} \tag{1.6}$$

is necessary to produce an ion in charge state n. Using the above–mentioned Lotz formula, Donets calculated the ionization factors required to ionize different elements into different charge states at given electron energies [7]. The results of these calculations for neon and argon are shown in fig. 1.3.

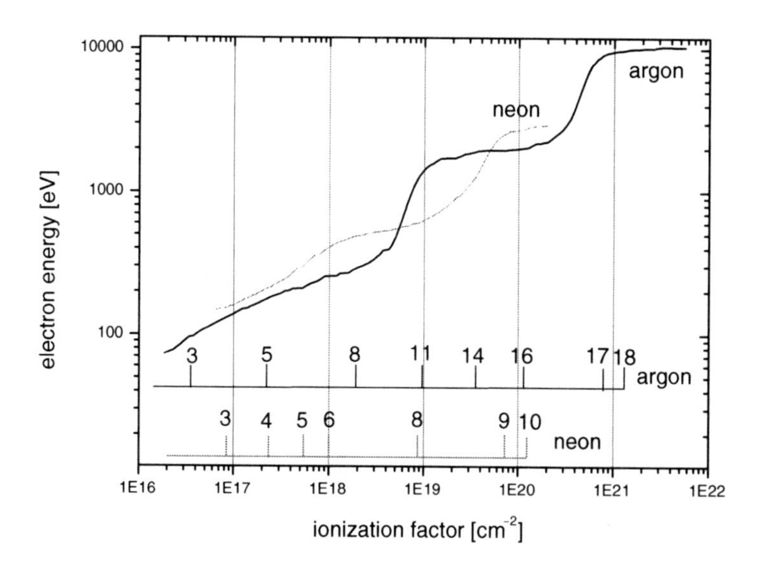

Figure 1.3. Calculation of ionization factors for the production of ions in the shown charge states at given electron energies [7].

From these considerations it is clear that, in order to get a high production rate of highly charged ions in the plasma, we have to achieve the following conditions:

- high electron energy
- high electron density
- long confinement time of the ions

Apart from electron–impact ionization there are also other processes in a plasma which have influence on the charge–state distribution in the plasma. Very important is the recombination of ions with free electrons:

$$A^{q+} + e^- \longrightarrow A^{(q-1)+} \tag{1.7}$$

An increase of the electron density leads to higher losses by recombination but the cross sections for this process are large at low electron energies and decrease rapidly with increasing energy. Therefore, recombination can only be suppressed when a high electron energy is achieved simultaneously. Another collision process that occurs in the plasma is charge exchange between ions and atoms:

$$A^{q+} + B^0 \longrightarrow (AB)^{q+} \longrightarrow A^{(q-1)+} + B^+ \tag{1.8}$$

The cross sections for these processes are usually much greater than those for the electron–impact ionization. Therefore, a high ionization degree of the plasma is required (charge–exchange cross sections in ion–ion collisions are much lower), which is usually true in an ECR plasma because the electron temperatures are high.

2.2 Electron Cyclotron Resonance – Production of High Electron Energies

The plasma electrons gain their energy from the microwaves that are radiated into the plasma. To understand the heating mechanism, some aspects of the interaction of electro–magnetic waves with a plasma are discussed in the following.

In a magnetic field the electrons gyrate around the magnetic field lines with the cyclotron frequency

$$\omega_{cyc} = \frac{e}{m} B \tag{1.9}$$

and the *Larmor radius*

$$r_L = \frac{v_\perp}{\omega_{cyc}} = \frac{mv_\perp}{eB} \tag{1.10}$$

Here, e and m denote the charge and mass of the electron, B the magnetic field and $v_\perp$ the velocity component perpendicular to B. Therefore, we have

$$\omega_{cyc} = 0.176GHz \cdot B \tag{1.11}$$

with B in [mT].

The ions can be regarded as fixed due to their large mass. Electrons that move in the plasma will be attracted by the positive charge of the ions which leads to an oscillation of the electrons around the ions. Quantitatively these oscillations are described by the *plasma frequency* ω_p which depends on the plasma density as shown in eq. 1.12:

$$\omega_p = \sqrt{\frac{ne^2}{\epsilon_0 m}} \simeq 0.178GHz \cdot \sqrt{n \cdot cm^3 \cdot 10^{-7}} \tag{1.12}$$

Here, n is the plasma density and ϵ_0 the dielectric constant. For a typical plasma density of $n = 10^{11} cm^{-3}$, the plasma frequency ω_p is 17.8 GHz.

In the following we will discuss under which conditions an energy transfer from the microwave to the electrons is possible. If an electro–magnetic wave is radiated into a plasma, we have to distinguish between different orientations of the field vectors ($\vec{E}_1, \vec{B}_1$), or the wave vector ($\vec{k}$) of the radiation, and the direction of the magnetic field ($\vec{B}_0$). The different wave modes interact differently with the plasma. The microwave is radiated into the ECR plasma under a fixed spatial orientation. However, the magnetic field is not homogeneous (see next paragraph) and therefore there are different orientations between the high frequency and the magnetic field vectors which is schematically shown in fig. 1.4.

The interaction of a plasma with an electro–magnetic wave can be described by the phase velocity v_Φ of the wave:

$$v_\Phi = \frac{\omega}{k} = \frac{c}{\tilde{n}} \tag{1.13}$$

with ω being the angular frequency of the electro–magnetic wave, k its propagation constant, c the vacuum velocity of light and $\tilde{n}$ the refractive index.

For v_Φ we have the following extreme cases: if the phase velocity becomes zero ($\tilde{n} \to \infty$) we speak of a *resonance*. The radiation cannot propagate in the plasma anymore, is absorbed and heats the plasma. A *cut–off* occurs if the refractive index becomes zero ($v_\Phi \to \infty$). Here, the wave is completely reflected and cannot transfer energy to the plasma. In table 1.1 the different orientations of the electro–magnetic waves with respect to the magnetic field are given together with their dispersion relations.

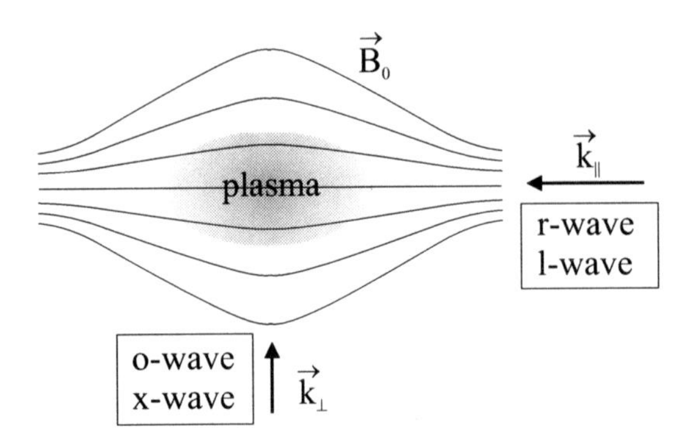

Figure 1.4. Electro–magnetic wave propagation perpendicular (ordinary and extra–ordinary wave) and parallel (r– and l–wave) to the magnetic field lines.

Table 1.1. Possible waves and their dispersion relation.

wave	orientation of the field vectors	dispersion relation
o–wave	$\vec{k}_1 \perp \vec{B}_0 \wedge \vec{E}_1 \parallel \vec{B}_0$	$\frac{v_\Phi^2}{c^2} = (1 - \frac{\omega_p^2}{\omega^2})^{-1}$
x–wave	$\vec{k}_1 \perp \vec{B}_0 \wedge \vec{E}_1 \perp \vec{B}_0$	$\frac{v_\Phi^2}{c^2} = (1 - \frac{\omega_p^2}{\omega^2}\frac{\omega^2-\omega_p^2}{\omega^2-\omega_h^2})^{-1}$
	$\omega_h^2 = \omega_p^2 + \omega_{cyc}^2$	upper hybrid frequency
r–wave	$\vec{k}_1 \parallel \vec{B}_0 \wedge \vec{E}_1$ righthanded circularly pol.	$\frac{v_\Phi^2}{c^2} = (1 - \frac{\omega_p^2/\omega^2}{1-\omega_{cyc}/\omega})^{-1}$
l–wave	$\vec{k}_1 \parallel \vec{B}_0 \wedge \vec{E}_1$ lefthanded circularly pol.	$\frac{v_\Phi^2}{c^2} = (1 - \frac{\omega_p^2/\omega^2}{1+\omega_{cyc}/\omega})^{-1}$

From the above relations it can be seen that an energy transfer from the electro–magnetic wave to the plasma (v_Φ=0) is only possible for the x– and r–wave. The dependence of their phase velocity on the angular frequency of the microwave is shown in fig. 1.5

From this diagram a possible energy transfer from the x–wave can be seen at the *upper hybrid frequency* ω_h. This is the frequency which arises from the superposition of the electron gyration around the magnetic field lines and the plasma oscillation. For a microwave frequency of 2.45 GHz this would happen for a plasma density of $n = 1.1 \cdot 10^{11} cm^{-3}$ at a magnetic field of 0.08 T.

Most important for the heating of the ECR plasma is the r–wave at $\omega = \omega_{cyc}$. At this frequency the electro–magnetic wave is in resonance with the gyrating motion of the plasma electrons. The field energy of the microwave is transformed into kinetic energy of the electrons. The conditions of electron heating are illustrated in fig. 1.6

When an electron enters the resonance region where $\omega = \omega_{cyc}$ it is accelerated or decelerated depending on the phase Φ between its transverse velocity

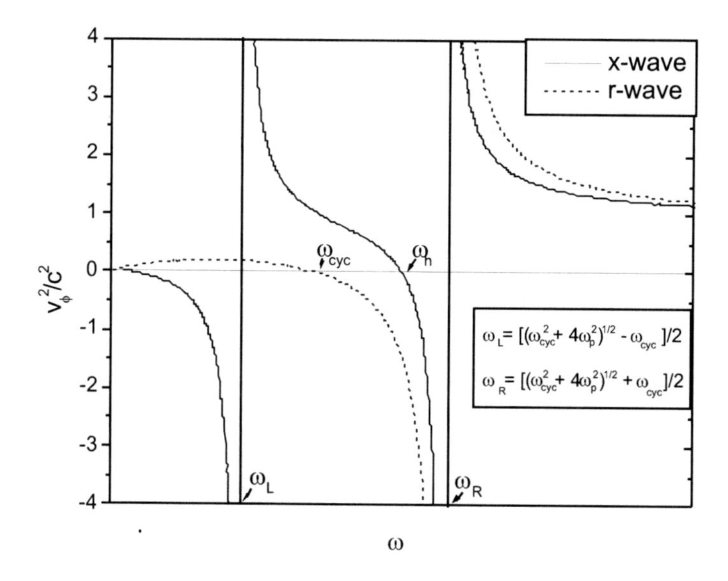

Figure 1.5. Dispersion relation for the x– and r–wave useful for plasma heating.

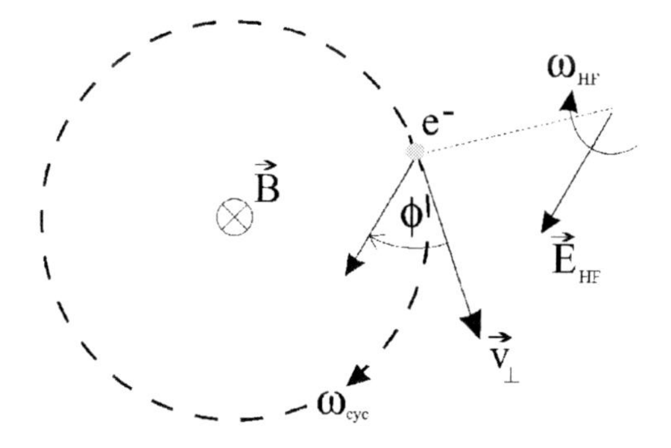

Figure 1.6. Phase relation Φ between the circularly polarized electric field vector $\vec{E}_{HF}$ and the transverse velocity vector $\vec{v}_\perp$ of the Larmor motion of the electron in the magnetic field $\vec{B}$ with the frequency ω_{cyc}.

component $\vec{v}_\perp$ and the electric field vector $\vec{E}_{hf}$ of the microwave. The change in velocity is given by:

$$\Delta v_\perp = \frac{e}{m} E_{hf} (\cos \Phi) \Delta t \tag{1.14}$$

where E_{hf} is the field strength of the microwave and Δt the time the electrons remain in the resonance region. Calculations done by Jongen showed a maximum velocity gain of about $0.5 \cdot 10^7 m/s$ when an electron passes the resonance region once [8]. Connected with that is a maximum energy gain of about 1.5 keV which is bigger than the maximum possible loss since the electrons have an initial velocity:

gain: $v_0 \to v_0 + \Delta v_{max} \Rightarrow E_{kin_0} \to \frac{m}{2}(v_0 + \Delta v)^2 \Rightarrow \Delta E_{kin} = \frac{m}{2}\Delta v(\Delta v + 2v_0)$

loss: $v_0 \to v_0 - \Delta v_{max} \Rightarrow E_{kin_0} \to \frac{m}{2}(v_0 - \Delta v)^2 \Rightarrow \Delta E_{kin} = \frac{m}{2}\Delta v(\Delta v - 2v_0)$

Spectroscopic investigations of ECR plasmas showed that there are electrons in the plasma with energies of several 100 keV [9]. Therefore, the electrons have to be well confined in the plasma (see next paragraph) to have the possibility to enter the resonance region many times.

The considerations above are essentially correct but only valid for the special cases of wave propagation perpendicular or parallel to the magnetic field lines as discussed. A real ECR plasma, however, is much more complicated and at present there is no general theory available which can explain the heating process in a satisfactory way. Usually plasma theories only take into account linear effects and they consider the plasma as isothermic and collision–free. These points, of course, would have to be included for a more detailed description.

Here, only an estimation of the maximum achievable electron energies shall be given using a theoretical description by Eldridge [10]. In this non–relativistic model which describes a plasma in a magnetic mirror field the transverse energy gain of the electrons is given by:

$$\frac{d}{dt}\left\langle n_e\, m_e\, v_\perp^2 \right\rangle = \frac{1}{3}\,\omega_P^2\,\frac{E^2}{L}\,\frac{d\omega_{cyc}}{ds}\bigg|_{\omega_{cyc}(s)=\omega} \tag{1.15}$$

Here, L denotes the distance between the two ECR regions in the mirror field and s the position on the axial field line. With eq. 1.15 and the assumption that 2/3 of the energy of the electric field is transferred into parts transverse to the magnetic field and 1/3 into longitudinal parts (equipartition of the energy to the three spatial degrees of freedom), the energy gain of a single electron can be estimated. For parameters of the Giessen 10 GHz ECRIS [11] (electrical field strength: 20 V/cm, a gradient of $\frac{d\omega_{cyc}}{ds} = 0.13cm^{-1}s^{-1}$ at the resonance region and plasma density being three times higher in the resonance region compared to the rest of the plasma) we get:

$$\frac{d}{dt}\left\langle m_e\, v_\perp^2/2 \right\rangle \simeq 6.3 \cdot 10^7\ eV/s \tag{1.16}$$

Assuming an electron confinement time of 5 ms this leads to an electron energy of 315 keV, which is in the same order of magnitude as the above mentioned measurements.

2.3 Plasma Confinement

A good plasma confinement is necessary for several reasons:

- the electrons have to pass the resonance region many times in order to gain high energies
- the ion confinement time should be as long as possible to achieve a high ionization factor
- the ions should be prevented from touching the plasma chamber wall in order to minimize losses due to recombination

However, the magnetic confinement must not be “perfect”. At least in the axial direction ions have to leave the plasma in order that an ion current may be extracted. Therefore, in an ECRIS a so–called open magnetic structure is used in the axial direction which is shown in fig. 1.7.

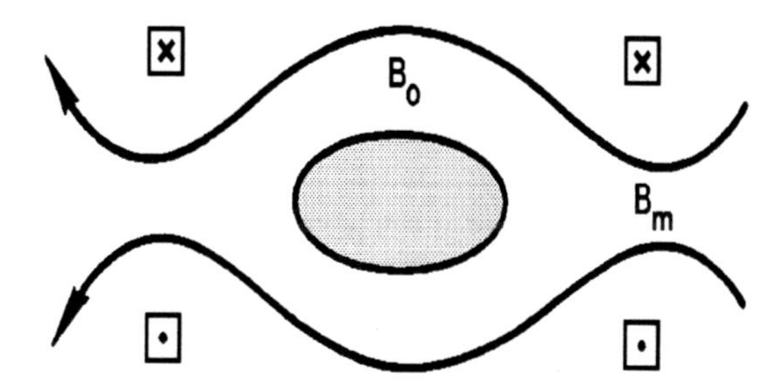

Figure 1.7. Open magnetic structure used in ECR ion sources for the axial plasma confinement.

The following considerations about the motion of charged particles in a magnetic field are valid for both electrons and ions. However, the ions have a very low cyclotron frequency due to their large mass. They are not heated by the microwave radiation, remain thermal and therefore their interaction with the magnetic field can be neglected. The confinement of the ions in the plasma is done by the space charge of the electrons and by the plasma potential, which will be discussed later. So, in the following we will only consider the aspects of the electron motion in a magnetic field, which is responsible for the electron confinement. The consideration of the other possible motions would exceed the framework of this article and can, for example, be found in [12].

2.3.1 Gyration. The equation of motion for an electron in an electric and magnetic field is given by the Lorentz force F_L:

$$m\frac{d\vec{v}}{dt} = \vec{F_L} = e(\vec{E} + \vec{v} \times \vec{B}) \tag{1.17}$$

Assuming $\vec{E} = 0$ and the the magnetic field as being purely axially $\vec{B} = B_z\vec{e}_z$, eq. 1.17 describes a harmonic oscillator with the *cyclotron frequency*

$$\omega_{cyc} = \frac{e}{m}B \tag{1.18}$$

At that frequency the electrons gyrate around the magnetic field lines with the *Larmor radius*

$$r_L = \frac{v_\perp}{\omega_{cyc}} = \frac{mv_\perp}{eB} \tag{1.19}$$

According to eq. 1.17, the velocity component parallel to the magnetic field does not interact with the magnetic field and so the resulting motion of the electrons is a helical trajectory tied to the magnetic field lines.

2.3.2 $\vec{\nabla}B \| \vec{B}$ Drift: The Magnetic Mirror. A magnetic structure, as shown in fig. 1.7, has a gradient which is in direction of the magnetic field. Such a configuration is called a *magnetic mirror* because it can reflect charged particles under certain circumstances as we will see in the following considerations.

As already shown, the electrons gyrate around the magnetic field lines. An electron circling around an area F can be regarded as a ring current I and related to that is a magnetic moment

$$\mu_e = I \cdot F = \frac{1}{2}\frac{m_e v_\perp^2}{B} \tag{1.20}$$

When the electron moves into a region with a different magnetic field strength the Larmor radius changes, but μ_e remains invariant. The interaction of a magnetic moment with the gradient of a magnetic field leads to a force F_e on the electrons

$$\vec{F}_e = -\mu_e \vec{\nabla} B \tag{1.21}$$

If an electron moves from a weak–field to a strong–field region, its velocity $v_\perp$ has to increase according to eq. 1.20 in order to keep μ_e constant. However, the total energy has to be conserved as well:

$$\frac{d}{dt}(E) = \frac{d}{dt}\left(\frac{1}{2}m_e(v_\|^2 + v_\perp^2)\right) = \frac{d}{dt}\left(\frac{m_e v_\|^2}{2} + \mu_e B\right) = 0 \tag{1.22}$$

When B increases, $v_{\|}$ has to decrease in order to conserve the energy and can eventually become zero if the magnetic field is strong enough. The electron is then reflected into a region with a weaker field.

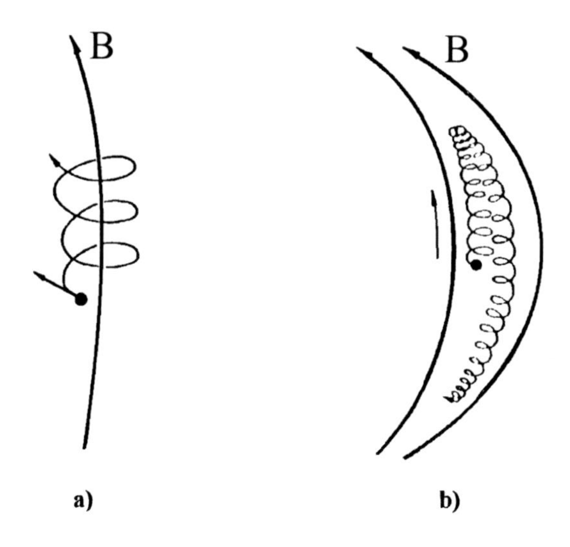

Figure 1.8. Motion of charged particles in a magnetic field: a) gyration; b)$\vec{\nabla}B\|\vec{B}$ drift (reflection in a mirror field)

The motions explained above are shown again in fig. 1.8. However, the electron confinement in a magnetic mirror is not perfect. For example, electrons with $v_{\perp} = 0$ have no magnetic moment and therefore do not interact with the magnetic field gradient. The condition for confinement is:

$$\frac{v_0^2}{v_{0,\perp}} \leq \frac{B_0}{B_m} \equiv R \tag{1.23}$$

Here it is assumed that the electron initially is in the middle of the mirror with a magnetic field B_0 (see also fig. 1.7) and has the velocity components $v_{0,\perp}$ and $v_{0,\|}$ perpendicular and parallel to the magnetic field. If we look at the conditions in velocity space, we can define the pitch angle θ of the so–called *loss cone* of a magnetic mirror [12]:

$$\sin^2\theta = \frac{v_{0,\perp}}{v_0^2} = \frac{1}{R} \tag{1.24}$$

Here R is the mirror ratio. The *loss cone* is shown in fig. 1.9. It can be seen that there has to be a minimum value θ_m in order to reflect the particle:

$$\sin^2\theta_m = \frac{B_0}{B_m} = \frac{1}{R} \tag{1.25}$$

Particles with smaller pitch angles in velocity space are not confined and can leave the magnetic mirror.

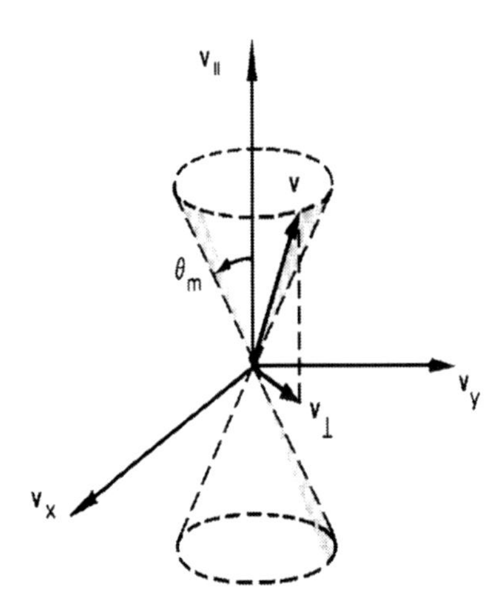

Figure 1.9. Loss cone in velocity space for charged particles confined in a magnetic mirror field.

The above considerations are only valid for charged particles in magnetic fields in the absence of electric fields. However, the presence of the resonance region where the electrons are heated by the microwave and drift or diffusion processes (which will not be regarded here) cause potentials in the plasma. Therefore, eq. 1.25 has to be modified. According to [13] the effective pitch angle θ_{eff} of the *loss cone* is then:

$$\sin^2\theta_{eff} = \frac{1}{R - 2e(U_0 - U_{B_m})/m_e v_\perp^2} \tag{1.26}$$

Here U_0 and U_{B_m} are the electrostatic potentials in the midplane of the mirror and in the region of the maximum magnetic field, respectively.

In order to achieve long confinement times, the mirror ratio should be as high as possible. As a general rule the mirror ratio should be about 2. The most powerful ECR ion sources operate with mirror ratios of 3 or better at the microwave injection side [14], where no electron loss is desired. This corresponds to a pitch angle θ_m of the *loss cone* of 35°. However, the ions are confined by the space charge of the electrons (see next chapter). So, if we want to extract ions from the plasma, electrons have to leave the magnetic mirror. The ions will then follow the electrons because the plasma tends to stay electrically neutral.

Up to now we only have considered the axial magnetic confinement. Such a system, however, is unstable. The curvature of the magnetic field lines results in a radial drift (see e.g. [12]) of the particles towards the plasma chamber and therefore leads to plasma instabilities. To minimize these, in ECR ion sources a radial magnetic field is superimposed between the maxima of the axial mirror field. This radial magnetic field is produced by a multipole magnet which is usually a hexapole. This superposition creates the already–mentioned *minimum–B–structure*. With such a configuration, the magnetic field lines are curved towards the plasma and therefore instabilities are strongly reduced [15] [16]. Confinement times of more than 10^{-3} s can be achieved which is a factor of 50 longer than with an axial mirror field only.

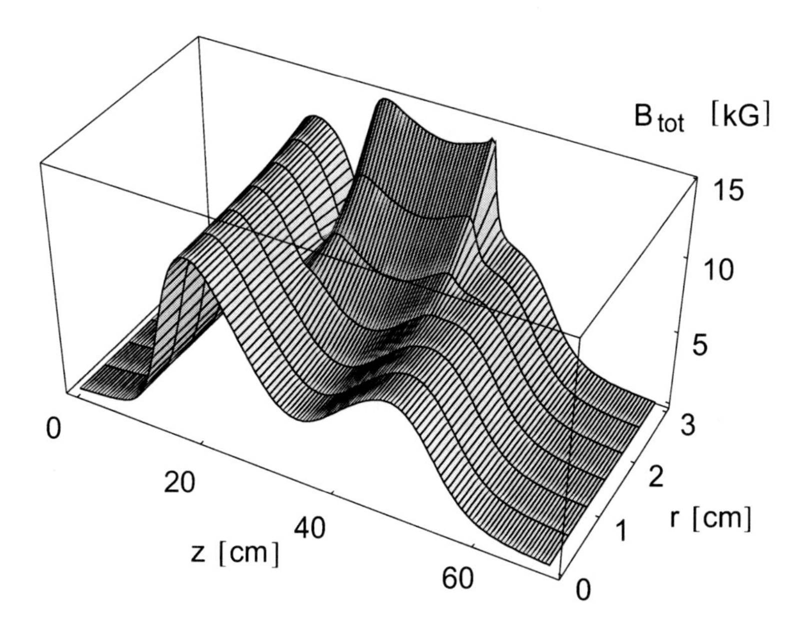

Figure 1.10. Total magnetic field of a minimum–B–configuration for the plasma confinement.

The resulting total magnetic field of such a *minimum–B–configuration* is shown in fig. 1.10. The minimum is in the middle of the plasma chamber and from there the magnetic field increases in all directions leading to a good electron confinement.

2.3.3 Electrostatic Confinement of the Plasma Ions. As already said before, the ions in an ECR plasma are thermal, so their interaction with the magnetic field can be neglected. The ions are confined by the electrostatic potentials created by the strong gradients of electron density and temperature. These are caused by the local microwave heating of the electrons on the ECR surface and by the shape of the magnetic field. Since some of the electrons are accelerated to very high energies through the high frequency, there are at least two different energy distributions for the electrons: a "cold" component with short confinement times and a "hot" component in the region of the ECR surface with a long confinement time. The plasma potential U_p is mainly determined by the electron – and therefore ion losses – from the plasma. A potential distribution, as shown in fig. 1.11, is created along the axis of symmetry.

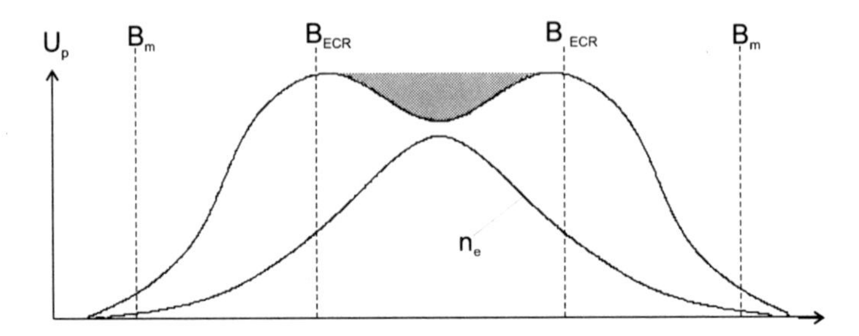

Figure 1.11. Distribution of the plasma potential U_p and the electron density n_e along the axis of a magnetic mirror.

In the middle, a dip in the potential can be seen in which the highly charged ions are well confined. Experimental investigations of the radial plasma potential show a similar distribution [17]. However, it is not clear yet whether the electron density in the centre of the discharge (*core plasma*) can be high enough to form a negative potential.

Apart from the potentials, the conditions for ion confinement also depend strongly on the ion energy and therefore on the heating processes in the plasma. These are mainly the following:

- Coulomb scattering with electrons
- charge exchange
- ion cooling by the loss of energetic ions from the plasma ($E_i >$ potential barrier U)

The time–development of the ion temperature $T_{i,q}$ due to Coulomb scattering can be written as [6]:

$$\frac{dT_{i,q}}{dt} = \frac{4\sqrt{2\pi}\, n_e\, q^2\, r_L^2\, m_e^2\, c^4 \sqrt{m_e}}{m_n\, A\, \sqrt{T_e}}\, L_C \qquad (1.27)$$

Here n_e denotes the electron density, q the charge state of the ion, r_L the Larmor radius, m_e the electron mass, m_n the nucleon mass, A the atomic mass number, T_e the electron temperature. The so–called *Coulomb logarithm* L_C is characteristic for a plasma. It contains the ratio of electron density and plasma temperature and usually has for ECR ion sources a value between 10 and 15 [6]. It can be seen from eq. 1.27 that even if the heating rate is low, highly charged heavy ions are heated faster than light ions in low charge states.

Charge exchange plays only a minor role for the ion temperature. However, the cooling of ions by losses from the plasma is important in an ECRIS. Ions in low charge states have a lower potential barrier and can escape from the plasma more easily. They can take away energy that has been transferred in collisions with the "hotter" heavy ions and therefore cool the ions remaining in the plasma. The time between collisions and hence the characteristic time until an equilibrium is reached is of the order of microseconds. This is short compared to the ion confinement time which is some milliseconds. Therefore, the ion energy distribution can be assumed to be a stationary Maxwell–Boltzmann distribution

$$f_q(E_q) = \frac{2}{\sqrt{\pi}} \frac{1}{T_q} \sqrt{\frac{E_q}{T_q}} \exp\left(-\frac{E_q}{T_q}\right) \tag{1.28}$$

as shown in the left side of fig. 1.12. In addition the potential barriers for the different charge states are displayed.

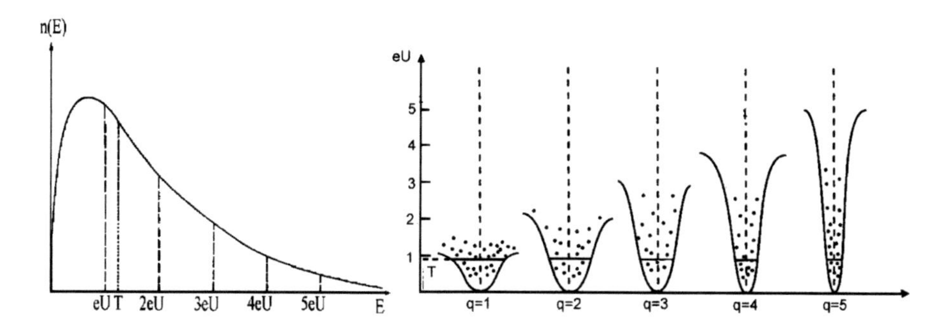

Figure 1.12. Left side: Maxwell–Boltzmann distribution of the ion temperature T with the potential barriers for different charge states q. Right side: ions of the same temperature T in different charge states q in a potential U.

The higher the charge state of the ions is the higher the potential barrier gets that confines them in the plasma (see also right side of the figure). Only ions with energies from the high–energy tail of the Maxwell–Boltzmann distribution can escape. The cooling of these ions further increases their confinement time and leads to higher charge states. Another consequence of the potential barrier is that the charge state distribution inside the plasma is higher than that of the extracted ions.

3. Increasing the Performance of an ECRIS

In the previous chapter we have seen that many plasma parameters influence the performance of an ECRIS. The most important ones are:

- high electron densities
- high electron energies
- long ion confinement times

In the following some methods shall be listed which can modify these parameters and increase the performance of an ECRIS.

3.1 Increase of the Electron Density

In order to reduce charge–exchange processes in collisions between ions and atoms, the flux of neutral gas into the plasma has to be minimized. However, this also reduces the number of particles which are available for the ionization and leads to a saturation of the electron density. The use of internal or external electron sources can overcome that limitation.

3.1.1 Wall Coating. Electrons and ions that leave the confinement radially hit the plasma chamber wall and can produce secondary electrons. If the wall is coated with a material with a high secondary electron coefficient, the electron density in the plasma can be increased. Materials often used in ECR ion sources are, for example, SiO_2 or Al_2O_3.

A similar effect can be seen with a plasma chamber made of aluminum. If the ion source is operated with oxygen in the plasma for a certain period, aluminum oxide is formed at the wall. Nowadays, almost all high–performance ECR ion sources use aluminum plasma chambers.

3.1.2 Biased Disk. A substantial enhancement of the extracted ion currents –especially for the high charge states– was observed with many ECR ion sources when a *biased disk* was used [18] [19] [20]. A *biased disk* is an electrode which is biased negatively with respect to the plasma and which is brought close to the plasma (mostly axially). Although the function of the *biased disk* is not yet fully understood, the following are possible explanations:

- Ions from the plasma are accelerated towards the surface of the electrode and secondary electrons are emitted. These propagate into the plasma and increase the electron density.
- The negative potential repels electrons that escaped from the plasma through the *loss cone*.
- The negative voltage at the electrode has a beneficial influence on the plasma potential, e.g. it could further increase the potential dip (see previous chapter) and therefore increase the confinement time for the ions.

First evidence for the last point was found in time–resolved measurements at the Frankfurt 14 GHz ECRIS [21].

The influence of the *biased disk* on the extractable ion currents depends mainly on the position of the electrode with respect to the plasma and on the applied voltage. Enhancement factors compared to the operation without a *biased disk* between 2 and 10 could be observed [22]. Similar results are obtained by the installation of an electron gun [23].

3.2 Improvement of the Magnetic Confinement

As already discussed in the previous chapter, the pitch angle of the loss cone of the magnetic mirror field depends on the mirror ratio B_m/B_0. Since the minimum magnetic field B_0 has to be at least as high as the resonance magnetic field, the mirror ratio can only be improved by increasing the magnetic field maximum B_m. This is, however, limited by the available currents through the solenoids or by the remanence of the permanent magnets unless a superconducting magnet system is used. The use of a soft–iron yoke can further increase the magnetic field and, when inserted into the plasma chamber, can also change the shape of the field as shown in fig. 1.13.

As can be seen in the figure, not only the maximum of the magnetic field increases by 25%, but also the gradient at the microwave injection side (left side) increases, leading to a better confinement (eq. 1.21). The magnetic field at the extraction side (right side) of the ion source is not affected.

R. Geller tried to describe the dependence of the performance of an ECRIS on the different parameters with empirical scaling laws [24] [25]. For the dependence on the magnetic field he found

$$q_{opt} \propto \log B_m \tag{1.29}$$

when only B is varied and all other parameters are kept constant. Here, q_{opt} denotes the highest q in a charge state distribution which is still the dominant peak.

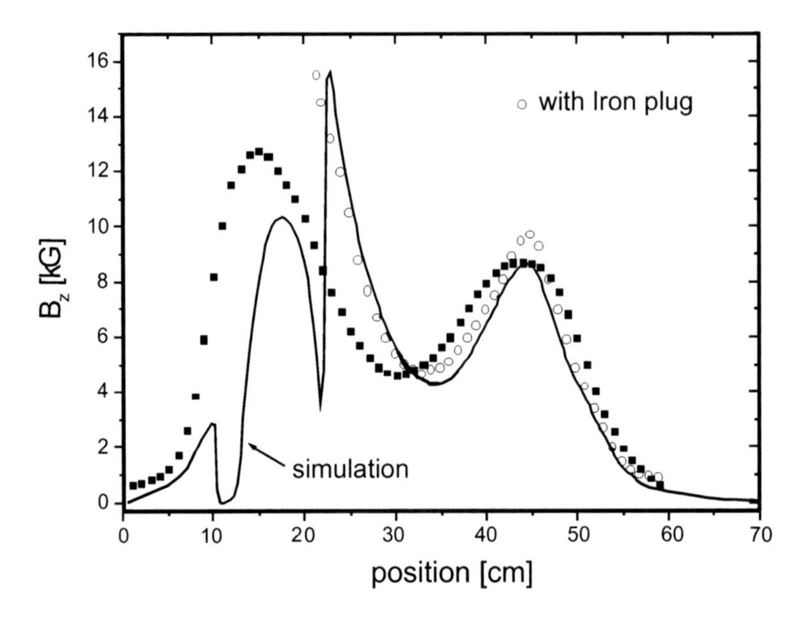

Figure 1.13. Improvement of the axial magnetic confinement by the insertion of a soft–iron plug.

3.3 Increase of the Microwave Frequency

An increase of the magnetic field strength also allows the use of higher microwave frequencies with a still sufficiently high mirror ratio. The comparison between different ion sources showed that with higher frequencies the maximum of the charge state distribution as well as the extracted ion currents could be improved. Geller found the following scaling laws:

$$\begin{aligned} q_{opt} &\propto \log \omega^{3.5} \\ I_q &\propto \omega^2 \end{aligned} \tag{1.30}$$

where I_q is the ion current of the charge state q_{opt}. One can see that an increase in the microwave frequency has a big effect on the performance of an ECRIS. However, since an increase in the microwave frequency also requires higher magnetic fields, there is a limit at about 18 GHz where the necessary mirror ratio can still be achieved with conventional solenoids. For higher frequencies superconducting structures have to be used. One example of this type is a 28 GHz ECRIS which is currently under construction at the Lawrence Berkeley National Laboratory, USA [26].

Generally it has to be said that the above mentioned scaling laws only show the correct tendency of the parameter dependence. They are, however, deduced from the comparison of different ion sources where it was certainly not possible to vary only one specific parameter. Therefore, their mathematical correctness has to be questioned.

3.4 Two–Frequency Heating

As described in chapter 1.2, the plasma electrons gain high energies in the resonance region. If a magnetic field configuration is chosen where the minimum field is sufficiently low, a second, lower frequency can be radiated into the plasma and heat the electrons on a second resonance surface. This method is shown schematically in fig. 1.14.

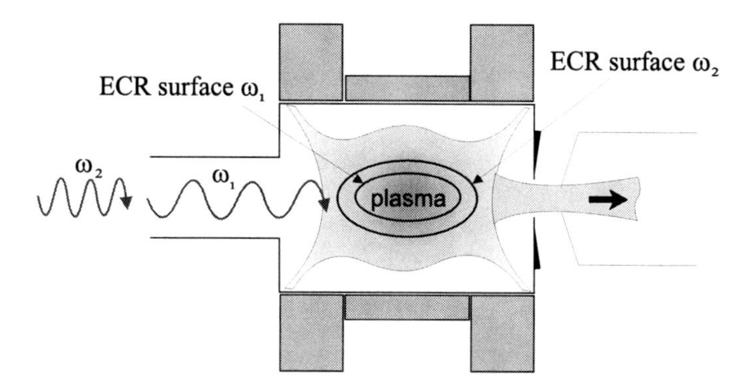

Figure 1.14. Principle of the two–frequency heating.

Two–frequency heating was first done at the AECR in Berkeley where a substantial improvement of both q_{opt} and the extracted ion currents was observed. This is shown in fig 1.15 for uranium ions [27]. The open circles are the best results obtained with single–frequency heating (14 GHz). In this distribution, the charge state with the maximum intensity is U^{33+} which was produced using a microwave power of 1540 W. With additional heating of the second frequency (14 GHz + 10 GHz), the power absorbed by the plasma could be increased to a total of 1770 W. The charge state distribution (full squares) shifts to higher values (q_{opt} is now U^{36+}) and the extracted ion currents increased as well.

3.5 Gas Mixing

In 1983 Drentje in Groningen discovered that the extracted ion currents of a certain element (e.g. argon) could be significantly increased when an additional light element (e.g. helium or oxygen) would be introduced into the ion source [28]. In fig 1.16 this effect is shown for nitrogen charge state distributions extracted from an all–permanent magnet 10 GHz ECRIS [29]. A significant shift of q_{opt} towards higher charge states can be seen when helium is added as

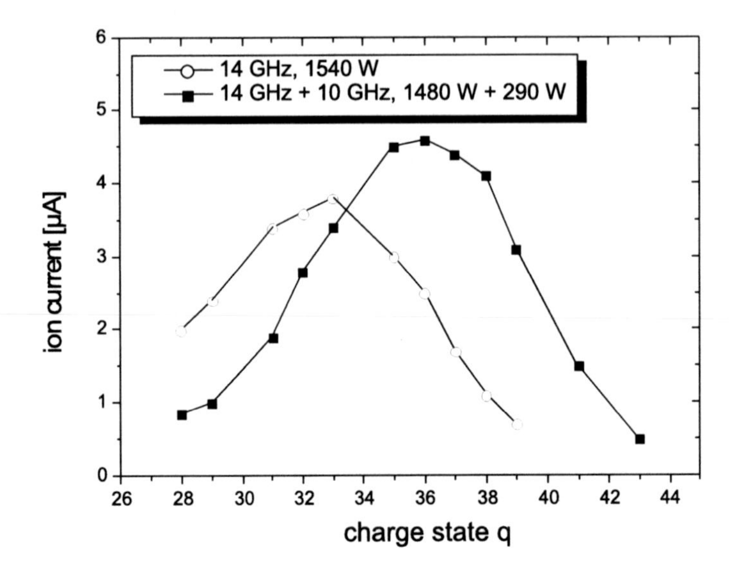

Figure 1.15. Charge state distributions for uranium with single–frequency heating (open circles) and two–frequency heating (full squares)[27].

a mixing gas. The ion currents increased as well, which was especially true for the higher charge states (a factor of 80 for N^{6+}).

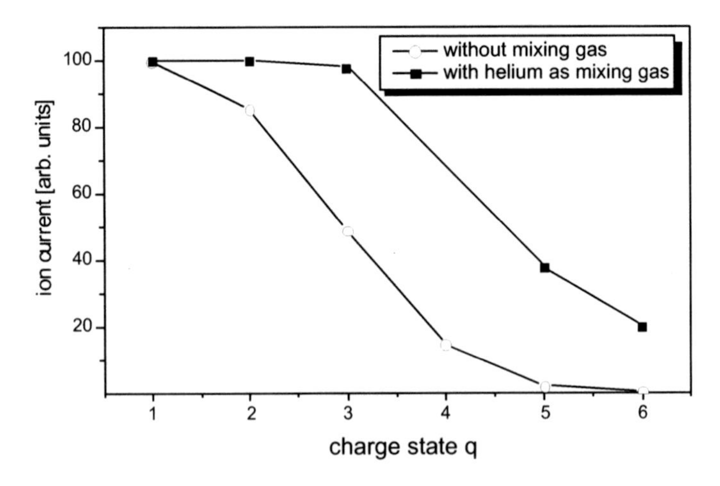

Figure 1.16. Gas mixing effect for nitrogen: without mixing gas (open circles) and with helium as mixing gas (full squares)[29].

Up to this day, there is no satisfactory explanation for this effect. However, there are several attempts of which the following two seem to be the most promising:

- The light element has a lower mass and a lower temperature. In collisions with the "hotter" heavier ions, energy is transferred to them. Since the light element is not as well confined as the heavier due to the lower plasma potential barrier for lower charge states, these ions can escape from the plasma quite easily. They take away energy from the plasma and therefore cool the heavier ions. This increases the confinement time of the heavy ions in high charge states and leads to a better *ionization factor* (see chapter 1.2). The decisive point for this effect is that the time until the plasma particles are in equilibrium is in the microsecond range whereas the confinement times in the plasma are several milliseconds [30].
- The light element is easy to ionize due to the low charge states and supplies electrons to the plasma which increases the electron density.

Typical mass ratios between process gas (A_p) and mixing gas (A_m) are A_p/A_m = 3-5. It turned out that the higher the desired charge state of an element is, the higher is the volume ratio V_m/V_p. Typical values for the volume ratio are between 5 and 10.

3.6 Afterglow Effect

The *afterglow effect* was discovered more than 10 years ago and one of the first reports was written by Geller [31]. He found out that after switching off the microwave frequency there is a drastic increase of the ion intensity in a short pulse (200 μs – 10 ms, depending on the ion source parameters). A typical afterglow pulse is shown on the left side of fig. 1.17 for Au^{27+} ions [32]. After switching on the microwave power at t=0, it takes some time for the plasma to reach equilibrium. At about 40 ms of plasma heating, there is a constant ion current. At 100 ms the microwave is switched off and then a very intense pulse in the output current can be observed. This effect becomes stronger with increasing charge state. The right side of the figure shows a comparison of the performances of an ECRIS in continuous operation (open circles) and in the *afterglow mode* (full squares) [33]. Not only is there an increase in the intensity of the highly charged ions but the charge state distribution is also shifted towards higher charge states.

An explanation of this effect can be given by regarding the confinement conditions for highly charged ions in the ECR plasma. As already mentioned in chapter 1.2, the charge state distribution inside the plasma is higher than that of the extracted ions due to the high potential barrier for highly charged

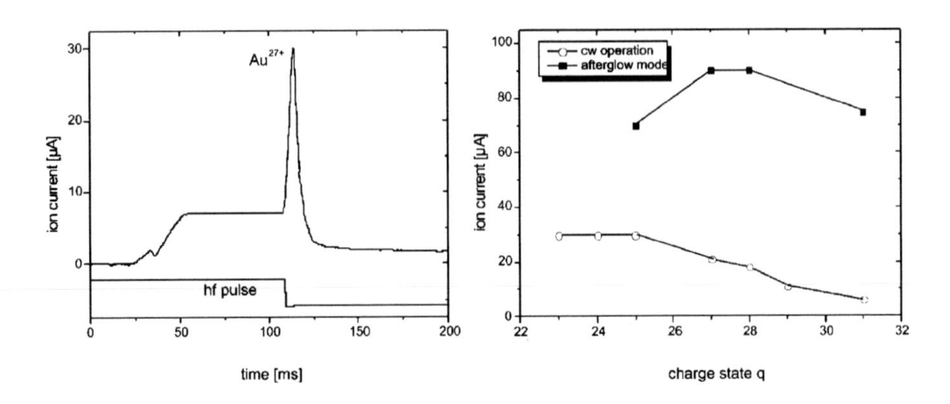

Figure 1.17. Left side: typical afterglow pulse for Au^{27+} ions [32]; right side: effect of the afterglow mode on a lead charge state distribution [33] (open circles: cw operation; full squares: afterglow mode).

ions. Only few ions in high charge states, which have energies in the tail of the Maxwell–Boltzmann distribution, can escape. If the microwave is switched off, the electrons are not accelerated anymore, their confinement conditions become worse and they leave the plasma. Since the plasma tends to stay electrically neutral, the ions that have been confined by the electron space charge, will leave the plasma as well. So, all the highly charged ions that have been trapped in the plasma can be extracted in a short pulse.

For many experiments the *afterglow effect* cannot be used since a continuous ion beam is required. However, it is perfectly suitable, e.g. for the injection into linear accelerator or synchrotrons where pulsed beams of high intensities are desired.

4. Extraction of Ions

Electrons which have velocity components in the *loss cone* of the magnetic mirror can escape from the plasma pulling ions after them. Usually, positive ions are extracted from an ion source by applying a positive high voltage to the ion source body. The insertion of an extraction aperture into the plasma chamber separates the plasma from the extraction region. Electrons and ions leaving the plasma drift through a hole on the axis of the ion source. Once outside the plasma, the charged particles interact with the high voltage. The electrons are pulled back towards the positive potential and hit the extraction aperture where they recombine; the ions are pushed away from the ion source. A second electrode on ground potential, the so–called *puller electrode*, accelerates the ions to the energy defined by the high voltage. The shape of the extracted beam depends on many parameters, but is mainly formed in the extraction region.

The extraction system should be variable in order to have the possibility to optimize the shape of the ion beam for various elements and charge states. The beam intensity and quality is determined by plasma parameters like ion temperature, density, etc., the magnetic field in the extraction region, the geometry of the extraction system and the applied electrical field strength. The ion beam is charcterized by its energy, intensity, shape and divergence. These parameters have to be optimized depending on the specific application.

4.1 Space–Charge Limited Ion Extraction

Since in an ion beam all particles have the same charge sign, there is an electric field from the space charge of the ions which will cause the beam to widen.

The maximum ion current density j_{sc} that can be extracted from a plane area under space–charge limiting conditions can be described by the law of Child and Langmuir [34] [35]:

$$j_{sc} = 1.72 \left(\frac{q}{A}\right)^{1/2} \frac{U_{acc}^{3/2}}{d^2} \qquad \left[\frac{\mathrm{mA}}{\mathrm{cm}^2}\right] \tag{1.31}$$

where U_{acc} denotes the acceleration voltage applied to the ion source and d the distance between extraction and puller electrode.

This relation, however, is only valid when the plasma is able to produce sufficient ions and therefore the ion density in the plasma plays an important role. The current density j_p that can be delivered by the plasma is given by [36]:

$$j_p = 4.91 \cdot 10^{-13} n_i \left(\frac{q}{A}\right)^{1/2} T_i^{1/2} \qquad \left[\frac{\mathrm{mA}}{\mathrm{cm}^2}\right] \tag{1.32}$$

Here, n_i and T_i are the ion density and temperature, respectively. If we neglect the space charge of the ion beam, a parallel ion beam can be extracted from a plasma when the emitting surface (plasma boundary or *meniscus*) is plane. This is true for the condition $j_{sc} \approx j_p$. If the space charge is taken into account the shape of the plasma meniscus is determined by the ratio between j_{sc} and j_p. The shape of the meniscus has a strong effect on the shape of the ion beam because it acts as a lens. The possible shapes of the meniscus are shown in fig. 1.18.

a)	$j_p > j_{sc}$	:	convex plasma meniscus
b)	$j_p \approx j_{sc}$	:	plane plasma meniscus
c)	$j_p < j_{sc}$	:	concave plasma meniscus

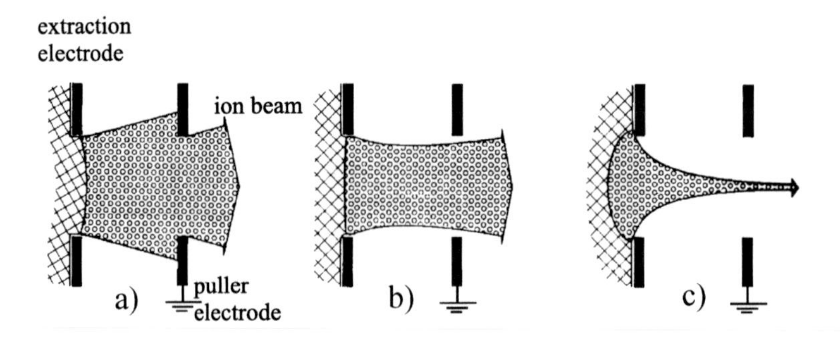

Figure 1.18. Possible shapes of the plasma meniscus.

For ECR ion sources we have usually $j_p < j_{sc}$, i.e. the extractable ion current is limited by the ion production in the plasma. The plasma meniscus is then concave and the shape of the beam can be adjusted to be slightly convergent by variation of the distance between extraction and puller electrode at a fixed acceleration voltage and by the geometry of the extraction system. Details about ion extraction and the beam quality can be found in [37]. On its path through a beam line the ion beam will widen again due to its space charge. Here, focusing elements like electrostatic Einzel lenses, magnetic lenses or electrostatic quadrupoles can be used to achieve a high transmission.

5. Production of Metallic Ions

The production of ions from gaseous elements is quite straight forward because the processing gas can be inserted into the plasma easily with a gas–inlet valve. For metal ions this is only possible for a few elements for which gaseous compounds exist. For other metals different techniques were developed. The methods that are most commonly used in ECR ion sources shall now be briefly discussed.

5.1 The Insertion Technique

With this method the element to be ionized is introduced into the ion source as a solid and brought close to the plasma. Usually a bundle of thin wires is used in order to get a surface as large as possible. Atoms are removed from the solid through sputter processes with energetic electrons from the plasma. This technique is mainly used for high–melting materials. The problem with this method is that the position of the bundle with respect to the plasma is very critical and therefore it is very difficult to get a stable ion beam for a long period. The metal sample is a kind of "antenna" and interacts with the microwave and the plasma. Furthermore, the position of the bundle has to be re–adjusted frequently because material is removed from the sample in order to keep the parameters constant. Another disadvantage is that the partial pressure of the sputtered metal atoms in the plasma is not high enough for a steady discharge. Therefore, an additional gas has to be introduced into the source. This also leads to relatively low ion currents of the desired element.

5.2 The Use of an Evaporation Oven

This is the method which is mostly used for low–melting materials. With this technique the metal is heated (which normally means melted) until the vapor pressure is sufficiently high to allow atoms to evaporate into the ECR plasma. With this method the vapour pressure can be adjusted by the oven temperature and usually no additional gas is needed leading to a "cleaner" charge state distribution. The disadvantage of this method is that not all elements can be evaporated because the oven temperature is limited to about 2000°C. Furthermore, the evaporation oven has to be refilled after a certain time, although the material consumption is rather low. Depending on the ion source parameters, this is in the order of a few mg/h, so that operation times of up to 1000 h can be achieved.

5.3 The MIVOC Technique

The MIVOC (Metal Ions from Volatile Compounds) technique [38] uses special metallic compounds which have a gas pressure in the range of 10^{-3} mbar at room temperature, i.e. *ferrocene* ($Fe(C_5H_5)_2$). These compounds can be used with an ordinary gas–inlet valve. With this technique very high ion intensities can be extracted for a very long period. Apart from the fact that these *metallocenes* are usually very toxic, there are two disadvantages. Only a few elements are available in these compounds (till today) and there are many impurity ions in the plasma (mainly carbon and hydrogen) which have much higher intensities than the desired metallic element.

6. Realization of an ECRIS

As already mentioned there are two major directions in the development of ECR ion sources. One is to build more and more powerful ion sources in order to produce higher and higher intensities of highly charged ions. We have seen in the previous chapters that this can be achieved by increasing the microwave frequency and the magnetic confinement field accordingly resulting in super–conducting magnetic structures for frequencies above 18 GHz. At these very powerful ion sources the magnetic fields (at least for the axial confinement) are produced by current–driven solenoids. Therefore their electrical power consumption is rather high (several 100 kW). On the other hand, there is a growing interest in compact ion sources where the necessary magnetic fields are produced by permanent magnets only. This type of an ECRIS has a very low electrical power consumption and is ideally suited for the use on high–voltage terminals or Van–de–Graaff generators. In the following a brief description of the realization of both types of ECR ion sources shall be given.

6.1 "Conventional" ECR Ion Sources

In fig. 1.19 a typical example for a "conventional" ECRIS [39] is shown schematically.

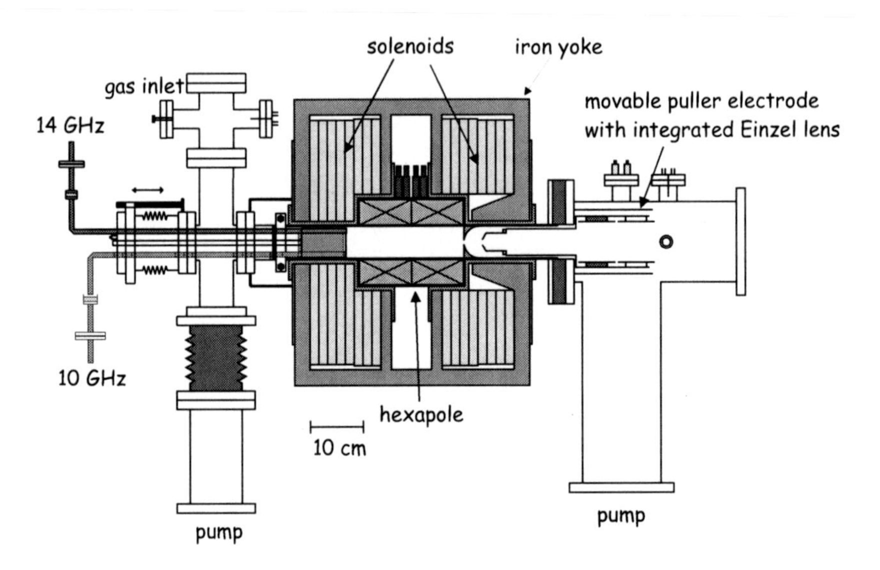

Figure 1.19. Schematic set–up of the Giessen 14 GHz ECRIS [39].

This ECRIS operates nominally at a microwave frequency of 14 GHz but 10 GHz can be additionally injected into the plasma in order to have *two–frequency heating*. The microwave power available is up to 2.2 kW, but usually during operation about 1 kW is needed depending on the charge state the ion source is optimized on. As a general rule one can say that the higher the desired charge state is, the higher is the necessary microwave power. For a description of the microwave system, refer to the next paragraph. The plasma chamber is pumped from both sides and a residual gas pressure of about $1 \cdot 10^{-7}$ $mbar$ is obtained. Although typical gas pressures during operation are in the range of about 10^{-4} $mbar$, the residual gas pressure should be as low as possible in order to avoid the extraction of impurity ions. The gas is introduced into the plasma by a thermally controlled valve. Apart from the microwave power and the magnetic field strength that determines the electron confinement the gas pressure is the most important parameter for the optimization of the ECRIS on different charge states. In general we can say that low charge states are produced at a high gas pressures whereas the pressure has to be reduced for the high charge states. A lower particle density increases the mean free path of the electrons between collisions and therefore they can gain higher energies.

The magnetic field for the longitudinal electron confinement is produced by current–driven solenoids surrounded by an iron yoke to increase the magnetic field strength. An additional iron plug can be inserted into the plasma chamber. The resulting longitudinal magnetic field has already been shown in fig. 1.13. The resonance magnetic field strength corresponding to a microwave frequency of 14 GHz is 5 kG (10^4G = 1T). So in this ECRIS there is a mirror ratio of 2 at the extraction side and 2.6 (without iron plug) at the injection side. The radial electron confinement is achieved by a Halbach–type hexapole [40] made of permanent magnets. The magnet configuration and the resulting magnetic field strength is shown in fig. 1.20

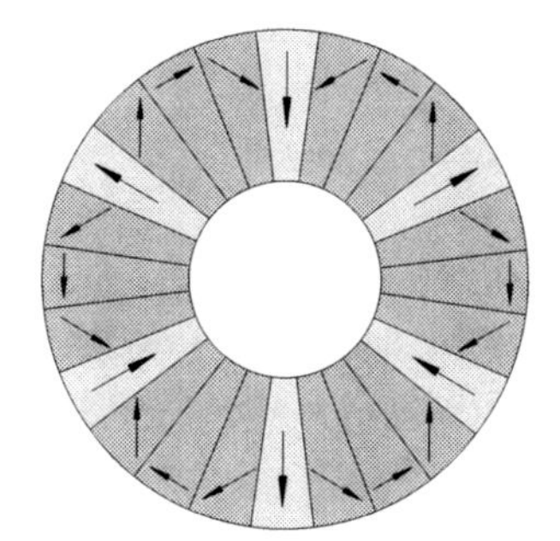

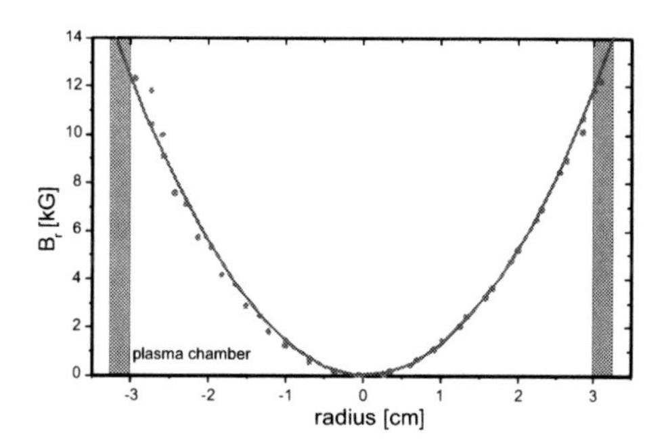

Figure 1.20. Radial magnetic field of the 14 GHz ECRIS. Left side: Halbach configuration [40]; Right side: resulting magnetic field strength.

The hexapole magnet is made of *NdFeB* which is the strongest permanent magnet material presently available (remanence $\approx$ 13 kG). With the present set–up a radial magnetic field strength of 12 kG is obtained at the inner wall of the plasma chamber.

The whole ion source body is at a positive high–voltage potential (for the extraction of positive ions). Ions that drift through the extraction aperture are accelerated by the grounded puller electrode to typical energies of 10–20 q$\cdot$ keV. Integrated into the puller electrode is an electrostatic Einzel lens to focus the ion beam. The position of the puller electrode with respect to the extraction aperture can be varied in order to allow the optimization on different charge states. Since the plasma electrons are not mono–energetic, not only one specific charge state can be extracted from the ion source but there is always a charge state distribution where q_{opt} can be varied depending on the ion source parameters. The desired charge state is usually selected using an analyzing magnet. In a

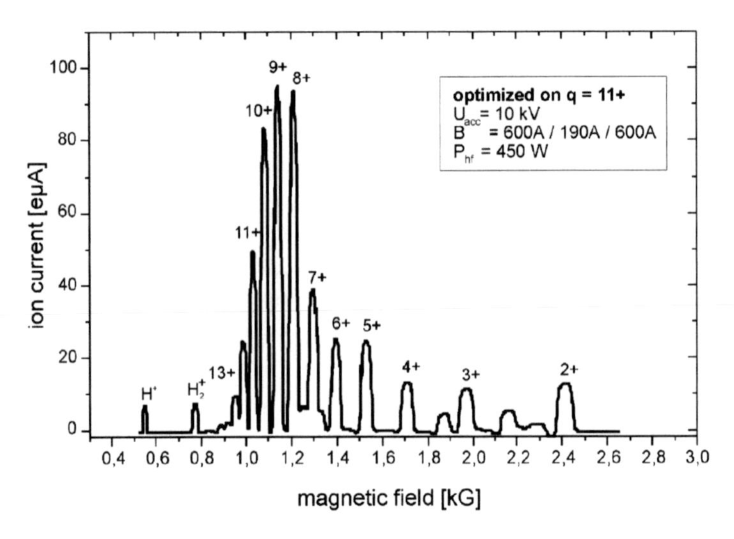

Figure 1.21. Argon charge state distribution extracted from the 14 GHz ECRIS.

magnetic field, ions with different m/q–values have different trajectories due to the Lorentz force (eq. 1.17). At the exit aperture of the analyzing magnet the desired charge state can be selected by adjusting the magnetic field strength of the analyzing magnet.

A typical charge state distribution for argon is shown in fig. 1.21 as a function of the magnetic field of the analyzing magnet. The ions were extracted with $U_{acc} = 10\ kV$ and the ion source was optimized on Ar^{11+} (P_{hf} = 450 W, electrical currents through solenoids: 600 A, 190 A, 600 A). Ar^{9+} is the charge state with the highest intensity (q_{opt}) and the ion currents are typically in the range of several 10 to several 100 microamps.

6.2 Compact All–Permanent Magnet ECR Ion Sources

At this type of an ECRIS the necessary magnetic fields are produced with permanent magnets only. This reduces the required electrical power consumption from several 100 kW for "conventional"ECR ion sources to several 100 W (mainly for the microwave system). Furthermore, these ion sources must have small dimensions in order to achieve sufficient magnetic field strengths with permanent magnets. An example of such an ion source is shown in fig. 1.22.

This ECRIS [41] operates at a microwave frequency of 10 GHz. In this ion source the microwave radiation is coupled to the plasma via a co–axial line to which an evaporation oven can be attached for the production of ions from metallic elements (see previous paragraph). The radial confinement of plasma electrons is again achieved with a Halbach–type hexapole magnet (see

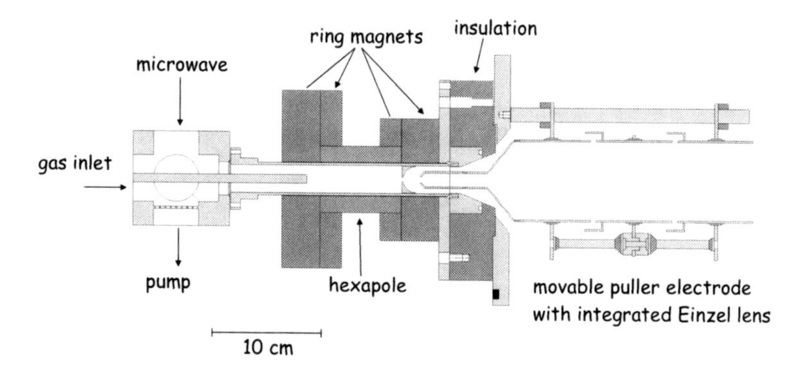

Figure 1.22. Schematic set–up of an all–permanent magnet 10 GHz ECRIS [41].

previous paragraph). The longitudinal magnetic field is produced by 2 ring magnets where one ring is magnetized towards the plasma chamber and the second away from the plasma chamber. Two additional ring magnets which are axially magnetized help to "close the field lines" and increase the field inside the plasma chamber. The longitudinal magnetic field configuration together with the resulting magnetic field strength is shown in fig. 1.23. The resonance magnetic field strength that corresponds to a frequency of 10 GHz is 3.6 kG, so with this configuration the mirror ratio is 2.2 at the injection side and 1.95 at the extraction side.

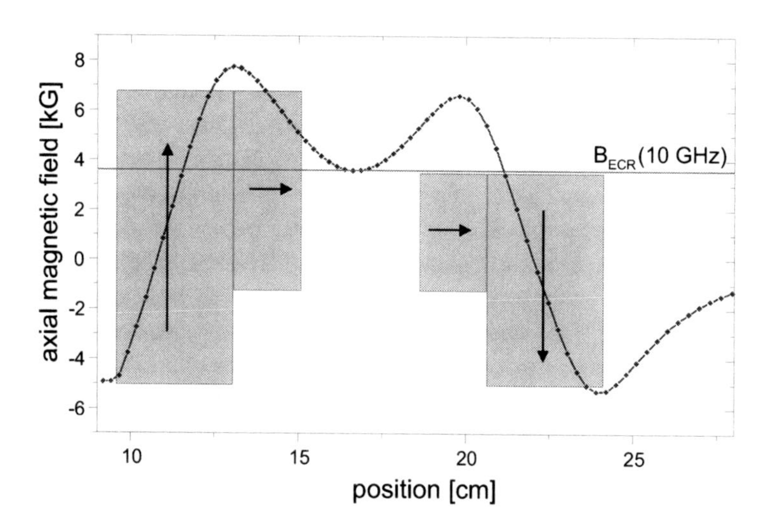

Figure 1.23. Longitudinal magnetic field configuration of the 10 GHz all–permanent magnet ECRIS.

When comparing the longitudinal magnetic field of the "conventional" ion source with the all–permanent type a difference can be seen. In the hexapole region, where the plasma is confined, both configurations form a magnetic mirror. In the extraction region, however, the field in the permanent magnet set–up becomes negative, which has some influence on the ion beam formation. This shall not be considered here.

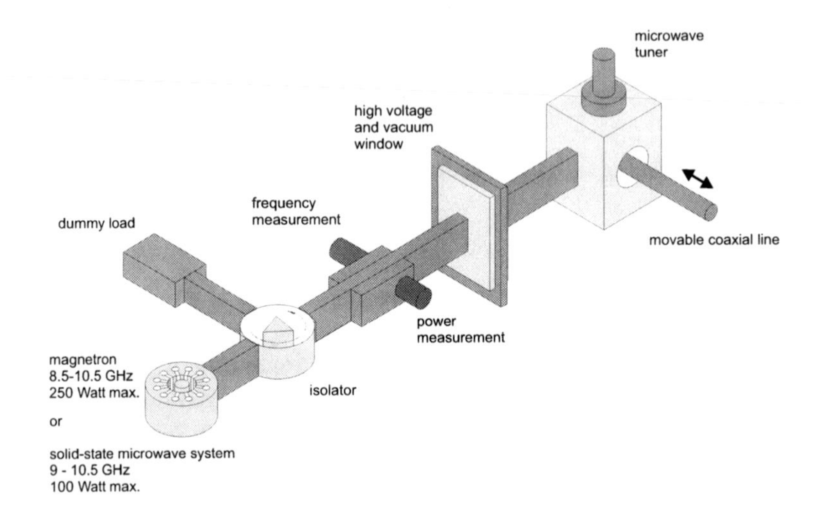

Figure 1.24. Microwave system of the 10 GHz all–permanent magnet ECRIS.

The high frequency that is necessary for the electron heating is produced in a microwave generator. Here, a magnetron or a solid–state microwave system (for microwave powers below 200 W) is used. "Conventional" ECR ion sources usually use klystron amplifiers which can deliver several kW of microwave power. When optimizing an ion source for different charge states the confinement conditions are usually varied by variation of the magnetic field. This can hardly be done when permanent magnets are used. Therefore, we use a frequency–tunable microwave generator and with which we vary the applied microwave frequency. The complete microwave system is shown in fig. 1.24.

The microwaves pass an isolator to protect the generator from reflected microwave power which could destroy it. The reflected power is guided into a dummy load where the microwave energy is converted to heat. After frequency– and power measurement there is a window which separates the ion source body, which is under vacuum and on high voltage, from the rest of the system. The microwave radiation is then coupled to the plasma via a co–axial line. A microwave tuner is installed to adapt the chamber volume to the applied microwave frequency.

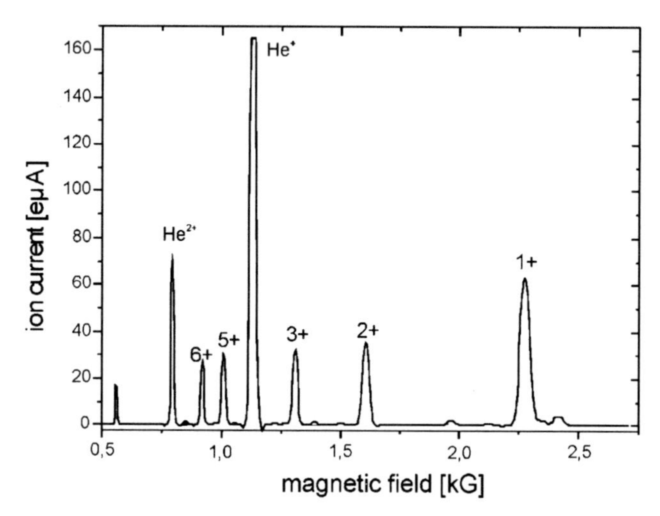

Figure 1.25. Oxygen charge state distribution from the all–permanent magnet 10 GHz ECRIS with helium as a mixing gas.

A typical charge state distribution for oxygen from this type of ion source is shown in fig. 1.25. In this spectrum helium was used as a mixing gas. The extracted ion currents from this type of ion source are usually lower than from the "conventional" ion sources due to the smaller plasma volume.

The two ECR ion sources mentioned above have been described in order to demonstrate how an ECRIS is built in general. Meanwhile there exists a large number of ECR ion sources throughout the world which have essentially the same set–up, but which may differ in many details and therefore also in their performance. For an overview as well as a more detailed description of this type of ion source see [2] or [42].

References

[1] P. Briand, R. Geller, B. Jacquot and C. Jacquot, Nucl. Inst. & Meth. **131**, 407 (1975)

[2] R. Geller, *Electron Cyclotron Resonance Ion Sources and ECR Plasmas.* Institute of Physics Publishing, Bristol (1996)

[3] W. R. Thompson, M. B. Shah and H. B. Gilbody, J. Phys. B **28**, 1321 (1995)

[4] K. Aichele, U. Hartenfeller, D. Hathiramani, G. Hofmann, V. Schäfer, M. Steidl, M. Stenke, E. Salzborn, T. Pattard and J. M. Rost, J. Phys. B **31**, 2369 (1998)

[5] W. Lotz, Zeitschr. f. Physik **232**, 101 (1970)

[6] G. D. Shirkov and G. Zschornack: *Electron Impact Ion Sources for Highly Charged Ions.* Vieweg Verlagsgesellschaft mbH, Braunschweig (1996)

[7] E. D. Donets, Soviet Journal of Particles and Nuclei **13**, 387 (1982)

[8] Y. Jongen, In: *Proc. of the 6th International Conference on ECR Ion Sources*, 238. Lawrence Berkeley National Laboratory, USA (1985)

[9] G. Gaudart and A. Girard, In: *Proc. of the 12th International Workshop on ECR Ion Sources*, 174. Riken, Japan (1995)

[10] O. Eldridge, Phys. Fluids **15**, 676 (1972)

[11] M. Liehr, M. Schlapp, R. Trassl, G. Hofmann, M. Stenke, R. Völpel and E. Salzborn, Nucl. Inst. & Meth. **B 79**, 697 (1993)

[12] E. F. Chen, *Introduction to Plasma Physics and Controlled Fusion.* Plenum Press, New York (1984)

[13] B. Lehnert, *Dynamics of Charged Particles.* North Holland Publishing Company (1964)

[14] Z. Q. Xie and C. M. Lyneis, In: *Proc. of the 13th International Workshop on ECR Ion Sources*, 16. Texas University, USA (1997)

[15] J. Taylor, Phys. Fluids **6**, 1529 (1963)

[16] J. B. Taylor, Phys. Fluids **7**, 767 (1964)

[17] K. S. Golovanivsky and G. Melin, In: *Proc. of the 10th International Workshop on ECR Ion Sources*, 63. Oak Ridge National Laboratory (1990)

[18] G. Melin, In: *Proc. of the 10th International Workshop on ECR Ion Sources*, 1. Oak Ridge National Laboratory, USA (1990)

[19] S. Cammino, J. Sijbring and A. Drentje, Rev. Sci. Inst. **63**, 2872 (1992)

[20] K. Matsumoto, Rev. Sci. Inst. **65**, 1116 (1994)

[21] K. E. Stiebing, O. Hohn, S. Runkel, L. Schmidt, V. Mironov, G. Shirkov and H. Schmidt-Böcking, Phys. Rev. Special Topics: Accelerators and Beams (submitted) (1999)

[22] Z. Q. Xie and C. M. Lyneis, Rev. Sci. Inst. **65**, 2947 (1994)

[23] C. M. Lyneis, Z. Q. Xie, D. J. Clark, R. S. Lam and S. A. Lundgren, In: *Proc. of the 10th International Conference on ECR Ion Sources*, 47. Oak Ridge National Laboratory, USA (1990)

[24] R. Geller, In: *Proc. of the 10th International Workshop on ECR Ion Sources*, 381. Oak Ridge National Laboratory, USA (1990)

[25] R. Geller, Ann. Rev. Nucl. Part. Sci. **40**, 15 (1990)

[26] M. A. Leitner, S. A. Lundgren, C. M. Lyneis, C. E. Taylor and D. C. Wutte, In: *Proc. of the 14th International Workshop on ECR Ion Sources*, 66. CERN (Geneve), Switzerland (1999)

[27] Z. Q. Xie and C. M. Lyneis, In: *Proc. of the 12th International Workshop on ECR Ion Sources*, 24. RIKEN (Tokyo), Japan (1995)

[28] A. G. Drentje and J. Sijbring, KVI Report 996 (1983)

[29] R. Trassl, Ph.D. thesis, Justus-Liebig-University Giessen (1999)

[30] G. D. Shirkov, Plasma Sources Sci. Technol. **2**, 250 (1993)

[31] R. Geller, In: *Atomic Physics of Highly Charged Ions*, E. Salzborn, P. H. Mokler and A. Müller, eds., 117. Springer Verlag (1991)

[32] G. M. et al, Rev. Sci. Inst. **65**, 1051 (1994)

[33] P. Sortais, Rev. Sci. Inst. **63**, 2801 (1991)

[34] C. D. Child, Phys. Rev. **32**, 492 (1911)

[35] I. Langmuir and K. T. Compton, Rev. Mod. Phys. **3**, 251 (1931)

[36] D. Bohm, *The Characterisctics of Electrical Discharges in Magnetic Fields*. McGraw Hill, New York (1949)

[37] I. G. Brown, *The Physics and Technology of Ion Sources*. John Wiley & Sons Inc., New York (1989)

[38] H. Koivisto, *The MIVOC Method for the Production of Metal Ion Beams*. Ph.D. thesis, University of Jyväskylä, Finland (1998)

[39] M. Schlapp, Ph.D. thesis, Justus–Liebig–University, Giessen (1995)

[40] K. Halbach, Nucl. Inst. & Meth. **169**, 1 (1980)

[41] R. Trassl, P. Hathiramani, F. Broetz, J. Greenwood, R. McCullough, M. Schlapp and E. Salzborn, Physica Scripta **T73**, 380 (1997)

[42] B. Wolf, *Handbook of Ion Sources*. CRC Press, New York (1995)

Chapter 2

ELECTRON BEAM ION TRAPS AND THEIR USE IN THE STUDY OF HIGHLY CHARGED IONS

F. J. Currell
School of Mathematics and Physics,
The Queen's University of Belfast,
Belfast BT7 1NN, UK.
f.j.currell@qub.ac.uk

Abstract This chapter presents the basic principles underlying the operation of electron beam ion traps, with the machine physics underlying their operation being described in some detail. Predictions arising from this description are compared with some diagnostic measurements. This is followed by a brief description of their uses in atomic physics.

Keywords: electron beam ion trap, EBIT, ion trap, ion source

1. Introduction

Electron beam ion traps (EBITs) are widely recognised as versatile instruments, capable of performing a wide range of measurements on highly charged ions. A beam of electrons typically as thin as a human hair, with 10^{18} electrons passing through it every second simultaneously creates, confines and probes highly charged ions. The first EBIT was developed by Levine *et al* [1] based on the electron beam ion source (EBIS) developed by Donets and co-workers [2, 3] and related developmental work by Schmieder *et al* [4]. EBITs make use of split Helmholtz coils along a short trap length (typ. 3cm) to facilitate detection of emitted radiation. In contrast EBISes have trap lengths of up to a meter, with the axial field being provided by a single solenoid. The shorter trap length of EBITs was used to ensure certain plasma instablities did not occur [5] although there has been some debate regarding this effect.

F.J. Currell (ed.), The Physics of Multiply and Highly Charged Ions, Vol. 1, 39-75.

Building on the success of the original EBIT, a second device was constructed at LLNL and the original EBIT was upgraded for higher energy operation, being dubbed super-EBIT [6]. Among the range of notable achievements made with the upgraded device, was the creation of bare uranium [7]. With this important milestone, EBIT technology came of age, demonstrating that any ion of any element of the periodic table can in principle be created and trapped with these machines.

It is a tribute to Levine and co-workers that this design is still producing fresh and interesting results more than ten years after its initial realization and has been closely copied elsewhere. Devices constructed along broadly similar lines to the original EBIT include installations in laboratories at Oxford [8], NIST [9] and Berlin [10]. The installations at Tokyo [11] and Frieburg [12] (now operating at Heidelburg) more closely resemble super-EBIT although both incorporate a number of new design features. There have also been a number of less conventional designs, including those which do not use an axial magnetic field [13] or those which use high residual field rare earth magnets [14] to create the axial magnetic field. New machines are currently being developed based along broadly similar principles to those outlined below.

In EBITs, ion creation occurs through electron impact ionization, whereby an electron from the beam is able to liberate a bound electron from one of the trapped ions. Since the ionization potential grows rapidly with increasing nuclear charge (becoming as high as 137keV for the final electron orbiting hydrogen-like uranium), a very high energy electron beam is required. To achieve the ionization rate required to form high charge states, the electron beam is magnetically compressed whilst being accelerated electrostatically to an energy ranging typically from 2keV to 200keV. This beam is launched from a high perveance electron gun sited in a region of zero magnetic field. From here it is accelerated towards the trap region, whilst being compressed by the rising magnetic field, which reaches a peak value of several Teslas at the trap region, where a pair of superconducting Helmholtz coils is situated.

Ion confinement is achieved by the space charge of the electron beam (radial trapping) and the potentials applied to a series of drift-tubes through which the beam passes (axial trapping). Again, a highly compressed electron beam is advantageous since it leads to a deeper axial well, more able to trap the ions.

Electrons can be captured, cause excitation or further ionization, each of which gives rise to characteristic signals. The excitation rate is sufficiently low that usually all of the probing reactions can be considered to start from the ground state although a mix of charge states (referred to as the charge balance) may be present. Usually closed shell, hydrogen-like or bare ions dominate this charge balance although it can to some extent be controlled by choice of the operating conditions. A particular closed shell configuration is usually achieved by setting the beam energy just below the value required to open the next shell.

The trapped ions form a cloud, which surrounds and penetrates the electron beam. Typically, this cloud has a 3cm long, 1mm diameter cigar-like shape, with the highest density along the electron beam axis. The 3cm length is the typical distance between the end drift-tubes both of which are usually positively biased (typ. 0-500V) with respect to the end drift tube. Within this cloud, individual ions are always in motion, some orbiting around the electron beam, others entering and leaving it about once every 10nsec. This motion is determined by a combination of the electron beams space charge and the Lorentz force on the ions due to the axial magnetic field. All these ions move up and down the cigar's length about $1/10^{th}$ as often. Still less frequently (typ. msec timescale), two ions collide, suddenly changing their trajectories. This in turn leads to cross-field difussion (typ. 10msec timescale). The overall effect of cross-field diffusion is to allow the ions to 'hop' across the magnetic field lines to which they would otherwise be pinned. Once these collisions are considered it can be shown [15, 16] that the magnetic field no longer plays any role in long-term trapping of the ions.

In the trap region of an EBIT, a bewildering array of processes occurs. Sometimes an ion will enter the electron beam to emerge having been stripped of one of its electrons. This is due to electron impact ionization, the ion creation process described above. Recombination reactions can occur, whereby an ion can gain an electron, emitting one or more photons of characteristic energy to stabilize the ion, which hence has its charge state reduced by one. In other interactions, no charge-changing will occur, only energy will be transferred; electron impact excitation and subsequent emission resulting in characteristic line radiation, non-excitational scattering resulting in continuum Bremsstrahlung light. Occasionally an ion might have a close enough encounter with a neutral atom or molecule when it can 'steal' at least one electron in a process know as charge exchange. These processes typically occur on time scales greater that 10msec.

The alert reader might have noticed that the timescales associated with the processes form a hierarchy as is shown in Table 2.1. It is through this hierarchy that much of the machine physics can be gleaned. Section 2.3 deals with this machine physics. This material is then used in section 2.4 which concentrates on uses of EBITs with reference to this machine physics.

After a short description of the main component parts of an EBIT, the electron gun and beam it produces are described in section 2.2.1. Some equations governing the properties of the electron beam and the space-charge potential it produces are given. As is described in section 2.3.1, the space-charge potential of the electron beam plays a significant role in determining the ion cloud shape of each trapped species (charge-state and type of ion). Section 2.3.2 describes the dynamics of creation of highly charged ions in EBITs. Associated with each species is a characteristic temperature. The dynamics of these characteristic temperatures is discussed in section 2.3.3. This section then completes a self-

Table 2.1. Heirarchy of typical timescales in an EBIT.

radial motion	axial motion	ion-ion collisions	cross-field diffusion	charge changing reactions	ion escape
10ns	100ns	0.1-1ms	1-10ms	10ms-10s	100ms-hours

consistent description of the machine physics of EBITs through a series of coupled differential equations will have been outlined. These equations can be solved to give the number densities, energy dynamics and electron beam overlap factors of the various species.

2. Device Description

The form of an EBIT can be considered as several separate assemblies, of approximately cylindrical symmetry about a beam axis. These sub-assemblies are listed below, together with a description of their functions.

- Electron gun, to produce a high current laminar flow electron beam suitable for acceleration into the trap region. The outside of the electron gun is encased in soft Iron, to shield the cathode from the magnetic field due to the superconducting magnet.

- Cryostat/trap region, to house the drift tubes and super-conducting coils held at a temperature of around 4 Kelvin by liquid Helium. The drift tubes are in good thermal contact with the cryostat, so they are cooled to a similar temperature, acting as an efficient cryopump for the trap region.

- Electron collector, to safely decelerate and collect the electron beam after it has passed through the trap. This region includes an extractor electrode through which ions can pass, having traveled along the beam axis, when using the machine as a source. Efficient collection is important to ensure that secondary electrons do not escape to be reaccelerated, subsequently colliding with parts of the cryostat/trap region. Such collisions would cause undesirable loading on the main power supply and electron heating, leading to excessive evaporation of the liquid Helium.

Additionally, super-EBITs have a set of transport coils to transport the electron beam over a greater distance, giving a sufficient space to ensure electrical breakdown does not occur and efficient transport of the high energy beam into the collector. Furthermore, to facilitate the high voltage isolation required between the trap and the gun and collector assemblies, accelerator tubes are normally used.

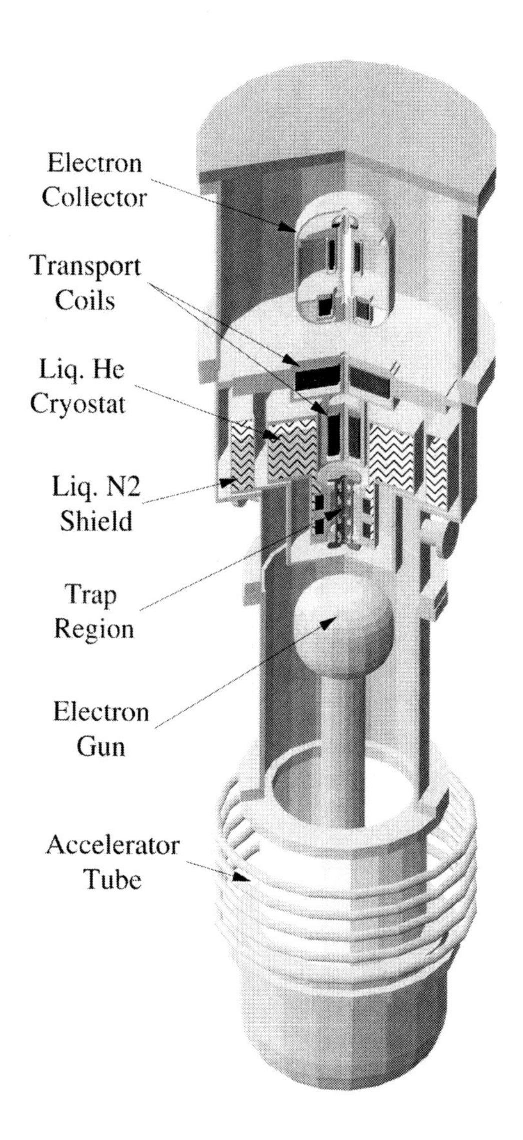

Figure 2.1. Schematic figure of a super-EBIT. Note, the collector support and associated accelerator tube are not shown. Magnetic coils are shown as the darkest regions of the figure

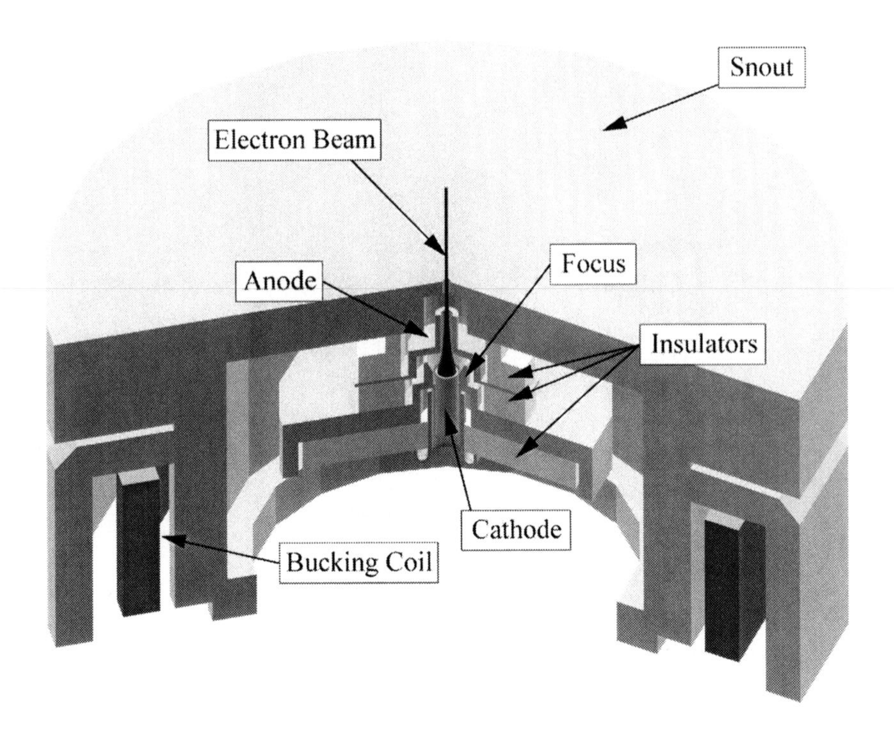

Figure 2.2. Schematic figure of an electron gun for use with an EBIT.

2.1 The Electron Gun and Beam

To achieve the required high current, high energy and high current density, the electron gun of any EBIT is designed carefully, to have a high perveance. The perveance of a particular gun design is a constant relating the space charge limited emission to the electrostatic field at the cathode, having dimensions $A/V^{3/2}$. The cathode is situated in a region where the magnetic field is zero, since this results in the maximum beam compression. In addition to the soft Iron shield surrounding the whole electron gun assembly, a bucking coil is used to ensure the zero field condition is satisfied. In practice, the current through this coil and voltages applied to electrodes in the gun assembly are usually tuned (i.e. empirically adjusted) to achieve a suitable beam condition, as judged by currents landing on various electrodes and the ionization balance created. Fig. 2.2 shows a diagram of the Tokyo-EBIT's electron gun, which consists of the following parts:

- Cathode, a convex spherical surface of a good electron emitter, doped with Barium Oxide which acts as a source of electrons.

- Heater, to heat the cathode and hence facilitate thermionic emission of electrons.
- Focus, to fine-tune the electrostatic field to compensate for edge effects of the spherical cathode, hence allowing the gun to produce a laminar beam over a wide range of electron energies.
- Anode, to create the initial field in the region of the cathode; the anode voltage controls the beam current.
- Snout, to match the electron beam to the field outside the gun.
- Soft Iron casing, to shield the cathode from the magnetic field due to the superconducting magnet as mentioned above.
- Bucking coil, to fine-tune the magnetic field at the cathode.

A zero-temperature plasma of electrons launched from such a cathode would be expected to form a beam with a characteristic radius equal to the Brillion radius

$$r_b[\mu\text{m}] = \frac{150}{B[\text{T}]}\sqrt{\frac{I_e[\text{A}]}{E_e[\text{keV}]}}, \tag{2.1}$$

where I_e is the electron current in the beam, B the magnetic field at the trap and E_e is the electron beam energy at the trap, with the dimensions of the various quantities being given in square barackets. Since the electrons are emitted thermionically, they have a finite temperature. In such a situation, Herrmann theory [17] has been found to give a good prediction for the radius [18, 19] through which 80% of the beam passes as

$$r_h = r_b\sqrt{\frac{1}{2} + \frac{1}{2}\sqrt{1 + 4\left(\frac{8kTr_c^2}{m\eta^2 r_b^4 B^2} + \frac{B_c^2 r_c^4}{B^2 r_b^4}\right)}} \tag{2.2}$$

where r_c is the cathode radius, kT_c is the characteristic electron energy at the cathode, m is the electron mass, η is the electron's charge to mass ratio and B_c is the magnetic field at the cathode. The beam radius can be seen to take on a minimum value when $B_c = 0$, hence care is taken to ensure this condition can be met through the use of a soft Iron shield and bucking coil described above. Any convenient unit system may be used for the quantities in eq. 2.2, provided that the ratios in round brackets are dimensionless. For any EBIT, the first term in round brackets is always much greater than unity. Hence, the equation can be considerably simplified by neglecting the small constants that are added to this term, under the square root signs and enforcing $B_c = 0$ Combining eq.s

2.1 and 2.2 and converting to suitable units then gives the minimum attainable beam radius (i.e. optimum bucking coil tuning) to be

$$r_h[\mu\text{m}] = 260\sqrt{\frac{r_c[\text{cm}]}{B[\text{T}]}\sqrt{kT_c[\text{eV}]}} \tag{2.3}$$

Typically this predicts a beam radius of about 30 μm, consistent with several measurements of this important parameter [18–20]

With an electron density typically between $1 \times 10^{11}/\text{cm}^3$ and $6 \times 10^{11}/\text{cm}^3$, the electron beam is responsible for radial trapping of the ions. Assuming the electron beam has a 'top-hat' like profile with all the charge bounded inside a radial region r_e (typically about equal to the Herrmann radius r_h) lying along the axis of cylindrical symmetry of the machine, the potential is given by Gauss' law to be

$$V_e(\rho) = V_0\left(\frac{\rho}{r_e}\right)^2 \quad if \;\; \rho < r_e \tag{2.4}$$

$$V_e(\rho) = V_0\left(2\ln\left(\frac{\rho}{r_e}\right) + 1\right) \quad if \;\; \rho > r_e. \tag{2.5}$$

Here V_0 is determined by the total charge per unit length of the electron beam and is given (in convenient units, denoted in square brackets) by

$$V_0[\text{V}] = \frac{30I[\text{A}]}{\sqrt{1 - \left(\frac{Ee[\text{keV}]}{511} + 1\right)^{-2}}} \tag{2.6}$$

where $I[\text{A}]$ is the total beam current in amps and $E_e[\text{keV}]$ is the electron beam energy. V_0 has a value of between 5V and 10V for a typical EBIT set up. It is noteworthy that since V_0 does not depend on the electron beam radius, neither does the potential energy required to move an ion from the centre of the beam to the edge. The form of eq. 2.5 however is such that the potential required for an ion to reach the drift tube wall (typically situated 5mm from the beam axis) does depend on the electron beam radius.

3. Machine Physics

It is usual to make several assumptions to give rise to a tractable and reasonably accurate description of the machine physics of EBITs. These assumptions are listed below and will be justified as the description is developed.

- Cylindrical symmetry of trap environment.
- Square shaped axial potential.

- Each charge state of trapped ions is characterised by a single temperature applicable to all degrees of freedom.
- Trap and ion distribution are much longer than they are wide.

The last of these assumptions is the only questionable one, being valid only when the depth of the axial potential is less than about 150V as is discussed later.

The central concepts underlying this machine physics and their inter-relations are summarised in fig. 2.3.

- Electron beam: already described above, it is responsible for ion creation (proportional to the current density, J_e) and radial trapping, through $V_e(\rho)$, which determines ion spatial distributions. An unavoidable but unwanted side effect related to the electron beam is ion heating, through the Ladau-Spitzer and ionization heating mechanisms described below.
- Ion spatial distributions: distributions describing the locations of ions of the trapped charge states. From these distributions, the overlap with the electron beam, $f_{e,i}$, or other ion species, $f_{i,j}$, are determined. Such overlap factors act to scale the rates of various processes which lead to change of charge or energy or both. Furthermore, work must be done to change the spatial distributions since this involves transport of ions from one potential to another. The energy to do this work is accounted for within the framework of the energy dynamics.
- Charge dynamics: processes which lead to trapped ions being created or changing their charge state through atomic physics reactions or escaping (usually axially) from the trap. Hence, these processes affect the total number of ions of any charge state present. As is described below, the number densities on the axis, N_i, are used to describe this dynamics and to normalize the corresponding ion spatial distribution. These become axial number densities per unit length once we invoke the assumption that the ion cloud is much longer than it is wide. Once the space-charge of the ions is considered, the N_i affect the ionic space-charge potential which is an extra term in the radial potential determining the spatial distributions. As ions change their charge states, there is also a corresponding (small) change to the temperature associated with each species. For example, if colder, more lowly charged ions are ionised to higher charge states, there will be a net reduction in the temperatures associated with the higher charge states, as the new (formerly lowly charged) ions form part of the ensemble.
- Energy dynamics: processes which change the characteristic temperature, kT_i, of ions of a particular charge state. The units of temperature

normally used (and used throughout this article) are eV as this simplifies expressions somewhat when comparing the temperature to the potential experienced by the ions. The temperature of the ions directly affects the spatial distribution, with hotter ions spending less of their time close to the beam axis. Dominant processes are the electron beam heating mechanisms and cooling through axial escape, which happens faster for higher kT_i. Ion-ion collisions give rise to energy sharing between the different charge states.

The physics underlying these processes is described in more detail below.

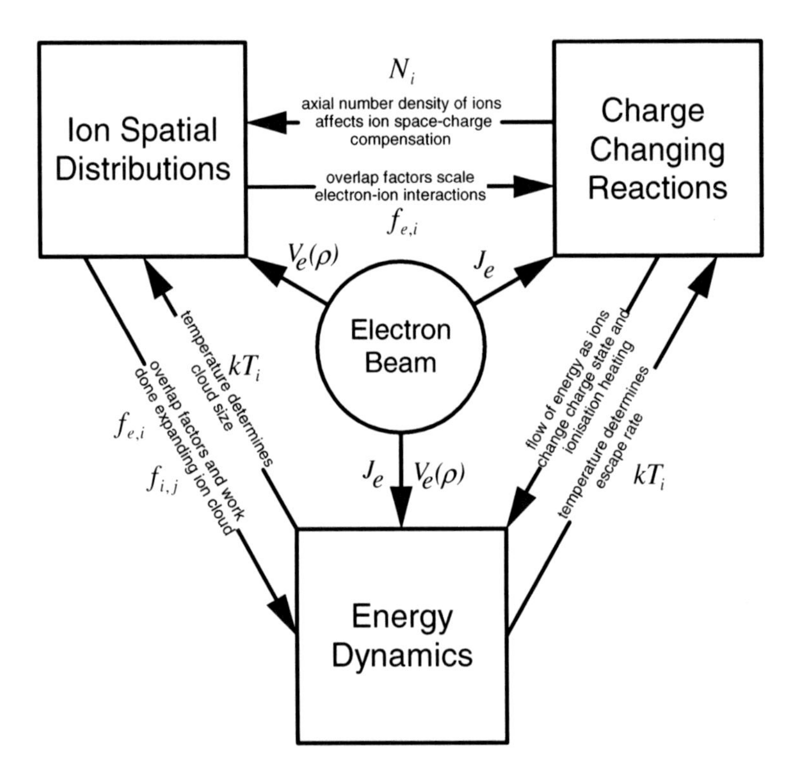

Figure 2.3. Summary of relationships between concepts underlying the machine physics of EBITs. Arrows show causal links between the various process concerned with the associated variables being shown alongside the arrows.

3.1 Trapping and the Ion Cloud Shape

The essentials of the trapping environment is shown in fig. 2.4. The motion of a single ion in the trap is shown schematically in fig. 2.5. The motion results from forces on the ion due to the axial magnetic field, the space charge of the electron beam and the charges residing on the drift-tube walls. These charges

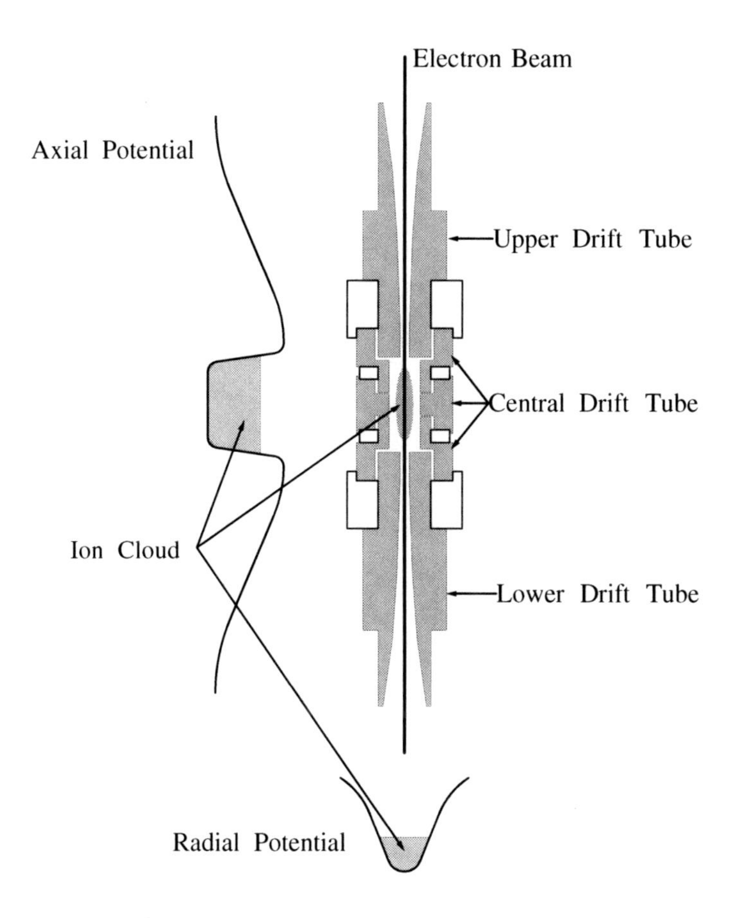

Figure 2.4. Schematic figure of the trapping environment of the Tokyo EBIT. Insulators are shown by the white regions of the figure, metal by the darkest regions. The trapped ions are schematically shown as the light gray regions. The whole trap has cylindrical symmetry. Usually, the three central metal electrode are shorted, to act as a single drift tube.

residing on the their walls are predominantly due to the voltages applied to the drift-tubes. The trapping effect due to these charges can be described in terms of an electrostatic potential which is the solution of Laplace's equation with boundary conditions determined by the shapes of and voltages applied to the drift-tubes.

Many ions undergo motion similar to that shown in fig. 2.5, occasionally colliding with one another to give rise to an ion cloud with an associated density distribution. In practice there is a different density distribution for each charge state because the ratio of charge to temperature (q_i/kT_i) can be different for different values of i.

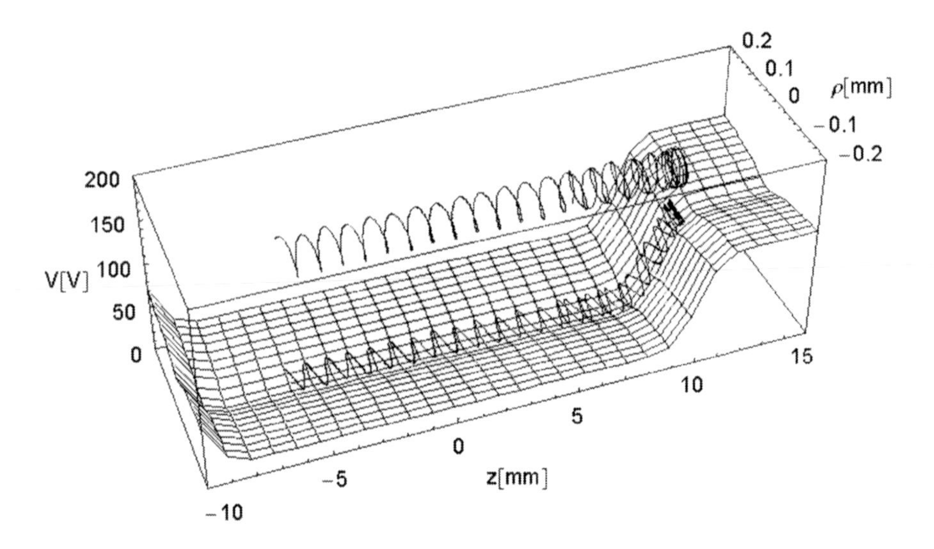

Figure 2.5. Schematic figure of the motion of a single trapped ion. The electrostatic trapping potential is indicated by the surface at the bottom of the figure. Note the different scales for the radial and axial axes. The small axial depression in the trapping potential due to the electron beam's space-charge. Many such ions are trapped, giving rise to an ion cloud and density distribution for each charge state present.

The ions are trapped axially by applying suitable voltages to a series of cylindrical drift tubes comprising the trap. This gives rise to the well shaped structure observed in fig. 2.5. Variations in the specific drift tube structure occur from device to device but the conventional mode of operation is where the central drift tube (comprised of three separate electrodes in fig. 2.4) is negatively biased by a few volts to a few hundred volts with respect to the outer drift tubes. Radial trapping occurs through the space charge potential of the electron beam, as given by eq.s 2.4 and 2.5 and is illustrated somewhat schematically at the bottom of fig. 2.4. Axial trapping occurs for all positive ions with a sufficiently low axial kinetic energy (i.e. $p_z^2/2m$ where p_z is the ions momentum component along the beam axis and m its mass) to be 'reflected' between the two drift-tube potentials. If a voltage V_{app} is applied to the outer drift tubes with respect to the central drift tube, the trap depth is approxiamtely given by

$$V_t \approx V_{app} + (V_e(r_{1,3}) - V_e(r_2)) \tag{2.7}$$

where $r_{1,3}$ is the radius of the inner wall of the two end drift tubes and r_2 is the same quantity for the central drift tube. The second term represents a small offset (typically $\sim V_0$) due to the fact that the electron beam induces an image charge on the drift tube walls. If the trap is configured with voltages applied asymmetrically (e.g. for ion extraction) then the lowest of the two V_t values is the one used since ions make many passes along the trap between energy-changing events (such as collisions) and so can reach this minimum barrier once they have sufficient axial kinetic energy (i.e. $p_z^2/2m > eqV_t$).

It is useful to assume that the ions of each trapped charge state can be assigned a single temperature but reasonable to treat the temperature of each charge state as distinct. This assumption can be justified in terms of table 2.1 as is discussed in section 2.3.3 below. The trapping occurs through three mechanisms discussed below.

- Trapping due to the magnetic field: Single particle Lagrangian analysis [21] shows the magnetic field contributes a term to the effective potential. However a more complete many particle analysis in terms of the Valasov equation [15, 16] shows that cross-field diffusion gives rise to ion-ion collisions on a time scale of typically 10msec. This cross-field diffusion negates the trapping effect of the magnetic field under usual EBIT operation. This trapping mechanism is however important for experiments where the electron beam is turned off suddenly [22], as the magnetic field acts to retard the expansion of the ion cloud somewhat.

- Potential due to voltages applied to the drift tubes: The ions are trapped axially by the potentials applied to the drift tubes (i.e. the central one is biased negatively with respect to the two ends). The drift-tubes are deliberately shaped so as to produce a nearly square-shaped potential along the beam axis so to a good approximation, the axial ion density distribution is uniform in the space between the two end drift tubes (3cm typically). The depth of this axial potential determines the rate of axial escape of the ions and hence controls the predominant cooling mechanism as is disussed in section 2.3.3 below.

- Space charge of the electron beam: Radial trapping is by the potential due to the space charge of the electron beam as given by eqs. 2.4 to 2.6. From this potential and the characteristic temperature of any trapped species it is possible to calculate the radial distribution and hence the proportion of ions of that species inside the electron beam.

Since only electrostatic terms have an effect, the equilibrium spatial distribution of the ions of charge state i and characteristic temperature kT_i is given by a Maxwellian distribution

$$n_i\left(\rho, z\right) d\rho dz = N_i 2\pi\rho e^{\left(-\frac{q_i V(\rho,z)}{kT_i}\right)} d\rho dz, \tag{2.8}$$

where $V\left(\rho, z\right)$ is the electrostatic trapping potential and N_i is the axial density (strictly a function of z). Conveniently for analysis of the spatial distribution of the trapped ions, the trapping potential is separable into a radial and axial term to a good approximation. The form of Laplace's equation means that there is a coupling term at the ends of the drift tubes. However, since the ion cloud is much longer than its characteristic radius (one of our assumptions), this coupling term is unimportant and will be neglected. The seperability of the trapping potential in turn implies seperability of the spatial distribution. As discussed above, the axial distribution is approximately uniform between the drift tubes. Hence the axial dimension can simply be integrated over in eqs. 2.8, which is equivalent to dropping the variable z from the analysis with a redefinition of terms. This leads to a first approximation (i.e. ignoring the effect of the space-charge of the trapped ions and end effects), for the radial spatial distribution

$$n_i(\rho)\,d\rho = N_i 2\pi\rho e^{\left(-\frac{q_i V(\rho)}{kT_i}\right)} d\rho, \tag{2.9}$$

where N_i is now the axial number density of ions of species i **per unit length** along the trap and $V(\rho)$ is the potential due to the space charge of the electron beam, as given by Eqs. 2.4 to 2.6.

Once the space charge of the trapped ions is considered, the description is no longer possible in a closed form. Once again, the fact that the ion cloud is long compared to its radius however gives rise to simplification through reduction of the problem's dimensionality and the use of Gauss's theorem. The potential used in Eq. 2.9 now contains a term describing the potential due to all of the trapped ions. This potential is determined in a self-consistent manner from the density distributions. The potential becomes

$$V(\rho) = V_e(\rho) + \sum_i V_i(\rho) \tag{2.10}$$

where $V_e(\rho)$ is the potential due to the space charge of the electron beam as as given by eqs. 2.4 to 2.6 and $V_i(\rho)$ is the potential due to the space charge of ions of charge state i. $V(\rho)$ must now satisfy Poisson's equation [16]

$$\frac{1}{\rho}\frac{\partial}{\partial\rho}\left(\rho\frac{\partial V(\rho)}{\partial\rho}\right) = \left(\{\rho < r_e\}\left(\frac{4V_0}{r_e^2}\right) - \frac{4\pi}{\varepsilon_0}\sum_i e q_i N_i \exp\left(-\frac{q_i V(\rho)}{kT_i}\right)\right), \tag{2.11}$$

where $\{\rho < r_e\}$ is interpreted to take the value 1 for $\rho < r_e$ and zero otherwise (i.e. a top-hat shape). Other forms of electron beam profile can be modeled by changing the term $\{\rho < r_e\}\left(\frac{4V_0}{r_e^2}\right)$. Notice this is the radial charge distribution of the electron beam. One simply replaces it with the appropriate distribution function to model a different electron beam distribution. For other realistic beam profiles (e.g. Gaussian) the results are similar to those reported here. By imaging visible light produced by the decay of long-lived states it is possible to directly observe the ion cloud shape [23].

For a known set of N_i and kT_i eq. 2.11 can be solved for $V(\rho)$ by integrating numerically away from the axis $\rho = 0$. These requirements for solving the right hand side of eq. 2.11 correspond to the two 'causal link' arrows pointing in to the box labeled 'Ion Spatial Distributions' in fig 2.3. The dominant effect in determining the ion cloud shapes is due to the space charge of the electron beam, denoted by the term containing $V_e(\rho)$ in fig 2.3 and represented in the right hand side of eq. 2.11 by the term $\{\rho < r_e\}\left(\frac{4V_0}{r_e^2}\right)$. Note that due to this term, if all the N_i are set to zero, then solving eq. 2.11 will lead to eq. 2.4 and eq. 2.5.

Once $V(\rho)$ is determined, the ion cloud radial distributions can be calculated from eq. 2.9 from which electron-ion overlap factors can be determined. For species i, the total number of ions in the trap N_i^{tot}, and the total number in the beam N_i^{in} are given by

$$N_i^{tot} = \int_0^{r_{dt}} 2\pi\rho \exp(-(\frac{q_i V(\rho)}{kT_i}))\, d\rho \tag{2.12}$$

and

$$N_i^{in} = \int_0^{r_e} 2\pi\rho \exp(-\left(\frac{q_i V(\rho)}{kT_i}\right))\, d\rho, \tag{2.13}$$

where r_{dt} is the central drift tube radius. Hence the electron-ion overlap factor is the ratio of ions inside the electron beam to ions in the trap and is given by

$$f_{e,i} = \frac{N_i^{in}}{N_i^{tot}} \tag{2.14}$$

In summary, hotter ions spend less time in the electron beam and hence are subject to lower rates for electron-ion interactions. The trapped ions tend to compensate for the space charge of the electron beam which acts to trap them. Provided the axial trap depth is shallow enough (typ. a few times V_0), the more highly charged ions are confined inside the electron beam and the space charge of the electron beam is only slightly compensated by the trapped ions.

3.2 Charge Evolution

Inside the trap, we find electrons, trapped ions in a range of charge states and neutral gas (either introduced deliberately or as an inevitable background) interacting with one another. The array of charge states and processes involved can be represented as a 'staircase' as shown in figure 2.6. The electron beam

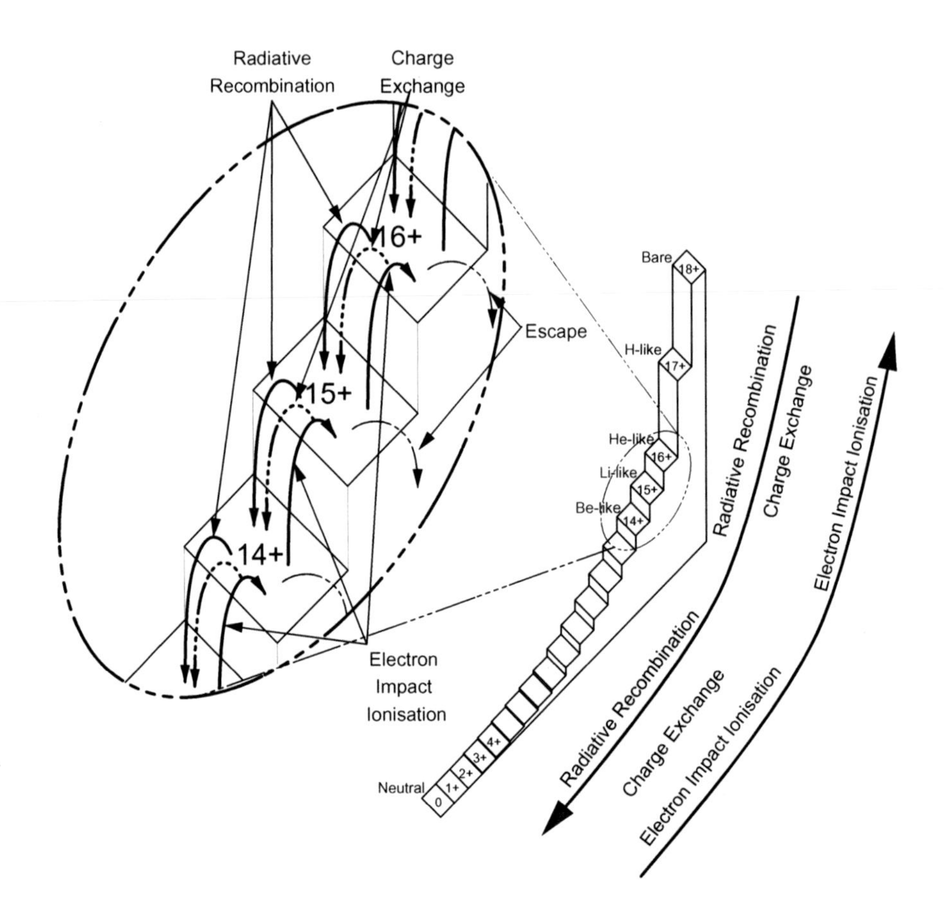

Figure 2.6. Schematic representation of the coupling of neighbouring charge states due to atomic physics processes. The height of each step is proportional to the ionization potential required to form that charge state to emphasise the increased difficulty in creating the higher charge states. The insert shows the couplings between neighbouring charge states and the escape process.

causes electron impact ionization (EI) (driving ions up the stairs) and radiative recombination (RR) (bringing them down). Further reduction of charge state occurs through charge-exchange (CX) caused by ion-neutral collisions, although this is accompanied by creation of a new singly charged ion. Double ionization and charge transfer generally have cross sections at least an order of magnitude less than there single counterparts so these processes can be neglected. In the resultant picture, neighbouring stairs on the staircase couple to each other through various processes and to creation of singly charged ions through charge exchange.

Additionally, ions from each charge state can escape from the trap as is described in section 2.3.3. Hence the time evolution of trapped ions can be described by a series of coupled differential equations, one for each charge state

$$\begin{aligned}\frac{dN_i}{dt} &= \frac{J_e}{e}\left(N_{i-1}\sigma_{i-1}^{EI} f_{e,i-1} + N_{i+1}\sigma_{i+1}^{RR} f_{e,i+1} - N_i(\sigma_i^{EI} + \sigma_i^{RR}) f_{e,i}\right) \\ &+ N_0(N_{i+1}\langle v\sigma_{i+1}^{CX}\rangle - N_i\langle v\sigma_i^{CX}\rangle) \\ &- N_i R_i^{ESC}.\end{aligned} \tag{2.15}$$

The first line of eq. 2.15 describes the interactions with the electron beam so all of the cross sections in this term are functions of the electron beam energy. Hence the term scales linearly with the current density $\frac{J_e}{e}$ and describes all of the interactions denoted by 'solid arrows', joining adjacent steps in fig. 2.6. Ions of the charge state i can in general be created by ionising charge state $i-1$ or recombination into charge state $i+1$, giving rise to the first two terms in the following bracket. Since these both lead to creation of charge state i, they are proceeded by a positive sign and correspond to 'solid arrows' leading onto step i from the two adjacent steps in fig. 2.6. The next two terms describe annihilation of charge state i, again through either EI or RR and accordingly are proceeded by a minus sign, corresponding to 'solid arrows' leading away from step i onto adjacent steps in fig. 2.6.

The second line of eq. 2.15 describes charge exchange (CX) processes through interactions with neutral species. Hence the term scales linearly with the neutral density N_0 and corresponds to the 'thick dashed arrows' in fig. 2.6. Again, in general two rates occur, to describe creation and annihilation of ions of charge state i. Charge exchange cross sections are averaged over the velocity distribution of ions of species i (hence the dependence on kT_i) to give a rate of the form $\langle v\sigma_i^{CX}(kT_i)\rangle$. The neutral background gas is considered to be uniform and is effectively stationary, having a much lower temperature that the trapped ions.

Eq. 2.15 describes the general case but special cases exist where some of the terms vanish or require modification. For example, a source term must be included to describe injection of either neutrals or ions of low charge state into the trap. Furthermore, ionization beyond bare ions or radiative recombination into bare or from neutrals can not occur. If the beam energy is tuned to be below the ionization potentials of some charge states, then some of the ionization cross sections are zero so higher charge states will not be created. Since opening a new shell gives rise to a large increase in the ionization potential, this usually means that the final charge state created corresponds to a closed shell configuration.

The final term of eq. 2.15 represents escape from the trap, corresponding to the solid 'thin dashed arrows' in fig. 2.6. Fussmann *et. al.* [16] give the axial rate of escape of ions to be

$$R_i^{ESC} = \frac{3}{\sqrt{2}}\nu_i \frac{e^{-\omega_i}}{\omega_i} \tag{2.16}$$

where ν_i is the Coulomb collision rate for ions of charge state q_i with all ions and ω_i is given by

$$\omega_i = \frac{q_i V_t}{kT_i}, \tag{2.17}$$

where V_t (as given by eq. 2.7) is the shallowest of the two end axial trap depths over which an ion must pass to escape .

Although the charge balance dynamics has been described briefly, it should be clear that in principle a series of differential equations can be set up to describe the time evolution of the ions in the trap. In conjunction with the equations governing the energetics of the ions (below) and those for the overlap factors (above), these equations can be solved to predict the time dependent behaviour of the device [24, 25].

Even without solving these differential equations fully, one can glean a lot of useful information from simplified analysis of their form. For example, suppose the trap is initially empty due to one of the outer drift-tubes being biased negative with respect to the central one (i.e. 'open') and then it is suddenly 'closed' by raising the outer drift-tube to a potential above the central one. Ions will start to accumulate, either sourced from a background of neutral gas or from (for example) singly charged ions injected just before the trap is closed. For times soon after the closure of the trap, the dominant process is ionization as the ions progressively climb the staircase of fig. 2.6. In this case, all of the other reactions can be neglected and the electron-ion overlaps $f_{e,i}$ can be considered to be unity (i.e. all the ions of all the charge states are inside the electron beam). In this case, the characteristic time for the appearance of charge state i is given by

$$t_i^{app} = \frac{1}{J/e}\sum_{j=1}^{i}\frac{1}{\sigma_i^{EI}}. \tag{2.18}$$

In practice, the ions may attain a high enough temperature before this time to reduce the overlap factor significantly below unity or the other reactions may play a role, particularly when very high charge states are being created. These effects all act to increase the characteristic appearance time t_i^{app} so eq. 2.18 can be considered to provide a lower bound.

If the trap is constantly filled due to ionization of a background of neutral gas (i.e the system has a constant source term) then eventually an equilibrium condition will be reached, where the left hand side of eq. 2.15 is zero for every charge state. Escape tends to occur predominantly for lower charge states as is outlined in section 2.3.3 below. Furthermore, for a shallow axial trap depth, all ions of the higher charge states tend to be inside the electron beam (i.e. $f_{e,i} \approx 1$). Furthermore, provided a sufficiently low neutral gas injection rate is used, the rate for charge exchange is much less than the rate for radiative recombination. Hence for the higher charge stages present, one can write

$$\tfrac{dN_i}{dt} = 0 \approx \tfrac{J_e}{e}\left(N_{i-1}\sigma_{i-1}^{EI} + N_{i+1}\sigma_{i+1}^{RR} - N_i(\sigma_i^{EI} + \sigma_i^{RR})\right). \quad (2.19)$$

Neglecting charge exchange and escape, this equation can be used to predict the relative abundance of two highest adjacent charge states at equilibrium (the highest being labelled i_{max}) to be related by

$$\tfrac{dN_{i_{max}}}{dt} = 0 \approx \tfrac{J_e}{e}\left(N_{i_{max}-1}\sigma_{i_{max}-1}^{EI} - N_{i_{max}}\sigma_{i_{max}}^{RR}\right), \quad (2.20)$$

since there are no ions of charge state $i_{max} + 1$ and no ionization of ions of charge state i_{max}, from which it follows that

$$\frac{N_{i_{max}-1}}{N_{i_{max}}} \approx \frac{\sigma_{i_{max}}^{RR}}{\sigma_{i_{max}-1}^{EI}}. \quad (2.21)$$

For the next lowest charge state, the equilibrium equation is

$$\begin{aligned} \tfrac{dN_{i_{max}-1}}{dt} = 0 \approx\ & \tfrac{J_e}{e}(N_{i_{max}-2}\sigma_{i_{max}-2}^{EI} + N_{i_{max}}\sigma_{i_{max}}^{RR}) \\ & -\tfrac{J_e}{e}(N_{i_{max}-1}(\sigma_{i_{max}-1}^{EI} + \sigma_{i_{max}-1}^{RR})). \end{aligned} \quad (2.22)$$

Rearranging this and substituting eq. 2.21 leads to

$$\frac{N_{i_{max}-2}}{N_{i_{max}-1}} \approx \frac{\sigma_{i_{max}-1}^{RR}}{\sigma_{i_{max}-2}^{EI}}. \quad (2.23)$$

This process can be repeated down the charge ladder, becoming ever more approximate, giving the general result

$$\frac{N_{i-1}}{N_i} \approx \frac{\sigma_i^{RR}}{\sigma_{i-1}^{EI}}. \quad (2.24)$$

It is emphasised that this result relies on the charge exchange and escape rates being negligible, which may not always be the case. Checks can be used to ensure these conditions are fulfilled when a closely related result is central to

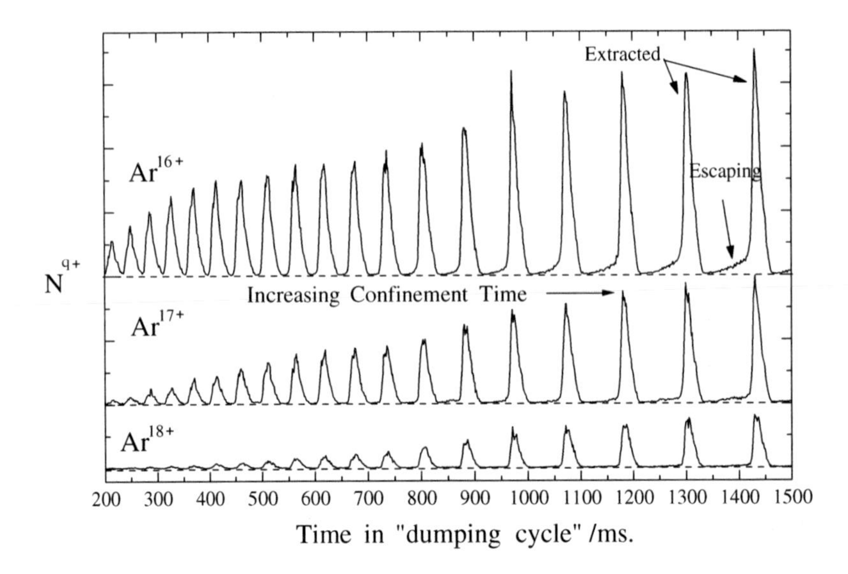

Figure 2.7. Detected yield of ions after charge analysis, during a sequence of successively less frequent dumps of the trap. Ions due to the dumping process form the combs of peaks. Ions detected between these peaks are due to axial escape. This data was taken with a 40.5keV, 100mA electron beam, with a constant background flux of neutral argon being injected throughout.

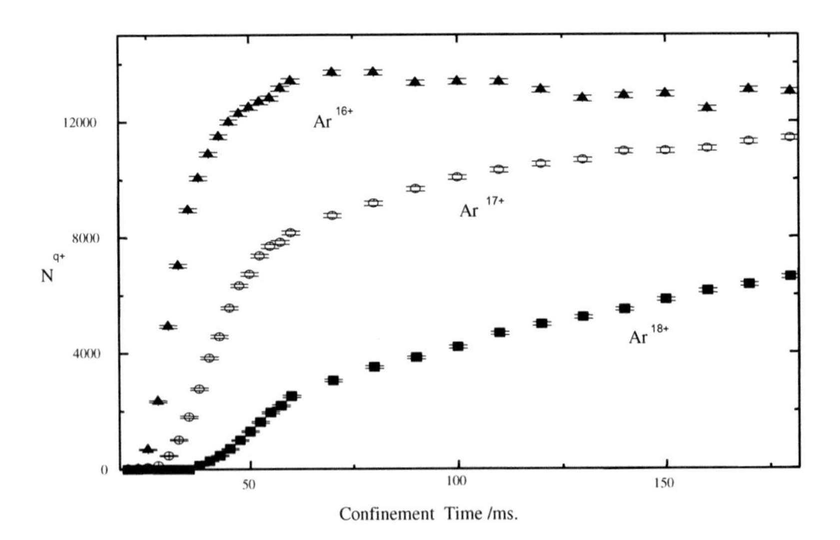

Figure 2.8. Relative abundance of trapped ions as a function of time from starting trapping for a 40.5keV, 100mA electron beam, with a constant background flux of neutral argon being injected throughout. The successive onsets of the various charge states and the approach to equilibrium can be seen.

a technique for measuring ionization cross sections as described elsewhere in this volume.

In summary, the charge balance is determined by a number of factors but essentially the charge states are successively created by stepwise ionization, each higher charge state appearing in turn. Eventually the creation rates (for a constant source term) are balanced by the escape and annihilation rates of each charge state, when the system is in equilibrium. It is possible to measure this charge balance dynamics by dumping the trap inventory into a beam line for subsequent detection. The beam line is equipped with a charge analysing magnet to separate the diferent charge states present. Doing this for progressively lengthening times between dumps gives rise to a data set as shown in fig. 2.7.

Integrating the areas under the peaks for the successively less frequent dumps gives rise to a data set showing the onset of ion creation as is shown in fig. 2.8. As expected the various charge states successively onset, with lower charges being created first. Eventually the system tends towards an equilibrium condition (not attained here for Ar^{17+} and Ar^{18+} until much later).

3.3 Energetics of the Trapped Ions

Previously, the existence of a single temperature for each charge state has been assumed although the temperatures for the various charge states have been considered distinct. It may not immediately be clear why this assumption is made until the relevant time scales are considered. As shown in table 2.1, transport of ions about the trap occurs more rapidly than any processes involving energy exchange. Hence all the ions of the same charge state should have the same energy distribution, regardless of where they are in the trap. Ion-ion collisions within the trap then give rise to Maxwellian energy distributions, characterised by a single temperature for each charge state. However, the rate of energy exchange between the various charge states (predominantly mediated by ion-ion collisions) is not high enough to equilbriate different change states so the temperature of each charge state can be distinct.

There are several processes involved in determining the temperature dynamics of each trapped ion species. The dominant processes are listed below.

- Electron beam heating: Called Landau-Spitzer or Spitzer heating in the plasma physics community, this is a process which occurs through numerous long-range collisions which occur between electrons from the beam and trapped ions. It has a rate given by

$$\frac{dkT_i}{dt} \approx f_{e,i} \frac{8\pi \left(J_e/e\right) q_i^2 e^2 ln\Lambda_i}{3m_i}. \tag{2.25}$$

Here kT_i is the temperature, $f_{e,i}$ is the electron-ion overlap and q_i is the charge of species i. J_e is the current density and $\ln \Lambda_i$ is the Coulomb logarithm of species i, a parameter describing the length scale over which the electron beam appears 'grainy' rather than as a smooth fluid. In practice, the logarithm takes on a value of approximately 10 but more exactly it can be evaluated using either the formula found in plasma physics texts [26] or the same formula with the Debye radius replaced with the electron beam radius, if it is smaller. Since this heating has a q_i^2 dependence, it is stronger for the more highly charged ions which hence are expected to be hotter.

- Charge-changing reaction heating: Inside the trap, there are reactions occuring such as electron impact ionization and radiative recombination, which lead to a change of the charge state of an ion. These reactions occur at a range of positions across the electron beam and hence at a range of different values of potential. The simplest example to understand and the most significant in the temperature dynamics is ionization heating. Imagine a low temperature neutral gas background uniformly penetrating the electron beam. After single ionization by the beam, each ion will find itself at a different electrostatic potential ranging from 0 volts to V_0 volts. Subsequent motion will cause this energy to be shared among the available degrees of freedom, leading to a net increase in temperature. Assuming that the ionising reactions occur uniformly across the whole cross section of the electron beam, each ionising reaction will (on average) result in an increase in the energy associated with each ion of

$$\Delta \mathrm{V_{av}} = \frac{1}{\pi \mathrm{r_e^2}} \int_0^{\mathrm{re}} \mathrm{V_0} \left(\frac{\rho}{\mathrm{r_e}} \right)^2 2\pi \rho d\rho = \frac{\mathrm{V_0}}{2}. \tag{2.26}$$

Due to ion-ion collisions, this $V_0/2$, will be shared among the available degrees of freedom, as dictated by the principle of equipatition. For cold ions, motion occurs in a simple harmonic potential (since $\rho < r_e$) for the two coordinates perpendicular to the beam axis. This is analogous to an ion in a crystal lattice, so for each of these coordinates there are two degrees of freedom for energy storage, corresponding to potential and kinetic energy. In contrast, the axial potential is like a square-well (analogous to a gas atom in a box) so there is one degree of freedom, associated with the kinetic energy. These five degrees of freedom each store $\Delta kT/2$ of the energy. Hence, the average change in temperature due to each ionization event is given by $\Delta kT/2 = \Delta V/5 = V_0/10$ or $\Delta kT = V_0/5$. Note this is per ion ionised, independently of how quickly the ionization occurs. Hence, this effect occurs even in the limit $t \to 0$ [27].

- Power supply heating: Power supplies can never be totally free from ripple. Some portion of this ripple will be resonant with frequencies associated with characteristic ion motions about the trap. The result is that the ions absorb radiation at these frequencies, gaining kinetic energy and hence increasing in temperature.

- Evaporative cooling: As discussed in the section on charge evolution above, the more energetic ions can escape along the beam axis. Since the hotter ions escape preferentially, there is a net lowering of the ion temperature. This process can be deliberately enhanced by introducing a light gas (typically Ne or N_2) which is rapidly ionised to its bare state when it acts to efficiently cool the more highly charged ions under study as given by eq. 2.16. This rate acts so as to limit the equilibrium temperatures of the ions to be typically 0.1 to 0.4 $q_i V_t$ (i.e $\omega_i \approx 2.5$ to 10). Since lower charge state ions can escape most easily, this process tends to cool them, with the cooling effect being transferred to higher charge states through ion-ion collisions (i.e. ion-ion energy exchange).

- Ion-ion energy exchange: Collisions between the trapped ions lead to energy exchange, much as in a gas. Since the particles are charged, they interact Coulombically, and hence the collisions have a longer range. The characteristic time for Coulomb collisions of ions of species i interacting with those of species j is given by [28]

$$\tau_{i,j} = \frac{25.8\sqrt{\pi}\varepsilon_0^2 m_j k T_j^{3/2}}{q_i^2 q_j^2 N_j \ln\Lambda_{i,j}} \tag{2.27}$$

  where $\Lambda_{i,j}$ is the ion-ion Coulomb logarithm. The entry for the ion-ion collision time given in table 2.1 was derived from this equation, using an ion density of 10^8/cm^3 (a reasonable number for an EBIT). When scattering occurs by more than one type of ion, the appropriate frequencies $\tau_{i,j}^{-1}$ must be summed to give a characteristic frequency from which the total collision time is determined. For species i and j, this frequency will have a $q_i^2 q_j^2$ dependence, meaning that the coupling between higher charge states is strongest. This means that they tend to share energy most rapidly, and hence have similar temperatures.

- Energy flow due to charge-changing reactions: Ions of neighbouring charge states have slightly different temperatures. Hence, when ions change charge state through the processes discussed in section 2.3.2, on average, they join the new ensemble with a new characteristic temperature. This energy flow acts to further couple the temperatures of neighbouring charge states slightly.

- Work done expanding the ion cloud: The average potential energy associated with ions of species i is given (in dimensions of eV, the same as the potential) by

$$\langle U_i \rangle = \frac{1}{N_i^{tot}} \int_0^{r_{dt}} 2\pi\rho\, q_i V(\rho) \exp(-(\frac{q_i V(\rho)}{kT_i}))\, d\rho \tag{2.28}$$

when the ion cloud expands, work must be done as the ions climb up the potential. The heat capacity of ions of species i is then given by

$$\begin{aligned} C_v &= \frac{d\langle U_i \rangle}{dkT_i} + 3/2 \\ &= \frac{1}{N_i^{tot}} \int_0^{r_{dt}} 2\pi\rho\, \frac{{q_i}^2 V(\rho)^2}{kT_i^2} \exp(-(\frac{q_i V(\rho)}{kT_i}))\, d\rho + 3/2 \end{aligned} \tag{2.29}$$

The first term in this expression corresponds to the work done in changing the size of the ion cloud, the second term corresponds to the change in kinetic energy.

In summary, the dominant effects in determining the temperature dynamics of the ions are electron beam heating, ionization heating (at short times after trapping is started), evaporative cooling and ion-ion energy exchange. Since ion-ion heating occurs with an approximately constant value per ionization reaction and electron beam heating is proportional to q_i^2, the ions with highest charge states are expected to be hottest. Since however ion-ion energy exchange occurs most efficiently between high charge states (frequency proportional to $q_i^2 q_j^2$), the difference in temperature between neighbouring charge states is expected to decrease as the charge increases. Eventually the effects of the heating mechanisms are balanced by evaporative cooling when the system attains a dynamic thermal equilibrium with different charge states being heated and escaping at different rates. Hence even in this equilibrium situation, they have different temperatures.

Since the temperature of the ions is reflected in the axial escape rate, it is possible to infer details of the temperature dynamics from data such as that shown in fig. 2.7 [27]. This is achieved by comparing the relative numbers and escape rates for two adjacent charge states and assuming they have a similar temperature. fig. 2.9 shows the measured temperature dependence of Ar^{17+} and Ar^{18+} ions as a function of time after the trap was closed. Even in the limit $t \to 0$ the measured temperature is not zero due to the presence of ionization heating [27]. Using the machine parameters, this temperature is predicted to be 23eV, in good agreement with the observation. The temperature of the ions

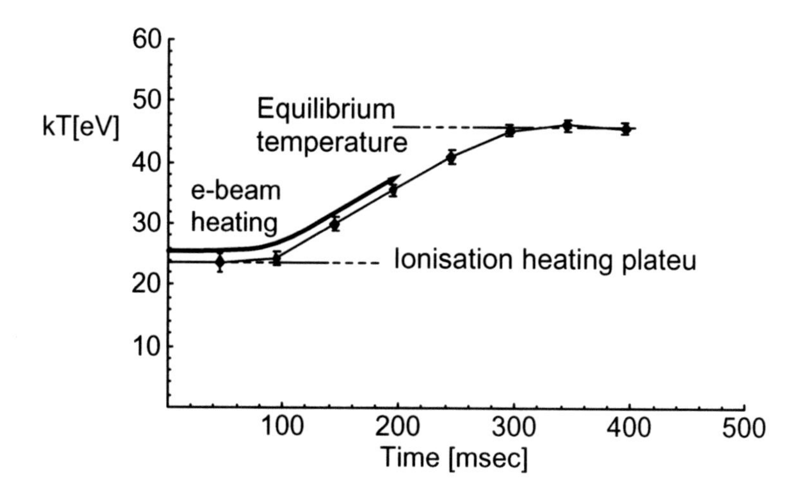

Figure 2.9. Measured temperature of Ar^{17+} and Ar^{18+} ions as a function of time after the trap was closed (i.e. ion creation initiated). This data was taken with a 20keV, 60mA beam compressed by an axial magnetic field of 4T at the trap and with a 10V trapping potential applied axially, corresponding to ($V_t \approx 16V$).

subsequently increases due to electron beam heating until it eventually attains an equilibrium value.

By further modeling the temperature as a function of charge state by

$$kT\,(t; q_i) = a\,(t) + b\,(t)\,q_i + c\,(t)\,q_i^2 \tag{2.30}$$

it is possible to deduce the full temperature dynamics of the trapped ions [27]. This was found to be the lowest order polynomial able to account for the measurements. In this way the time dependence of temperature for a range of charge states was deduced as shown in fig. 2.10. The conditions were the same as for fig. 2.9. Notice the temperatures increase with time and with charge state as expected. Furthermore, the increase with charge state becomes less for higher charge states, again in line with expectations.

From this data, we can justify our final assumption that the ion cloud is much longer than it is wide. Such data or simulations of EBITs generally confirms the assumption that the equilibrium temperatures of the ions to be typically 0.1 to 0.4 $q_i V_t$ (i.e $\omega_i \approx 2.5$ to 10). A 'characteristic radius' can be defined as the radius at which the radial potential is equal to the ion temperature. Ignoring the space charge of the trapped ions (i.e. the 'uncompensated limit') the characteristic radius is given by

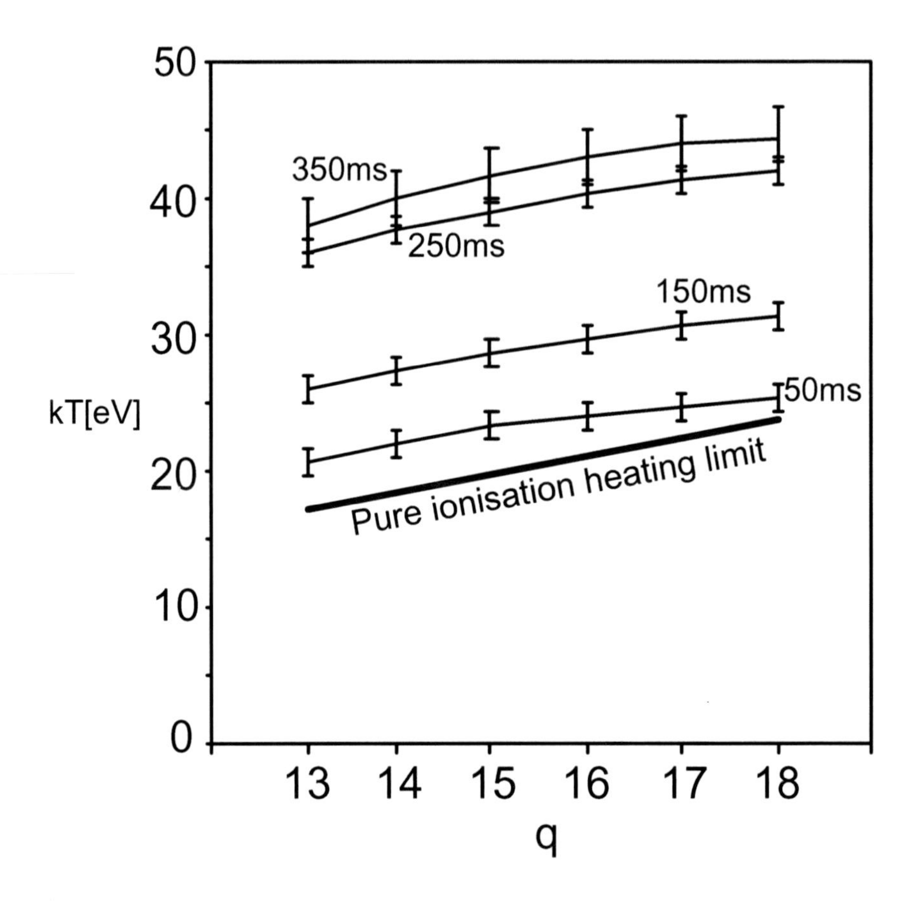

Figure 2.10. Measured temperature distribution of Ar ions as a function of time after the trap was closed (i.e. ion creation initiated). This data was taken with a 20keV, 60mA beam compressed by an axial magnetic field of 4T at the trap and with a 10V trapping potential applied axially, corresponding to ($V_t \approx 16V$).

$$r_{ch} = r_e \exp(\frac{kT_i}{2V_0 q_i} - \frac{1}{2}). \tag{2.31}$$

This result is obtained by equating the temperature to the expression for the potential (eq. 2.4) and solving for r. For a trap length l_{trap}, the 'aspect ratio' of the ion cloud is r_{ch}/l_{trap} which is usually much less than 1. Constraining (somewhat arbitrarily) $r_{ch}/l_{trap} \lesssim 0.05$ we then get a bound on the ion temperature below which the ion cloud is much longer than it is wide of

$$kT_i \lesssim q_i V_0 (2\ln(\frac{0.05 l_{trap}}{r_e} + 1)). \tag{2.32}$$

Noting that the $\frac{l_{trap}}{r_e} \approx 2000$ for a typical EBIT (30mm trap length, 30μm beam diameter), the logarithm takes a value of about 5, hence giving a temperature below which the ion cloud is much longer than it is wide of

$$kT_i \lesssim 10 q_i V_0. \tag{2.33}$$

The temperature will be related to the trap depth, taking on a value (very approximately) of $kT_i \approx 0.3 q_i V_t$, giving rise to a very rough rule to thumb as to when the ion cloud is much longer than it is wide of $V_t \lesssim 30V_0$. Recalling that V_0 is usually about 5V, one then gets a maximum trapping voltage of 150V up to which the ion cloud can be expected to be much longer than it is wide and accordingly the above analysis should hold.

4. Uses of EBITs

EBITs are versatile experimental tools capable of making a wide range of physical measurements when coupled to other apparatus. Broadly speaking these uses can be grouped into three categories

- A spectroscopic source of photons. Various interactions (predominantly with the electron beam) can lead to excitation of the trapped ions followed by decay via photoemission. A small portion of the emitted photons are able to leave the trap through small holes in the central drift tube. Spectroscopic measurements of these photons give details about the energy level differences of the trapped highly charged ions. These energy level differences are then used to learn about atomic structure of highly charged ions, deduce information about the nucleus or perform test of quantum electrodynamics [29].

- A source of ions. Many chapters later in this textbook and its accompanying volume (The Physics of Multiply and Highly Charged Ions: Interactions with Matter) deal with the interactions of multiply and highly charged ions with a range of forms of matter, particularly gases and surfaces. EBITs tend to provide moderately low currents but are able to provide the highest charge states. As outlined in the previous chapter, electron cyclotron resonance sources (ECRs) are preferable when higher currents are required and lower charge states.

- A tool for measuring cross sections and rates for various processes. The intensities of photons emitted due to electron ion interactions can be used to infer information about the rates for various processes, from which cross sections can also be inferred. Sometimes this is done by controlling various machine parameters of the EBIT and correlating the observed signals with the values of the machine parameters at the time of detection.

The remainder of this chapter will discuss these last two categories of uses as illustrations of the machine physics presented earlier.

4.1 A Source of Ions

Once ions leave the trap axially, passing over the potential barrier provided by the final drift tube, they are quickly accelerated electrostatically towards the collector, traveling along the beam axis. The collector assembly includes a final 'extractor' electrode through which some of the ions pass to enter a beam optics system such as that shown in fig. 2.11. Once the desired ions are selected by passing through the beam optics they are then brought into an interaction region where the physics processes of interest are studied.

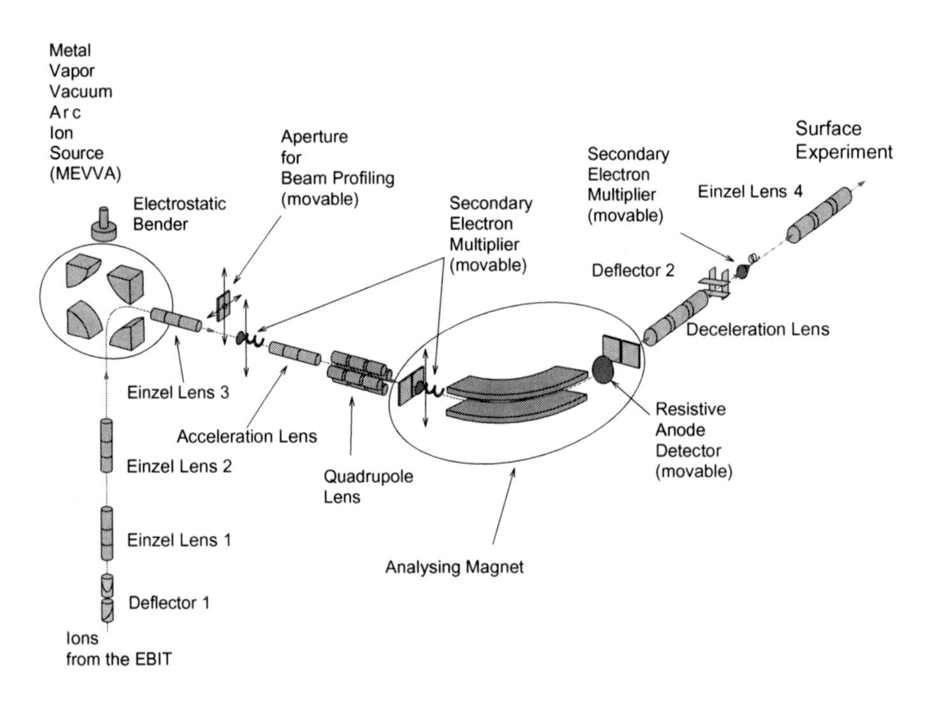

Figure 2.11. Beam optics and diagnostic systems used with the Tokyo EBIT. The resistive anode detector was used to measure the extracted ion spectra shown in fig. 2.8. This detector and the secondary electron multipliers are used to tune the extracted ion beam through the beam line.

Ions can be deliberately extracted from the trap (pulsed mode) or allowed to leak away naturally (leaky mode) through the axial escape process described in sec. 2.3.3. When ions are extracted in pulsed mode, the extraction is usually achieved either through raising the potential of the central drift-tube above that of the two end drift tubes or through lowering the potential of the final drift tube (each occurring typically in a few milliseconds). In either case, ions are expelled from the trap to travel along the beam axis into the collector and out through the extractor electrode. Although the two techniques of expelling the ions are similar, raising the potential of the central drift-tube has the advantage that the ions all leave the trap from the same potential throughout the extraction process and hence are equally well matched to the ion optics. The observed ion distribution after charge state analysis tend to be more peaked at lower charges state for leaky mode than for pulsed mode (i.e. a greater proportion of lower charge states are observed in leaky mode) since the lower charge states can escape preferentially. However, the energy distribution of the leaky mode ions is narrower.

Other extraction modes are possible, such as the 'pulsed evaporative cooling' scheme [30] which can be viewed as a hybrid of leaky and pulsed mode extraction. By gradually lowering the electrostatic trap potential in a programmed manner, the evaporation rate is controlled. In this manner, hotter ions are sacrificially evaporated leading to cooling (see sec. 2.3.3) which is accompanied by a contraction of the ion cloud (see sec. 2.3.1). The contraction of the ion cloud leads to an increase of the number of ions in the electron beam even through the total number of ions trapped has decreased. Furthermore, simulations suggest that the ions leaving the trap in this manner (particularly the latest ones to leave) have a very narrow energy spread whilst retaining the higher charge states in the extracted ion distribution.

4.2 Cross Section Measurement

EBITs have a successful history in the measurement of cross sections and rates for various types of electron highly charged ion interaction [31]. Either ions or photons leaving the trap can be used to infer information about processes occurring in the trap. There are two regimes where the machine physics outlined above simplifies considerably. In one regime, effects are measured at early times in the charge state evolution and the second regime the equilibrium condition is used. Normalisation of the cross sections is usually achieved through reference to calculated cross sections for other process. This section concentrates on measurement of dielectronic recombination rates because they are illustrative of the range of techniques (both photons and ions can be used) and show how the machine physics outlined in sec. 2.3 supports these measurements. There is also a discussion of measurement of electron impact ionization cross sections elsewhere in this volume, which includes a description of how EBITs are used for such measurements. Essentially, a more rigorous version of eq. 2.21 is used, with charge exchange being explicitly accounted for through an extrapolation procedure. The relative abundances of $N_{i_{max}-1}$ and $N_{i_{max}}$ are inferred from the detection rates of radiative recombination photons produced through capture into these ions.

Dielectronic recombination cross sections have been measured using extracted ions [32, 33] and photons [34]. This process involves resonant capture of an electron (in an inverse Auger process) followed by radiative stabilisation. For example for helium-like ions this process can be represented

$$\begin{aligned} I^{q+}(1s^2) + e^- \;\rightarrow\; I^{(q-1)+}(1s2lnl') &\rightarrow I^{(q-1)+}(1s^2nl') + hv \\ &\rightarrow I^{(q-1)+}(1s^2 2l) + hv. \end{aligned} \tag{2.34}$$

For ion based measurements, the equilibrium balance of the ions is monitored as a function of beam energy. On resonance the recombination rate increases which is reflected in either the observed equilibrium charge balance or the initial rate at which that balance is attained. In contrast for photon based measurements, the energy range of interest is rapidly swept after an initial charge balance has been prepared with the yield of detected photons being correlated with the electron beam energy.

For example, for studies of dielectronic recombination into hydrogen-like ions based on ion detection, the EBIT is programmed in three distinct phases: ionization (cooking), probe and dump. During the ionization phase, ions predominantly of the desired charge state are created using an appropriate beam energy and a high beam current. During the probe phase, the beam energy E_p is held at a particular value for some period of time t_p, possibly with the electron beam current lowered to reduce the electron beam's energy spread. Finally in the dump phase the beam energy and current are set to values suitable for extracting the ions and then the trapping potential is lowered so the ions are expelled from the trap to be detected. This process is repeated for a few values of t_p at many different values of E_p. The rate of disappearance of hydrogen-like ions is then measured as a function of E_p to determine the dielectronic recombination rates.

During the probe time, the rate of change of hydrogen-like ion population is given by

$$\begin{aligned} \frac{dN_H}{dt} = {} & \frac{J_e}{e}\left(+N_B\sigma_B^{RR} f_{e,B} - N_H(\sigma_H^{DR}(E_p) + \sigma_H^{RR}) f_{e,H}\right) \\ & + N_0(N_B\langle v\sigma_B^{CX}\rangle - N_H\langle v\sigma_H^{CX}\rangle) \\ & - N_H R_H^{ESC}. \end{aligned} \tag{2.35}$$

Note this is essentially eq. 2.15 with the subscripts referring to the various charge states being replaced by labels referring to the number of electrons attached to the species, i.e. N_B refers to the number of bare ions, N_H the number of hydrogen-like and N_{He} the number of helium-like etc. The expressions accounting for electron-impact ionization have been removed since the main dielectronic recombination resonances occur below (converging to) the helium-like ionization threshold so E_p is too low for this process to occur. Additionally the radiative recombination cross section into hydrogen-like σ_H^{RR} has been replaced by the sum of this and a dielectronic cross section $\sigma_H^{DR}(E_p) + \sigma_H^{RR}$ to include the resonance term. Note the energy dependence is only included for $\sigma_H^{DR}(E_p)$ to emphasize this is a resonant process having a rapidly varying energy dependence in contrast with the other cross sections which only vary slowly with energy.

Strictly our subsequent analysis should be referring to the total number of trapped ions of a given charge-state, not the axial density (as we are doing here). However this distinction does not affect the analysis as the temperature of the ions of a given species does not vary as E_p is varied since it is determined by the much longer cooking phase. If we were to work with the total number of trapped ions in the analysis this procedure would simply introduce an arbitrary multiplicative factor into the analysis. Several other such multiplicative factors exist such as detector sensitivity but these are accounted for by the normalisation process used. Hence we can absorb the extra multiplicative factor ignored by working with axial densities into the normalisation constant.

For short probe times t_p, the terms $N_B \sigma_B^{RR} f_{e,B}$ and $N_0 N_B \langle v \sigma_B^{CX} \rangle$ can be treated as approximately constant. Hence the differential equation 2.35 takes the form

$$\frac{dN_H}{dt} = \alpha(E_p) N_H + \beta \tag{2.36}$$

where the slowly varying energy dependence of σ_B^{RR} is neglected. Hence, for each value of E_p, the rate of disappearance of the hydrogen-like ions can be fitted to an exponential $A \exp[-\gamma(E_p)t] + B$ where A and B are slowly varying functions of energy, related to the initial density of hydrogen-like ions and the rates of various processes. Their slow variation as E_p passes through resonances can be used as a check that the experimental system has been set up correctly, and factors such as detector dead time have been correctly accounted for. The slowly varying part of $\gamma(E_p)$ can be removed numerically after which a series of peaks remain corresponding to the dielectronic recombination resonances (since this is the only term which varies rapidly with E_p) under study. These are then normalised at a single energy to theoretical calculations of the same resonance strengths.

In contrast, dielectronic recombination into helium-like ions can be studied through the equilibrium ratio of helium-like and lithium-like ions. The beam energy is set to a value E_e for sufficient time to establish the equilibrium charge balance after which the beam energy and current are set to values suitable for ion extraction. The ions are extracted and detected to infer the equilibrium charge balance. At equilibrium eq. 2.15 takes the form

$$\begin{aligned}\frac{dN_{He}}{dt} = 0 = & \frac{J_e}{e}(N_{Li}\sigma_{Li}^{EI} f_{e,Li} - N_{He}(\sigma_{He}^{RR} + \sigma_{He}^{DR}(E_e))f_{e,He}) \\ & - N_0 N_{He}\langle v\sigma_{He}^{CX}\rangle - N_{He}R_{He}^{ESC}\end{aligned} \tag{2.37}$$

where again an additional term has been added to account for the dielectronic recombination and again terms due to hydrogen-like ions have been removed since E_e is too low for these ions to be created. Assuming the overlap factors are equal, this can be rearranged to give the dielectronic recombination cross section as

$$\sigma_{He}^{DR}(E_e) = \frac{N_{Li}}{N_{He}}\sigma_{Li}^{EI} - K \tag{2.38}$$

where K is a term, accounting for reduction of the helium-like population due to radiative recombination, escape and charge exchange and hence is slowly varying with E_e. The rapidly varying dielectronic recombination contribution is mirrored by the rapidly varying part of the ratio N_{Li}/N_{He} which can be extracted numerically. A theoretical value of σ_{Li}^{EI} can then be used to normalise this rapidly varying contribution to give resonance strengths.

An alternative to measuring the ions is to measure photons produced as the doubly excited state formed through dielectronic recombination decays [34]. The EBIT is first tuned to a constant beam energy chosen to be just below the ionization potential required to ionize beyond the charge-state of interest, which is usually a closed shell. After a cooking period a single probing period is used to probe all the resonances of interest. During this probe period the beam energy is swept rapidly (typ. a few milliseconds) down across the energy range of interest and back up again. During this sweeping process the detection of photons by a solid state detector is correlated with the electron beam energy. Many such sweeps are performed in order to obtain the required statistical quality. The detector has an intrinsic energy resolution so different decay channels can be distinguished. The data set produced is a map of photon yield as a function of photon energy and electron beam energy as is shown in fig. 2.12.

The rapid sweeping is performed to ensure the whole energy range is covered with a constant charge balance (i.e. the enhanced recombination process does not significantly degrade the charge balance). The constancy of charge balance can be verified by comparing the spectra produced from the down sweep and the subsequent up sweep. If the charge balance changes during the down sweep, this will be reflected in different appearances of the measured data for up and down sweeping. Referring back to eq. 2.34 we see that the resonant state can stabilize in one of two ways, each giving rise to a photon of characteristic energy. These give rise to different photon energies observed for each

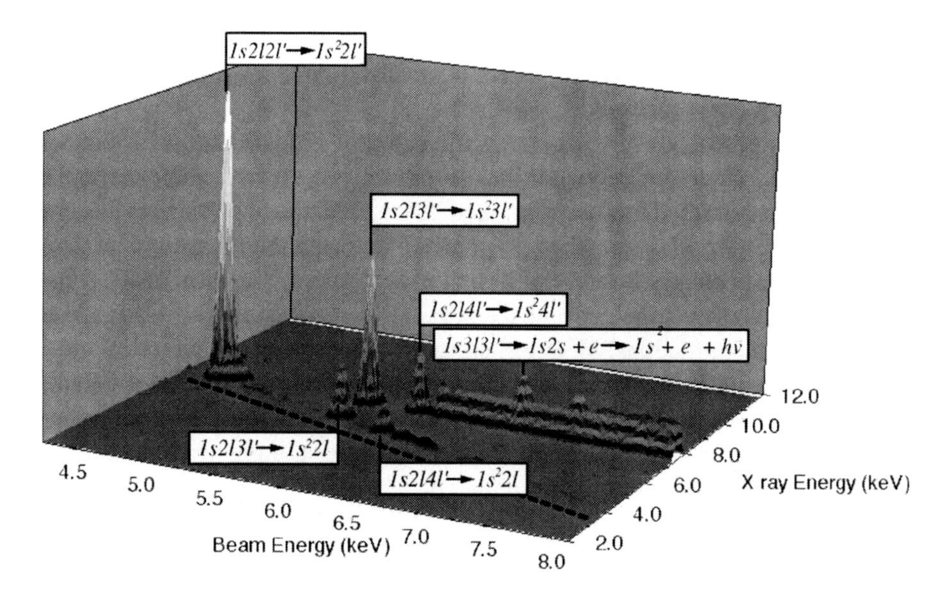

Figure 2.12. Map showing the yield of photons detected as a function of beam energy and photon energy as the beam energy is swept across a series of dielectronic recombination resonances for Fe^{24+}. The labels indicate the various decay routes. Note, evidence for resonant excitation can also be seen at beam energies of about 7 (labelled) and 7.4keV. Weaker radiative recombination features are also observed such as the feature due to $1s^2 + e^- \rightarrow 1s^2 2l + hv$ lying just above the dotted line in the figure.

dielectronic recombination resonance. Off the resonance energies, photons can still be created due to radiative recombination although the yield is too weak to see clearly in fig. 2.12. However the cross section for radiative recombination can be calculated precisely so this yield can be used to normalise the data from which resonance strengths can be obtained although this process was not performed for the data shown in fig. 2.12 [35]. Instead the data was normalised to a previous measurement of the resonance strength of the lowest lying resonance manifold [36], which in turn used a normalisation to a theoretical electron impact excitation cross section.

Even if one does normalise directly to a radiative recombination cross section, since the observations occur at 90° to the electron beam, the polarisations associated with photons produced by radiative recombination and dielectronic recombination must be used to deduce resonance strengths during the normalisation process. These polarisations in turn must be calculated theoretically so again effectively the resonance strengths are normalised to theoretical calculations of dielectronic recombination. A new electron beam ion trap [37] is currently under construction which will overcome this limitation by facilitating a wide range of angles for photon detection with respect to the electron beam. Then, positioning the detector at the magic angle will remove the polarisation dependence and hence allowing direct normalisation to the simpler radiative recombination cross sections.

The two approaches to measuring dielectronic recombination resonance strengths (i.e. detecting extracted ions or photons) each have their particular merits. When extracted ions are measured, the data obtained pertains to just one single charge state whereas when photons are detected the measurement does not discriminate clearly between different charge states. In some cases, it has been possible to detect dielectronic recombination resonances due to a range of charge states simultaneously [38, 39] although comparison to theoretical cross sections is required. In contrast, detection of photons has the advantage that the decay route is determined. Also, because the whole data set is covered during each probing period, the data is insensitive to changes in the absolute number of ions traped from measurement cycle to measurement cycle. This means that this technique can be used with injection schemes such as MEtal Vapor Vacuum Arc source (MEVVA) injection, where the loading of the trap may differ from shot to shot.

This section has only briefly reviewed some of the techniques through which EBITs can be used to measure cross sections or resonance strengths for various processes. This field is still developing and more techniques or refinements of the present techniques are sure to occur in the near future. The common theme underlying these approaches, and one likely to also be common to future developments is the idea of 'programming' the EBIT in some cycle designed with reference to a knowledge of the machine physics so as to highlight the particular process of interest and hence to learn about it. In a sense the rich range of processes occurring inside an EBIT gives rise to a wealth of possibilities whereby the physics of these processes can be studied.

5. Conclusion

The general form of EBITs has been described briefly with the various constituent parts being described in turn. Following this, the machine physics underlying EBITs has been discussed with reference to a wide range of atomic and plasma physics processes which occur inside the trap. Some diagnostic measurements have been presented which broadly support this model of the machine physics. Two main categories of uses of EBITs have been discussed, namely use as an ion source and as a tool to study atomic physics cross sections. Various techniques related to these uses have been outlined in terms of the machine physics developed.

Acknowledgments

The author is deeply indebted to many members of the EBIT-users community with whom he has had fruitful discussions, in particular Toru Kinugawa, Nobuyuki Nakamura, Shunsuke Ohtani, Brian O'Rourke, Emma Sokell and Hirofumi Watanabe.

References

[1] M. A. Levine, R. E. Marrs, J. R. Henderson, D. A. Knapp and M. B. Schneider, Phys. Scr. **T22**, 157 (1988)

[2] E. D. Donets and V. P. Ovsyannikov, Sov. Phys. –JETP **53**, 466 (1981)

[3] E. D. Donets V. I. Ilushenko and V. A. Alpert, "Ultrahigh vacuum electron beam ion source of highly stripped ions" in *Proceedings of the First International Conference on Ion Sources* (Saclay, France) 635 (1969)

[4] R. W. Schmeider, C. L. Bisson, S. Haney, N. Toly, A. R. Van Hook and J. Weeks, Rev. Sci. Instrum. **61**, 259 (1990)

[5] M.A. Levine, R.E. Marrs and R.W. Schmieder Nucl.Inst.Meth. **A237**, 429 (1985)

[6] D. A. Knapp, R. E. Marrs, S. R. Elliot, E. W. Magee and R. Zasadinski, Nucl. Instrum. Methods A **334**, 305 (1993)

[7] R. E. Marrs, S. R. Elliot and D. A. Knapp, Phys. Rev. Lett. **72**, 4082 (1994)

[8] J. D. Silver, A. J. Varney, H. S. Margolis, P. G. E. Barid, I. P. Grant, P. D. Groves, W. A. Hallet, A. T. Handford, P. J. Hirst, A. R. Holmes, D. J. H. Howie, R. A. Hunt, K. A. Nobbs, M. Roberts, W. Studholme, J. S. Wark, M. T. Williams, M. A. Levine, D. D. Dietrich, W. G. Graham, I. D. Williams, R. O'Neil and S. J. Rose, Rev. Sci. Instrum. **65**, 1072 (1994)

[9] C. A. Morgan, F. G. Serpa, E. Takacs, E. S. Meyer, J. D. Gillaspy, J. Sugar, J. R. Roberts, C. M. Brown and U. Feldman, Phys. Rev. Lett. **74**, 1072 (1994)

[10] C. Biedermann, A Forster, G Fuβmann and R. Radtke, Phys. Scr. **T73**, 360 (1997)

[11] F. J. Currell, J. Asada, K. Ishii, A. Minoh, K. Motohashi, N. Nakamura, K. Nishizawa, S. Ohtani, K. Okazaki, M. Sakurai, H. Shiraishi, S. Tsurubuchi and H. Watanabe, J. Phys. Soc. Jpn. **65**, 3186 (1996)

[12] J. Crespo López-Urrutta, B. Bapat, I. Draganic, A. Werdich and J. Ullrich, Phys. Scr. **T92** 110 (2001)

[13] M. Kleinod, R. Becker, H. Bongers, M Weildenmuller, B. Zipfel and E. D. Donets, Rev. Sci. Instrum. **67**, 986 (1996)

[14] V. P. Ovsyannikov and G. Zschornack Rev. Sci. Instrum. **70**, 2646 (1999)

[15] F. J. Currell "The physics of electron beam ions traps, section 5" in *"Trapping highly charged ions: fundamentals and applications"*, ed. John Gillaspy, Nova Science Publishers, ISBN 1-56072-725-X. (1999)

[16] G. Fussmann, C. Biedermann and R. Radke, NATO ASI Series (Proceedings Summer School, Sozopol 1998)

[17] G. J. Herrmann, Appl. Phys. **29**, 127 (1958)

[18] M. A. Levine, R. E. Marrs, J. N. Bardsley, P. Beiersdorfer, C. L. Bennet, M. H. Chen, T. Cowan, D. Dietrich, J. R. Henderson, D. A. Knapp, A. Osterheld, B. M. Penetrante, M. B. Schneider and J. H. Schofield, Nucl. Instrum. Meth. **B43**, (1989) 431

[19] R. E. Marrs, A. Schach von Wittenau, "EBIT Electron Beam Ion Trap, N-Division Experimental Physics Bi-Annual report 1996-1997" LLNL P107 (1998)

[20] H. Kuramoto, T. Kinugawa, H.Watanabe, C. Yamada and S. Ohtani, I. Yamada, F. J. Currell 2002 Rev. Sci. Instrum. **73**, 42

[21] F. J. Currell "The physics of electron beam ions traps, section 4" in *"Trapping highly charged ions: fundamentals and applications"*, ed. John Gillaspy, Nova Science Publishers, ISBN 1-56072-725-X. (1999)

[22] P. Beiersdorfer, L. Schweikhard, J. Crespo López-Urrutta and K. Widmann, Rev. Sci. Instrum. **67**, 3818 (1996)

[23] J. V. Proto, I. Kink and J. G. Gillaspy, Rev. Sci. Instrum. **71**, 3050 (2000)

[24] B. M. Penetrante, J. N. Bardsley, D. DeWitt, M. W. Clark and D. Schneider, Phys. Rev. A **43**, 4861 (1991)

[25] "Studies of Highly Ionized Atoms using an Electron Beam Ion Trap", Ph.D. Thesis H. S. Margolis, Pembroke College Oxford (1994)

[26] C. L. Longmire and M. N. Rosenbluth, Phys. Rev. **103**, 507 (1957)

[27] F.J.Currell, H.Kuramoto, S.Ohtani, C.Scullion, E.J.Sokell and H.Watanabe, Phys. Scr. **T92** 147 (2001)

[28] L. Spitzer, Jr., "Physics of fully ionized gases" (1956) Interscience publishers, inc. New York. (Library of Congress Catalog Card no. 55-11452)

[29] For a bibliography containing examples see: D. Schneider "The highly charged ion physics programme at the LLNL EBIT facilities", Ch. 19 in *"Trapping highly charged ions: fundamentals and applications"*, ed. John Gillaspy, Nova Science Publishers, ISBN 1-56072-725-X. (1999)

[30] T.Kinugawa, F.J.Currell and S.Ohtani, Phys. Scr. **T92** 102 (2001)

[31] R. E. Marrs, Comments At. Mol. Phys. **27** 57 (1991)

[32] D.R. DeWitt, D. Schneider, M.H. Chen, M.W. Clark, J.W. McDonald and M.B. Schneider, Phys.Rev.Lett. **68**, 1694 (1992)

[33] R. Ali, C. P. Bhalla, C. L. Cocke and M. Stockli, Phys. Rev. Lett. **64**, 633 (1990)

[34] D. A. Knapp, R. E. Marrs, M. B. Schineider, M. H. Chen and J. H. Schofield, Phys. Rev. A **47** 2039 (1993)

[35] H. Watanabe, F. Currell, H. Kuramoto, Y. M. Li, S. Ohtani and B. O'Rourke, J. Phys. B **34**, 5095 (2001)

[36] P. Beiersdorfer, T. W. Phillips, K. L. Wong, R. E. Marrs and D. A. Vogel, Phys. Rev. A **46** 3812 (1992)

[37] F. J. Currell "The Belfast EBIT source", LEIF Workshop on Ion Sources, Aarhus, Denmark (2001)

[38] T. Fuchs, C. Biedermann, R. Radtke, E. Behar and R. Doron, Phys. Rev. A **58** 4518 (1998)

[39] D. A. Knapp, Supp. to Z. Phys. D. At. Mol. and Clust. **21** S143 (1991)

II

APPLICATIONS

Chapter 3

COLLISION PHENOMENA INVOLVING HIGHLY-CHARGED IONS IN ASTRONOMICAL OBJECTS

A. Chutjian
Jet Propulsion Laboratory,
California Institute of Technology,
Pasadena, CA 91109 USA
ara.chutjian@jpl.nasa.gov

Abstract Recent advances in the atomic collision physics of highly-charged ions (HCIs) relevant to new astronomical observations are reviewed. The atomic phenomena include electron-impact excitation, charge-exchange, direct and indirect recombination, emission from HCI metastable levels, and X-ray emission in HCI-neutral collisions. Comparisons with theory are given, where available, to help establish "ground truth" for theory. The experiments and theories can then be applied with good confidence to the modeling of the rich body of high-resolution absorption and emission spectra observed in the interstellar medium, and from solar, stellar, and comet atmospheres by spectrometers aboard the SOHO, Chandra, Newton, Hubble and other NASA-ESA spacecraft. The sophisticated space-borne data require equally high-quality laboratory data to work out conditions of electron density and temperature, ion density, ionization fraction, and E,B fields present in the astronomical plasmas.

Keywords: electron energy loss, R-matrix, highly-charged ions, electron excitation, lifetimes, charge-exchange, recombination, X-ray emission, comets, solar wind, solar and stellar atmospheres

1. Introduction

This decade has seen a dramatic and international growth in space observations from an impressive array of ground and space-borne instruments. These efforts have been led by the U.S. National Aeronautics and Space Administration and the European Space Agency. The observational platforms include the Infrared Space Observatory, the Hubble Space Telescope, the Extreme Ul-

F.J. Currell (ed.), The Physics of Multiply and Highly Charged Ions, Vol. 1, 79-101.

traviolet Explorer (*EUVE*), *SOHO*, the Roentgen Satellite *(ROSAT)*, the Far Ultraviolet Spectroscopic Explorer (*FUSE*), *Chandra*, XMM-*Newton*, *TRACE*; as well as upcoming missions such as *SOFIA*, the Space Infrared Telescope Facility (*SIRTF)* and Constellation-X. Each platform has one or more spectrometers covering a range of wavelengths, from the infrared (*ISO*) to the X-ray regions (XMM-*Newton, Chandra*, Constellation-X). These spectrometers are *remote sensing*, as opposed to *in-situ* instruments which measure neutrals and charged particles from the Sun, a planet, or in the interstellar medium using neutral imagers, ion analyzers, electron analyzers, and mass spectrometers. As such, the purpose of these instruments is detection of photons, either through imaging instruments or spectrometers. One can obtain, for example, *images* of the Sun in the Fe IX ^{1}P^o $\rightarrow$ ^{1}S λ171Å, Fe XII ^{4}P $\rightarrow$ S^o λ195Å, Fe XV ^{1}P^o $\rightarrow$ ^{1}S λ284Å, and He II ^{2}P^o $\rightarrow$ ^{3}S λ304Å emission lines using the Extreme Ultraviolet Imaging Telescope (*EIT*) aboard *SOHO* [1]. One can also obtain *spectra* of the astronomical object, such as those returned by the *SUMER/SOHO* high-resolution spherical concave grating instrument [2]. Imaging information is useful for monitoring the morphology of the object, such as changes in the solar coronal and transition region due to coronal loops and mass ejections. Spectral data are critical for obtaining ionic species, densities, and electron temperatures from stars, the Sun, active galactic nebulae (AGNs), *etc.*[3]

In the following summaries the role of highly charged ions (HCIs) in the various astronomical objects will be summarized. Included will be the use of critical quantities such as cross sections for excitation, charge-exchange, X-ray emission, radiative recombination (RR) and dielectronic recombination (DR); and lifetimes, branching ratios, and A-values. These data, experimental and calculated, are required for interpretation of the spectral observations [4]. The astronomical objects include the Sun and stars, circumstellar clouds, the interstellar medium, planetary ionospheres, planetary magnetospheres ($e.g.$, the Io-Jupiter torus), and comets. The processes and their interpretation parallel in many respects the phenomena occurring in confined fusion plasmas where stellar-like temperatures are achieved. Unfortunately, space does not permit a treatment of these plasmas. The approaches summarized in Refs. [5–8] afford an entry into this parallel universe of the tokamak, JET, ITER, the National Ignition Facility, and the LaserMegajoule.

2. Basic Collision Phenomena and Atomic-Physics Parameters

2.1 Electron Excitation and Recombination in the HCIs

The spectrally-resolved photon emissions from the hot solar or stellar plasma are analyzed in terms of spectral line ratios to obtain the plasma properties. With increasing resolution of the spacecraft spectrometers, from the infrared to X-ray spectral regions, many once-blended emission lines can now be resolved, and hence many more useful line intensities and ratios are being established.

One of the most basic and important phenomena occurring in the high electron-temperature plasmas is the electron-impact (collisional) excitation of emitting spectral lines in the HCI. Electron temperatures and densities may be deduced from the line intensity ratios through the standard expressions for the statistical equilibrium. One treats the rate of change of the population of the emitting upper (i or j) state to the ground (g) state, taking into account all other levels. This process involves a large range of atomic data, such as level lifetimes and excitation cross sections (collision strengths).

For a simple three-level case the ratio of line intensities is given by [9, 10],

$$\frac{I(i \to g)}{I(k \to g)} = \frac{C(g \to i)}{C(g \to k)} \left(1 + \frac{N_e C(k \to m)}{A(k \to g)}\right). \tag{3.1}$$

Here, the C's are thermally-averaged collision strengths for which standard expressions are given [11], and $A(k \to g)$ is the Einstein A-coefficient for spontaneous radiation in the $k \to g$ transition. For small collision strengths $C(k \to m)$, low electron density N_e, and for an optically-allowed transition one has $A(k \to g) >> N_e C(k \to m)$, so that the line ratio is density insensitive, but may be T_e-sensitive *via* the behavior of the collision-strength ratio in eq. 3.1. This is the so-called coronal equilibrium limit. To qualify as useful diagnostics, the candidate line ratios should be monotonically increasing or decreasing in the range of desired plasma electron temperature T_e or electron density N_e. Otherwise, if a particular ratio is flat, then the astrophysical measurement of the ratio can lead to a wide and not helpful range for T_e or N_e. Many examples exist of the use of ratios, and recent calculations have been published for Fe XXI [12], Ca XVI [13], and Si VIII, S X [14]. Other ratios can be established, as when one member of the ratio is a forbidden transition to give a diagnostic for Ne [15]. By way of example, shown in fig. 3.1 are two calculated line ratios R_1 and R_2 as a function of N_e in Fe XXI. These ratios are given by,

$$R_1 = I(2\mathrm{s}^2 2\mathrm{p}^2\ {}^3\mathrm{P}_1 \to 2\mathrm{s}2\mathrm{p}^3\ {}^5\mathrm{S})/I(2\mathrm{s}^2 2\mathrm{p}^2\ {}^3\mathrm{P}_0 \to 2\mathrm{s}^2 2\mathrm{p}^2\ {}^3\mathrm{P}_1), \tag{3.2}$$

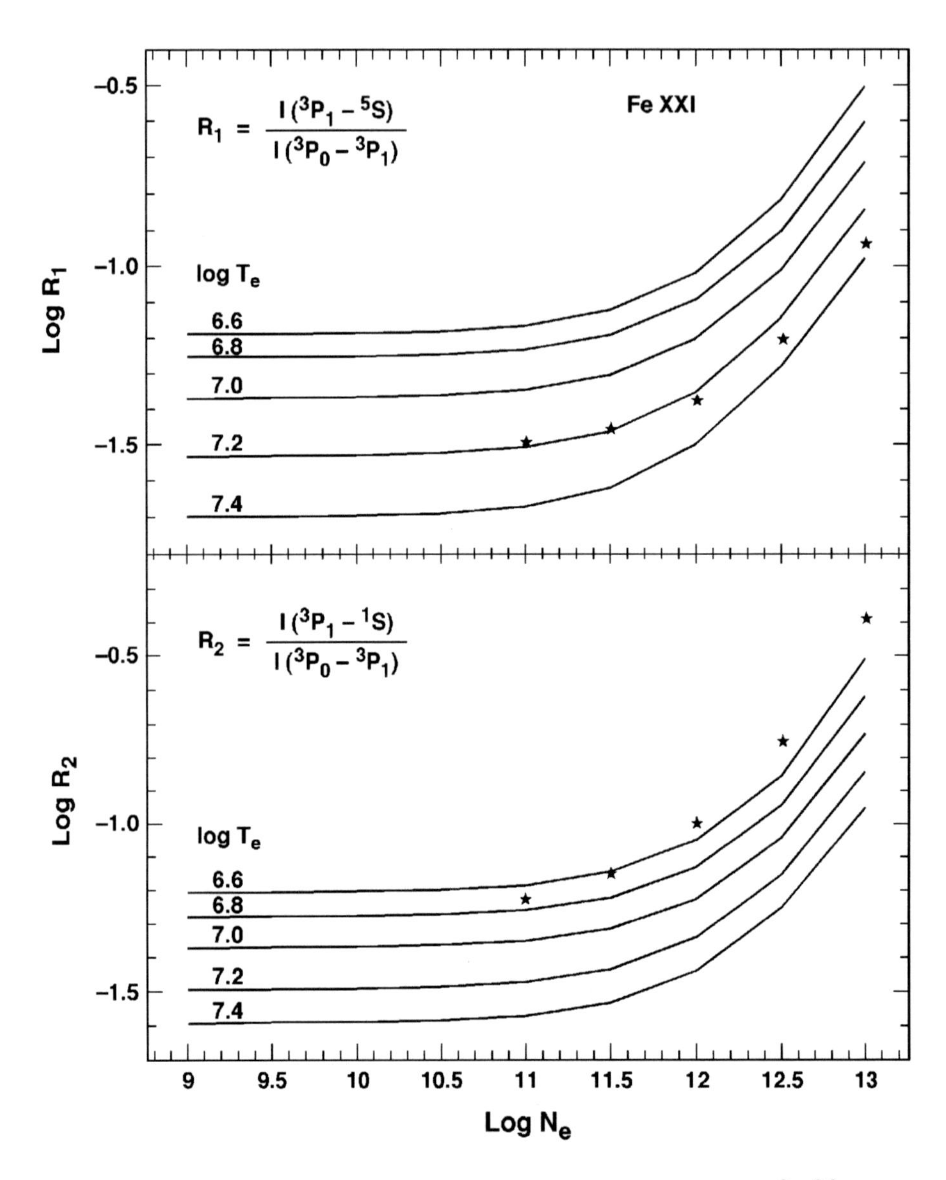

Figure 3.1. Theoretical Fe XXI line emission ratios $R_1 = I(2s^2 2p^2\,{}^3P_1 \rightarrow 2s2p^3\,{}^5S)/I(2s^2 2p^2\,{}^3P_0 \rightarrow 2s^2 2p^2\,{}^3P_1)$ (top), and $R_2 = I(2s^2 2p^2\,{}^3P_1 \rightarrow 2s^2 2p^2\,{}^1S)/I(2s^2 2p^2\,{}^3P_0 \rightarrow 2s^2 2p^2\,{}^3P_1)$ (bottom) in a 15-state ***R***-matrix calculation [12]. Stars are results in a distorted-wave calculation of Ref. [16]. The ratios make suitable diagnostics of N_e in the region above about $N_e \approx 10^{11}\mathrm{cm}^{-3}$, a density encountered in strong solar or stellar flares.

and

$$R_2 = I(2s^22p^2\,{}^3P_1 \to 2s^22p^2\,{}^1S)/I(2s^22p^2\,{}^3P_0 \to 2s^22p^2\,{}^3P_1). \quad (3.3)$$

One sees from fig. 3.1 that the ratio is a relatively flat function of N_e at $N_e \leq 10^{11}$ cm^{-3}, hence there is no accurate measure of electron density at any T_e. Above 10^{11} cm^{-3} one can, from knowledge of the measured ratio, pick off values of N_e and T_e using the family of calculated curves. As noted in Ref. [12], the ratios are useful in analyzing the plasma arising from an energetic solar or stellar flare; and emission from Fe impurities in a fusion plasma. The data of fig. 3.1 are all calculated results. There are at the moment no experimental data to provide a cross check of the accuracy of the theoretical approximations.

The use of eq. 3.1 requires knowledge of not only excitation cross sections *via* the C's, but also of level lifetimes and branching ratios to obtain the A's. Since excitation can occur from the kth level (which may be metastable) to all other levels m, a wide range of collision strengths is required, covering a multitude of transitions. Moreover, a respectable degree of accuracy in the atomic parameters is needed, almost always to better than 10%. As pointed out by Malinovsky *et al.* [17], shortcomings in the calculation of cross sections or A-values can detract from the usefulness of a line ratio. Hence the calculations, carried out in an appropriate theory, should address the following: the required size of the configuration-interaction expansion; resonances at the excitation threshold, especially for spin- and symmetry-forbidden transitions; a proper wavefunction description of the target to give the spectroscopic energy levels and oscillator strengths (length and velocity); use of continuum orbitals to describe short-range electron correlations, and pseudostates to describe resonances; sufficient number of partial waves; relativistic effects, *etc.* By the same token, experimental measurements of absolute excitation cross sections in HCIs must access both spin- and symmetry-allowed and -forbidden transitions; be able to access the threshold region, with extension to higher energies; account for metastable levels in the target beam current; and be free of (or corrected for) electron-ion elastic scattering. Moreover, when optical-emission and detection methods are used, the cross sections become *effective* excitation cross sections, as cascading contributions from excited electronic states will contribute to the measured emission.

The electron energy-loss approach, in both crossed beams [18–21] and merged beams [22–25] geometries affords many of the experimental advantages. In addition, those measurements are free of cascading contributions, since only a single energy-loss electron can result from each excitation of the upper state. Finally, the merged-beams geometry allows one to measure the integral excitation cross section directly, rather than an angle-by-angle differential cross section, which is then integrated. Angular differential measurements provide

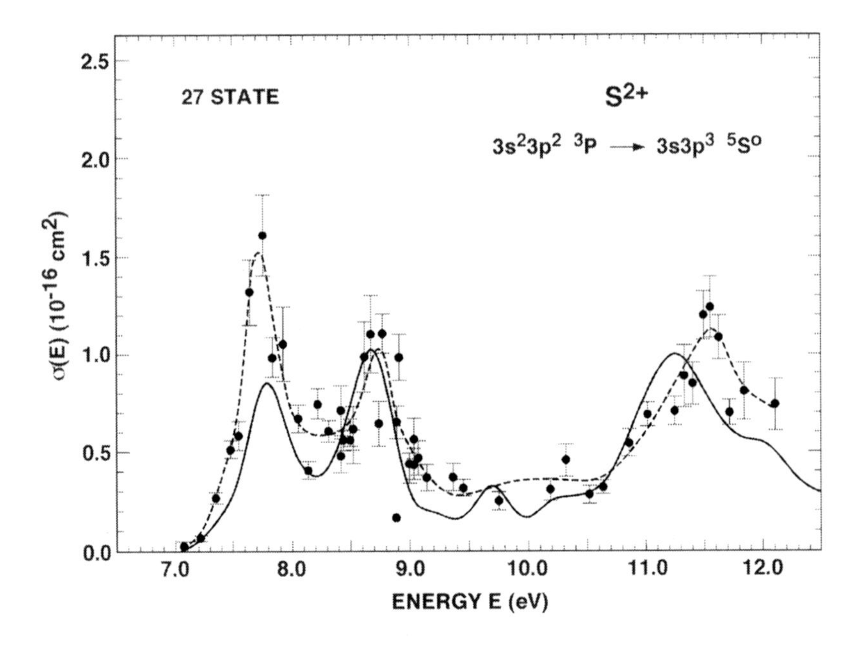

Figure 3.2. Experimental results (filled circles) and results of a 27-state ***R***-matrix theory (solid line) for the $^3P \rightarrow {}^5S^o$ transition in S^{2+} [30]. The theoretical results have been convoluted with an experimental, CM-variable Gaussian electron-energy width. The dashed line through the experimental data is meant to guide the eye.

a more critical test of physics theory, whereas only integral cross sections (at least at the moment) are used in the astrophysics models.

Comparisons of theory and experiment are steadily being published. The results to date show that the multistate ***R***-matrix approach to the calculation of HCI excitation cross sections in the threshold region generally agrees with experiment, to within the combined uncertainties. Such comparisons have been made for transitions in the HCI systems e-C^{3+} [26], e-O^{2+} [27], e-O^{5+} [22, 28], e-Si^{2+} [29], e-Si^{3+} [24], e-S^{2+} [30], e-Ar^{6+} [31], and e-Ar^{7+} [32]. In both e-O^{2+} and e-S^{2+} experiment and theory were able to account for the rich resonance structure at threshold. An example of the resonance structure in e-S^{2+} excitation is shown in fig. 3.2 for the spin-forbidden $3s^23p^2\,{}^3P \rightarrow 3s3p^3\,{}^5S^o$ transition. The experimental resolution was 80 meV in the center-of-mass frame, sufficient in many cases to resolve the envelope of overlapping, narrow resonances. Excitation rates for charge states of O and S are needed to understand the T_e, N_e, and energy balance in the Io-Jupiter plasma torus [33–35]. Long path length absorption by O^{5+} is commonly seen in our own

Milky Way gas [36]. Emission lines in O^{5+} and S^{5+}, as well as lines in C^{2+}, Ne^{4+}, and Ne^{5+}, have been detected in supernova remnant N49 by the *FUSE* satellite [37]. Higher charge states such as Si^{13+}, S^{15+}, and Fe^{24+} are seen by XMM-*Newton* at the cataclysmic variable *OY Carinae* [38], and Mg^{11+}, Ca^{18+}, and Ar^{16+} have been recorded by *Chandra* at *Capella* and at the binary HR 1099 [39].

Another important plasma property is the ionization fraction for each species; $i.e.$, the fraction of ions in a particular charge state as a function of solar/stellar temperature. These calculations require a large set of electron ionization and recombination cross sections for both direct and indirect processes (such as inner-shell excitation followed by autoionization). For example, high charge states of iron are common to the sun and stars. Charge states up to Fe^{25+} are seen in the more violent events, such as solar flares as observed by the Bragg Crystal Spectrometer (BCS) aboard the Japanese *Yohkoh* satellite, and in the *Chandra* X-ray observations of the Seyfert galaxy *Markarian 3* [40]. In the BCS measurements, the flare temperature was derived from the intensity ratio of a line formed by dielectronic capture to Fe XXV, relative to the intensity of a collisionally-excited resonance line in Fe XXV [41]. For general plasma modeling, a complete data set of ionization cross sections was used in a calculation of ionization and recombination rates as a function of T_e in plasmas containing charge states Fe^{14+} to Fe^{25+} [42]. The accurate measurement [43] and calculation [44] of ionization cross sections in HCIs not only reveal the physics of ionization but also provide important inputs to the ionization models.

Accurate DR cross sections are needed to establish diagnostic intensity ratios. It is well known that plasma microfields **E** of less than about 100 V/m, and motional electric fields can have a large effect on both the RR [45] and the DR cross sections [46–49]. Electric fields of the order of 100 V m^{-1} are sufficient to change DR rates by factors of 3-5 through mixing of ℓ levels [48–50]. Moreover, DR rate enhancements of up to 2 have been detected for Ti^{19+} in crossed $\mathbf{E} \times \mathbf{B}$ fields, with **B** in the range 30-80 mT [47]. This effect arises through the fact that the magnetic quantum number m is no longer a good quantum number in the crossed-fields geometry, and influences the ℓ mixing brought on by **E**. Our Sun's magnetic fields range from about 1 mT in the coronal region to 500 mT at the center of a sunspot umbra [51]. Hence *both* **E** (Stark) and **B** (Zeeman) mixing can effect DR rates in the Sun and stars. It is clear that as observations become more spatially localized, and the plasma models more sophisticated, one will have to take into account field effects on the cross sections in localized regions of high **E** and **B**.

The need in astrophysics for accurate data on all iron charge states is considerable. The IRON project is dedicated to the calculation of collision strengths, photoionization cross sections, oscillator strengths, radiative and DR cross sections in mainly Fe ions [52]. Because of the inherent difficulty of generating

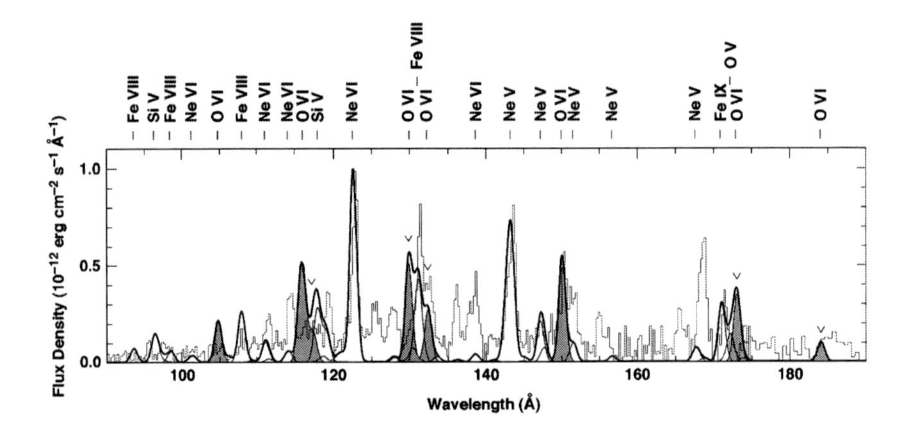

Figure 3.3. Coupling of database information (*CHIANTI*) with satellite observations (*EUVE*) to model the emissions from the dwarf nova *OY Carinae* in a superoutburst mode [58]. Emitting species are indicated at the top. The observed spectrogram is shown as gray, dotted histogram; the net model spectrum by the thick solid line, and the individual ion contributions by the thin solid line. The O VI model spectrum is the shaded area, and carets indicate nonresonance O VI lines.

high Fe charge states in the laboratory practically all the plasma parameters are calculated quantities, with almost no comparisons to experiment. The ion storage ring TSR in Heidelberg has been used to measure DR rate coefficients in Fe XVIII and Fe XIX [53]. Line emission in the 10.5-11.5 Å wavelength range corresponding to direct, resonant, and dielectronic excitations in the e-Fe XXIV system has been reported using the electron beam ion trap (EBIT) at Lawrence Livermore National Laboratory [54]. There is good agreement with distorted-wave and ***R***-matrix calculations. Data on collisional excitation of *any* Fe charge state is non-existent. As one part of a larger program involving Fe and Mg charge states, excitation cross sections have been reported for the $3s^2 3p^5\ ^2P^o_{3/2} \rightarrow 3s^2 3p^5\ ^2P^o_{1/2}$ λ6376 Å coronal red line in Fe^{9+} using the JPL electron cyclotron resonance ion source [55].

In order to make data available to modelers, experimentalists and theoreticians, the experimental and theoretical data are submitted to central databases. One frequently-used center is the *CHIANTI* atomic database [56].

2.2 Optical Absorption and Emission in the HCIs

Elemental abundances in the interstellar medium (ISM) are obtained through measurement of absorption intensities (equivalent widths) at the atomic and molecular wavelengths, using a bright star as the background continuum source. For example, recent Goddard High Resolution Spectrograph observations from the Hubble Space Telescope (GHRS/HST) reveal absorptions due to Si^{3+}, C^{3+},N^{4+}, and O^{5+} towards the Seyfert galaxy ESO 141-055 [57]. The wavelength range here is $\lambda = 1030 - 1550$ Å. An example of the use of the *CHIANTI* database in conjunction with a stellar superoutburst is shown in fig. 3.3 [58]. Rich emissions in O^{5+}, $Ne^{5,6+}$, Si^{4+}, and $Fe^{8,9+}$ are evident.

Emissions in HCIs are also detected by observation of solar/stellar winds, and of our Sun. *FUSE* has recorded emissions from stellar winds emanating from two supergiants in the Small Magellanic Cloud (SMC) and the LMC [59]. Emissions in C^{2+}, N^{3+}, O^{5+}, S^{3+}, S^{5+}, and P^{4+} are detected. Because of the greater sensitivity of *FUSE*, one is able to detect weaker absorptions and emissions. Hence absorption lines will tend to be unsaturated. Weak features will lie on the linear part of the curve of growth, leading to reliable measurements *via* the less-abundant elements or *via* optically-forbidden absorptions.

Column densities, or the integrated intensity of absorption along the line of sight between object and spectrometer, may be obtained from the expression for the so-called equivalent width of the absorption line given by [60],

$$W_\lambda = \int \left[1 - \frac{I_\nu}{I_\nu(0)} \right] d\lambda. \tag{3.4}$$

W_λ is the integrated area (in cm) at the wavelength λ (frequency ν) absorbed out of a background continuum by the absorbing species. The quantity $I_\nu(0)$ is the transmitted intensity in the absence of the line, and W_λ is related to the particle density by (in *cgs* units),

$$\frac{W_\lambda}{\lambda} = \frac{\pi e^2}{m_e c^2} N_j \lambda f_{jk}, \tag{3.5}$$

where e is the electron charge, m_e the electron mass, c the speed of light, and f_{jk} the absorption oscillator strength for transition between lower level j and upper level k. The quantity N_j is the number of absorbing molecules/cm^2 which is the desired elemental abundance in that region of the ISM.

Use of eq. 3.5 requires f_{jk} for the relevant transition in the neutral or ionized species along the line-of-sight. Since the accuracy of N_j is directly proportional to that of the f_{jk}, one would like oscillator strengths accurate to better than 10%. As is the case with collision strengths, experiment can only provide a handful of measured oscillator strengths for some HCIs and some transitions. The remainder has to be calculated through benchmarked theoretical methods. Many accurate measurements of lifetimes in HCIs are obtained in ion storage rings [61], the EBIT [62], and the Kingdon ion trap [63]. Representative examples are lifetimes for N^{5+} [64], Be-like N^{3+} and O^{4+} [65], intermediate$-Z$, He-like ions [66], and lifetimes for the coronal transitions in Fe^{9+} and Fe^{13+} [67].

Results of recent calculations of level lifetimes and oscillator strengths may be found in representative publications using the Breit-Pauli Hamiltonian in the *CIV3* atomic structure code, and using the ***R***-matrix method [68–74]. The order of 10^4 radiative transitions can be required to model stellar atmospheres and stellar spectra [75]!

Both the need for atomic A-values and collisions strengths in astrophysical observations, and the power of present-day orbiting X-ray spectrometers is best illustrated by recent data from *Chandra's* Low-Energy Transmission Grating Spectrometer (LETGS) [76] and XMM-*Newton's* Reflection Grating Spectrometer (RGS) [77]. The LETGS is a diffraction grating spectrometer with a resolution of $\approx 0.06\mathring{A}$ over the wavelength range 2-175 $\mathring{A}$ (0.07-6 keV). Shown in fig. 3.4 is the LETGS X-ray spectrum at the star *Capella* of the He-like triplets in C V, N VI, and O VII. These are the so-called forbidden (f) 1s2s ^{3}S $\rightarrow$ 1s^2 ^{1}S, intersystem (i) (1s2p ^{3}P^o + 1s2s ^{1}S) $\rightarrow$ 1s^2 ^{1}S, and resonance (r) 1s2p ^{1}P^o $\rightarrow$ 1s^2 ^{1}S transitions. The observed intensity ratios f/i and $(f+i)/r$ are used as density diagnostics. Due to the absence of experimental data the collision strengths for these ratios (needed to back out the electron density) were calculated in an early distorted-wave theory [78].

2.3 Highly-Charged Ions in Solar and Stellar Winds

Exciting and unexpected interactions of HCIs with neutrals were observed *via* resulting X-rays detected at the comet C/Hyakutake 1996 B2 [79] using *ROSAT*; and most recently at comet C/Linear 1999 S4 using the *Chandra* X-ray Observatory [80]. X-ray observations have also been reported from EUVE for the comets d'Arrest, Borrelly, Bradfield, Encke, Hyakutake, Hale-Bopp, and Mueller [81–85]. Further studies of the X-rays from Hale-Bopp [81–83] and Levy, Tsuchiya-Kiuchi, Honda-Mrkos-Pajdušáková and Arai [86] have also been reported.

It appears that X-ray emission from comets is a general phenomenon. Furthermore, the phenomena would appear to extend to *other* astrophysical objects where any "wind" of highly-charged ions (HCIs) can scatter from a neutral cloud, whether the cloud be from a comet, a planetary atmosphere [87], or a circumstellar neutral cloud or the ISM impinged by a stellar wind [88]. Predictions have been made of X-rays emitted by charge-transfer collisions of heavy solar-wind HCIs and interstellar/interplanetary neutral clouds [89].

As to the origin of the comet X-rays, it was initially suggested that the emissions could be explained by bremsstrahlung associated with hot electrons in the solar wind interacting with the cometary plasma [90]. These and other mechanisms, such as scattering of solar X-rays, and the interaction of the HCIs with cometary dust particles, were studied [81, 91]. Several proposed mechanisms

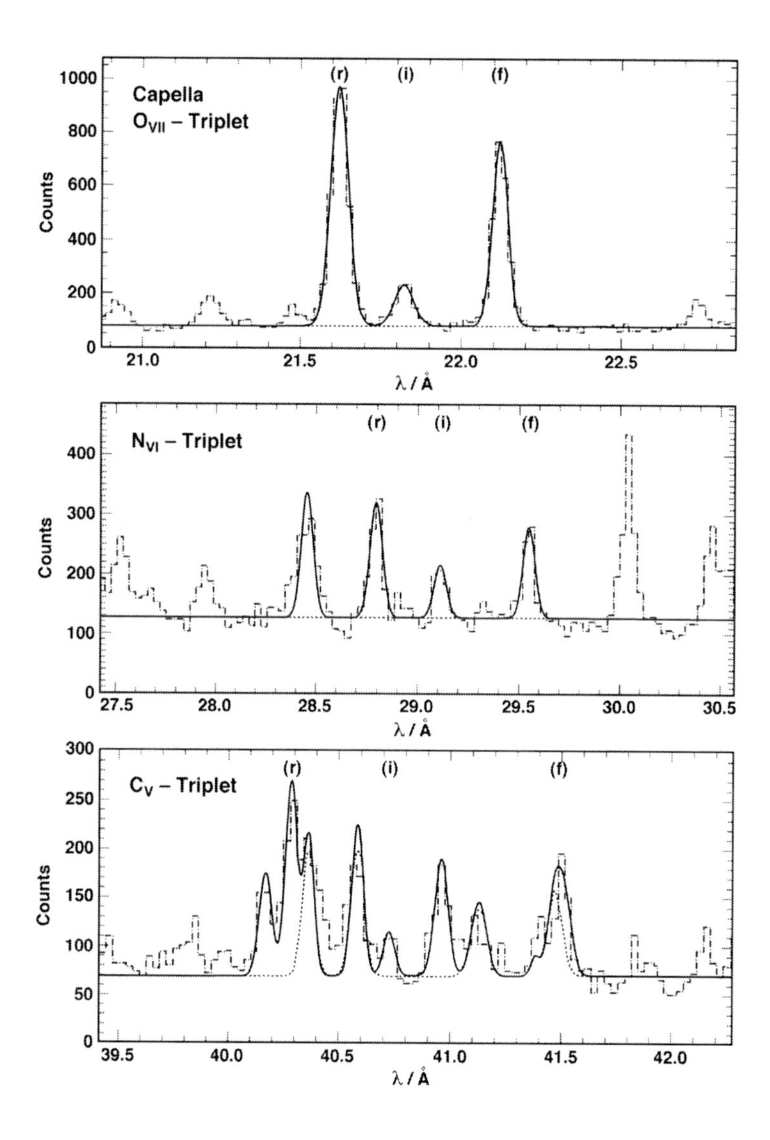

Figure 3.4. Spectra (dot-dash line), fit (solid line), and background (dotted line) for the resonance (r), intercombination (i), and forbidden (f) transitions in the He-like C V, N VI, and O VII triplets. Observations were made at the star *Capella* using the LETGS aboard *Chandra* [76].

are: (a) production of X-rays from charge-exchange with the cometary neutral gases, (b) bremsstrahlung between the solar-wind electrons and the cometary neutrals [90] and (c) combined bremsstrahlung and collisional excitation of highly-charged C, N, and O ions by electrons accelerated in the lower hybrid waves of the solar wind-photoion plasma [92]. The charge-exchange mechanism successfully accounts for the X-ray spectral energies and intensities based on best knowledge of ion fluxes and neutral densities [80–84, 86, 91, 93–96]. In this scenario highly-stripped solar-wind ions interact with the cometary neutral species. The excited, recombined solar-wind HCIs emits X-rays as they cascade to their ground electronic states. A list of the solar HCI abundances used in recent modeling [97] is given in Table 3.1.

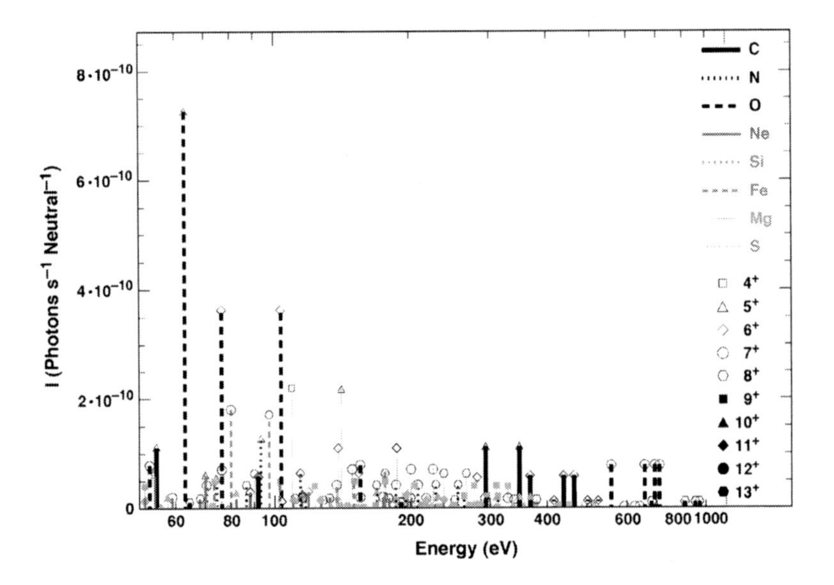

Figure 3.5. Charge-exchange, X-ray photon count rate per cometary neutral, for the condition of fast solar wind and optically thin target [97]. The contributing solar-wind species are listed at the right. A flat cross section *vs* energy is assumed, and multiple charge-exchanges ($\Delta q \geq 2$) are neglected.

Using O^{8+} as an example of a solar-wind ion, and neutral H_2O as a cometary species, the overall mechanism is,

$$O^{8+} + H_2O \rightarrow O^{7+*} + H_2O^+, \tag{3.6}$$

Table 3.1. Highly-charged heavy ions present in the solar wind, and their abundance relative to the total oxygen-ion abundance (taken as unity). Abundances in the slow (~4400 km s^{-1}) and fast (~4750 km s^{-1}) solar-wind distributions are listed, as well as total ion energies in each (adapted from Ref. [97]).

HCI, X^{q+}	[X^{q+}]/[O] fast	[X^{q+}]/[O] slow	Energy (keV) fast	Energy (keV) slow
C^{6+}	0.085	0.318	35.0	9.95
C^{5+}	0.440	0.210		
N^{7+}	0.000	0.006	40.8	11.6
N^{6+}	0.011	0.058		
N^{5+}	0.127	0.065		
O^{8+}	0.000	0.070	46.6	13.3
O^{7+}	0.030	0.200		
O^{6+}	0.970	0.730		
Ne^{8+}	0.102	0.084	58.3	16.6
Ne^{7+}	0.005	0.004		
Mg^{10+}	0.029	0.098	70.0	19.9
Mg^{9+}	0.044	0.052		
Mg^{8+}	0.028	0.041		
Mg^{7+}	0.007	0.017		
Mg^{6+}	0.003	0.009		
Si^{10+}	0.024	0.021	81.6	23.2
Si^{9+}	0.045	0.049		
Si^{8+}	0.022	0.057		
Si^{7+}	0.002	0.000		
S^{11+}	0.001	0.000	93.3	26.5
S^{10+}	0.008	0.005		
S^{9+}	0.027	0.016		
S^{8+}	0.023	0.019		
S^{7+}	0.005	0.006		
S^{6+}	0.001	0.002		
Fe^{13+}	0.005	0.002	163	46.4
Fe^{12+}	0.017	0.007		
Fe^{11+}	0.025	0.023		
Fe^{10+}	0.025	0.031		
Fe^{9+}	0.015	0.041		
Fe^{8+}	0.005	0.034		
Fe^{7+}	0.001	0.007		

where the excited O^{7+*} ions emit X-ray photons. Detailed X-ray spectra and absolute charge-exchange cross sections are required for modeling the spectra and intensity in the *ROSAT*, EUVE, *Chandra*, and XMM-*Newton* observations.

Using contributions from the many solar wind species and charge states, the expected changes of X-ray intensity have been calculated [96] as a function of fast (750 km/s) and slow (400 km/s) solar winds, and optically thick and thin scattering [97]. One result of Ref. [97] is shown in fig. 3.5. In the absence of data, assumptions had to be made about (a) the energy dependence of the single charge-exchange cross sections (assumed flat with ion energy), and about the magnitude of the multiple charge-exchange cross sections (neglected). The charge-exchange cross sections were calculated [98] from the expression given by the over-barrier model [99],

$$n \leq q\left(2I_p\left(1+\frac{q-1}{2\sqrt{q}+1}\right)\right)^{-1/2}, \tag{3.7}$$

where n is the largest integer satisfying the inequality, q is the ionic charge and I_p is the ionization potential (au). The charge-exchange cross sections $\sigma_{qq'}$ between ionic states q and q' are given as $\sigma_{qq'} = 0.88 \times 10^{-16}\ \mathrm{R}_c^2\ \mathrm{cm}^2$, where the crossing radius R_c is given as,

$$R_c = \frac{q-1}{q^2/2n^2 - I_p} \tag{3.8}$$

Where comparisons can be made between actual measurements and eq. 3.8, factors-of-two differences exist. For example, the single-exchange cross sections *assumed* [82] *vs measured* [100] for O^{5+} + H_2O collisions are (in units of 10^{-15} cm^2) 2 *vs* 4.3; and for O^{7+} + H_2O, 12 *vs* 5.3. Accurate single and multiple charge-exchange cross sections [101] and X-ray emission cross sections [102] must be measured for the various ionization states of the solar-wind species C, N, O, Ne, Mg, Si, Fe in the relevant solar-wind velocity range, and in collisions with neutral cometary species (table 3.1).

Shortly after the first ROSAT X-ray observations were reported, laboratory measurements were started on X-ray emission spectra for collisions of some of the partially- and fully-stripped ions in table 3.1 interacting with the major comet gases. These gases were CO, CO_2, and H_2O (and will also include NH_3). X-ray spectra for the system of fully-stripped oxygen O^{8+} + H_2O (eq. 3.6) and O^{8+} + CO_2 are shown in fig. 3.6 [103]. X-ray emission cross sections were obtained from these spectra by fitting the peaks to the underlying Lyman transitions *np* $\rightarrow$ *1s* for the H-like O^{7+} ion, then normalizing the area to the

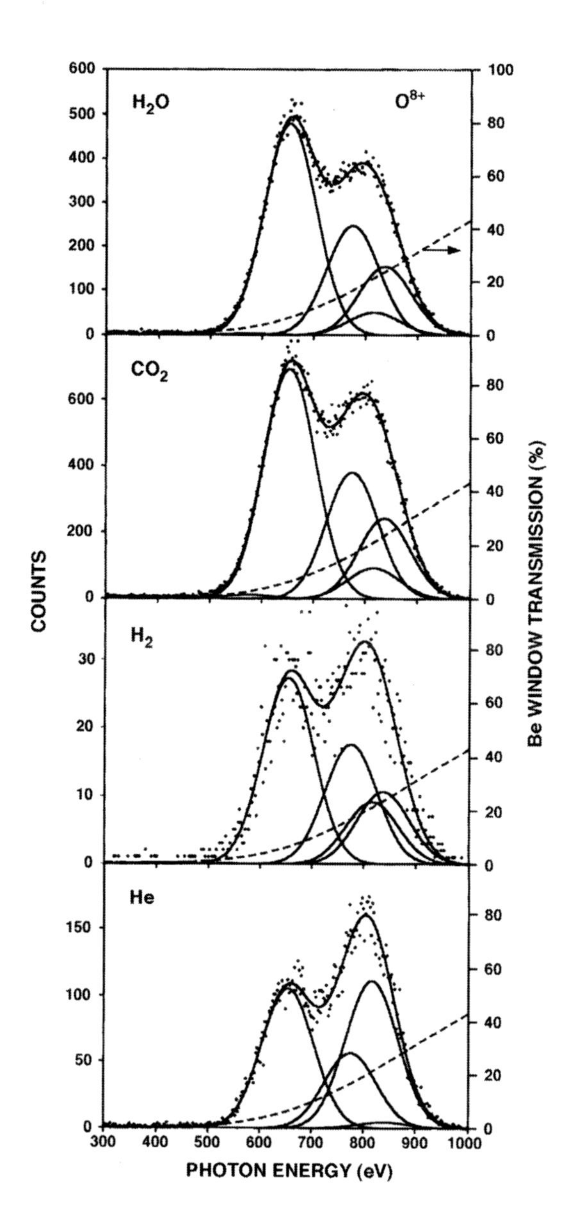

Figure 3.6. X-ray spectra for collisions of O^{8+} in He, H_2, CO_2, and H_2O at a total ion energy of 56 keV. Underlying curves are the Lyman contributions $np \rightarrow 1s$, and dashed line is the X-ray detector response curve [103].

Table 3.2. Compendium of home pages for satellite operations.

Satellite	Home Page
Hubble Space Telescope	http://www.stsci.edu/
SOHO	http://sohowww.nascom.nasa.gov/
FUSE	http://fuse.pha.jhu.edu/
EUVE (mission ended)	http://archive.stsci.edu/ euve/simple_euve.html
ROSAT	http://wave.xray.mpe.mpg.de/rosat
TRACE	http://vestige.lmsal.com/TRACE
Chandra X-Ray Observatory	http://asc.harvard.edu/
Constellation X Mission	http://constellation.gsfc.nasa.gov/
XMM-Newton	http://xmm.vilspa.esa.es/
SOFIA	http://sofia.arc.nasa.gov/
Infrared Space Observatory	http://www.iso.vilspa.esa.es/

total charge-exchange cross section, and correcting for cascades which populate metastable levels that decay outside the field of view of the detector [100, 101].

One sees that the data of fig. 3.6 are a subset of the emission channels addressed in fig. 3.5. There is a continuing need for emission cross sections for the system parameters listed in table 3.1, with attention to relevant species in the comet, planetary atmosphere [87], the ISM [88, 89], *etc.* Moreover, the energy range of required data is not limited to that in table 3.1, but can extend to several MeV/amu for charge transfer collisions with H_2 in the Jovian magnetosphere [87].

3. Summary

A list of useful web based resources are given in tables 3.2 and 3.3.

Laboratory collision physics involving highly-charged ions has direct contact with astrophysical phenomena. The surprising observation of cometary X-rays was quickly addressed both theoretically and (thanks to earlier investments

Table 3.3. Compendium of several useful atomic databases.

Atomic Database	Home Page
CHIANTI	http://www.solar.nrl.navy.mil/chianti.html
NIST Atomic Database	http://physics.nist.gov/cgi-bin /AtData/levels_form
Atomic Data for Astrophysics	http://www.pa.uky.edu/~verner/atom.html
The IRON Project	http://www.am.qub.ac.uk/projects/iron/
Redbook data for fusion, charge-exchange, *etc.*	http://www-cfadc.phy.ornl.gov/astro/ps/data/
Comet-Related Collision Research	http://www.qub.ac.uk/mp/ampr /networks/cometxrays.htm

in laboratory infrastructure by funding agencies) through laboratory measurements. A full range of atomic collision data is required to obtain accurate photon-emission intensities. These intensities can then be related back to electron and ion temperatures, densities, and ionization fractions in the astronomical plasma. A large body of data is needed for many phenomena; and for combinations of many transitions in a variety of ions and their charge states. Almost certainly the lion's share of the data must be calculated within an accurate theory. Experiment must be used to provide missing data and be required to provide critical tests of theory, as "ground truth." Those benchmarked theories can then, with higher confidence, be used to calculate collision phenomena for ions/charge-states/transitions where comparisons have not been made, usually for reasons of time and resolution.

In summary, the required data and their application are:

1. collision strengths for obtaining collisional excitation rates, assuming usually a Maxwellian electron energy distribution function in the astrophysical plasma,

2. lifetimes, branching ratios, and Einstein A-values for the coronal models,

3. direct ionization and indirect ionization cross sections, for both outer- and inner-shell electrons, to calculate ionization fractions in a plasma,

4. radiative and dielectronic recombination cross sections for calculation of ionization fractions,

5 dependencies of cross sections and lifetimes on ℓ, m level mixing by external **E** and **B** fields, such as encountered in a stellar object and within sunspots or high-velocity explosive events,

6 single and multiple charge-exchange cross sections from about 1-10^5 keV/amu for understanding solar- and stellar-wind interactions with comets and circumstellar neutral clouds; and for neutral collisions with magnetospherically accelerated ions,

7 X-ray emission cross sections in 6, above, with the relevant neutral atomic or molecular target.

There are other no less-important atomic parameters which go into a complete model. These include accurate HCI energy levels (wavelengths) [104], damping constants for atomic lines, photoionization cross sections, isotopic splittings, and hyperfine splittings. It is clear that space-based and ground-based instruments are becoming more sophisticated, sensitive, and able to cover a wider wavelength range at resolutions and sensitivities that are even challenging to laboratory instruments (*e.g.*, the GHRS and STIS spectrometers on *HST*). With increasing resolution, for example, more spectral lines can be identified, resulting in a wider range of detected species requiring more of (1)-(7) above. This trend will clearly continue with SOFIA and Constellation-X.

Acknowledgments

This work was carried out at the Jet Propulsion Laboratory/California Institute of Technology, and was supported under contract with the National Aeronautics and Space Administration.

References

[1] D. Moses and 34 co-authors, Solar Physics **175**, 571 (1997);
U. Feldman, I. E. Dammasch, and K. Wilhelm, Space Science Revs. **93**, 411 (2000)

[2] K. Stucki, S. K. Solanki, U. Schühle, I. Rüedi, K. Wilhelm, J. O. Stenflo, A. Brković, and M. C. E. Huber, Astronom. Astrophys. **363**, 1145 (2000)

[3] U. Feldman, W. Curdt, E. Landi, and K. Wilhelm, Astrophys. J. **544**, 508 (2000)

[4] See the issue, *Electron Excitation Data for Analysis of Spectral Line Radiation from Infrared to X-Ray Wavelengths: Reviews and Recommendations*, Atom. Data Nucl. Data Tables **57** (1994)

[5] See, for example, *Atomic and Molecular Processes in Fusion Edge Plasmas* Plenum NY, (1995)

[6] I. H. Coffey, R. Barnsley, F. P. Keenan, and N. J. Peacock, J. Phys. B **27**, 1011 (1994)

[7] S. H. Glenzer and 12 co-authors, Phys. Plasmas **6**, 2117 (1999)

[8] R. A. Phaneuf, Physica Scripta **T47**, 124 (1993)

[9] A. H. Gabriel and C. Jordan, Ch. 4 in *Case Studies in Atomic Collision Physics II* ed. E. W. McDaniel and M. R. C. McDowell, American Elsevier NY, (1972)

[10] A. H. Gabriel and H. E. Mason, Ch. 10 in *Applied Atomic Collision Physics, Vol.1* eds. H. Massey, E. W. McDaniel, and B. Bederson, Academic NY, (1982)

[11] S. S. Tayal, A. K. Pradhan, and M. S. Pindzola, Ch. 6 in *Atomic and Molecular Processes in Fusion Edge Plasmas* Plenum NY, (1995)

[12] F. P. Keenan, V. J. Foster, K. M. Aggarwal, and K. G. Widing, Solar Physics **169**, 47 (1996)

[13] F. P. Keenan, D. J. Pinfield, V. J. Woods, R. H. G. Reid, E. S. Conlon, A. K. Pradhan, H. L. Zhang, and K. G. Widing, Astrophys. J. **503**, 953 (1998)

[14] G. A. Doschek, H. P. Warren, J. M. Laming, J. T. Mariska, K. Wilhelm, P. Lemaire, U. Schühle, and T. G. Moran, Astrophys. J. **482**, L109 (1997)

[15] F. P. Keenan, P.44 in *UV and X-ray Spectroscopy of Laboratory and Astrophysical Plasmas* ed. E. Silver and S. Khan, Cambridge Univ. Press, UK, (1993)

[16] H. E. Mason, G. A. Doschek, U. Feldman, and A. K. Bhatia, Astron. Astrophys. **73**, 74 (1979)

[17] M. Malinovsky, Astron. Astrophys. **43**, 101 (1975); M. Malinovsky, L. Heroux, and S. Sahal-Bréchot, Astron. Astrophys. **23**, 391 (1973)

[18] A. Chutjian and W. R. Newell, Phys. Rev. A **26**, 2271 (1982)

[19] I. D. Williams, A. Chutjian and R. J. Mawhorter, J. Phys. B **19**, 2189 (1986)

[20] B. A. Huber, C. Ristori, C. Guet, D. Küchler, and W. R. Johnson, Phys. Rev. Lett. **73**, 2301 (1994)

[21] I. D. Williams, B. Srigengan, A. Platzer, J. B. Greenwood, W. R. Newell, and L. O'Hagan, Phys. Scripta **T73**, 121 (1997)

[22] J. A. Lozano, M. Niimura, S. J. Smith, A. Chutjian, and S. S. Tayal, Phys. Rev. A **63**, 042713 (2001)

[23] S. J. Smith, K–F. Man, R. J. Mawhorter, I. D. Williams and A. Chutjian, Phys. Rev. Lett. **67**, 30 (1991)

[24] E. K. Wåhlin, J. S. Thompson, G. H. Dunn, R. A. Phaneuf, D. C. Gregory, and A. C. H. Smith, Phys. Rev. Lett. **66**, 157 (1991)

[25] O. Woitke, N. Djurić, G. H. Dunn, M. E. Bannister, A. C. H. Smith, B. Wallbank, N. R. Badnell, and M. S. Pindzola, Phys. Rev. A **58**, 4512 (1998)

[26] J. B. Greenwood, S. J. Smith, A. Chutjian, and E. Pollack, Phys. Rev. A **59**, 1348 (1999)

[27] M. Niimura, S. J. Smith, and A. Chutjian, Astrophys. J. **565**, 645 (2002)

[28] E. W. Bell and 15 co-authors, Phys. Rev. A **49**, 4585 (1994)

[29] B. Wallbank, N. Djurić, O. Woitke, S. Zhou, G. H. Dunn, A. C. H. Smith, and M. E. Bannister, Phys. Rev. A **56**, 3714 (1997)

[30] S. J. Smith, J. B. Greenwood, A. Chutjian, and S. S. Tayal, Astrophys. J. **541**, 501 (2000)

[31] Y-S. Chung, N. Djurić, B. Wallbank, G. H. Dunn, M. E. Bannister, A. C. H. Smith, Phys. Rev. A **55**, 2044 (1997)

[32] X. Q. Guo, E. W. Bell, J. S. Thompson, G. H. Dunn, M. E. Bannister, R. A. Phaneuf, and A. C. H. Smith, Phys. Rev. A **47**, R9 (1993)

[33] F. Herbert and D. T. Hall, J. Geophys. Res. **103**, 19 915 (1998)

[34] D. T. Hall, G. R. Gladstone, H. W. Moos, F. Bagnal, J. T. Clarke, P. D. Feldman, M. A. McGrath, N. M. Schneider, D. E. Shemansky, D. F. Strobel, and J. H. Waite, Astrophys. J. Lett. **426**, L51 (1994)

[35] D. E. Shemansky, J. Geophys. Res. **93**, 1773 (1988)

[36] B. D. Savage, K. R. Sembach, E. B. Jenkins, J. M. Shull, D. G. York, G. Sonneborn, H. W. Moos, S. D. Friedman, J. C. Green, W. R. Oegerle, W. P. Blair, J. W. Kruk, and E. M. Murphy, Astrophys. J. Lett. **538**, L27 (2000)

[37] W. P. Blair, R. Sankrit, R. Shelton, K. R. Sembach, H. W. Moos, J. C. Raymond, D. G. York, P. D. Feldman, P. Chayer, E. M. Murphy, D. J. Sahnow, and E. Wilkinson, Astrophys. J. Lett. **538**, L61 (2000)

[38] G. Ramsay, F. Córdova, J. Cottam, K. Mason, R. Much, J. Osborne, D. Pandel, T. Poole, and P. Wheatley, Astron. Astrophys. **365**, L294 (2001)

[39] T. R. Ayres, A. Brown, R. A. Osten, D. P. Huenemoerder, J. J. Drake, N. S. Brickhouse, and J. L. Linsky, Astrophys. J. **549**, 554 (2001)

[40] M. Sako, S. M. Kahn, F. Paerels, and D. A. Liedahl, Astrophys. J. Lett. **543**, L115 (2000)

[41] G. A. Doschek, Astrophys. J. **527**, 426 (1999); G. A. Doschek and U. Feldman, Astrophys. J. **313**, 883 (1987)

[42] M. Arnaud and J. Raymond, Astrophys. J. **398** 394 (1992)

[43] R. Rejoub and R. A. Phaneuf, Phys. Rev. A **61**, 032706 (2000)

[44] J. Colgan, D. M. Mitnick, and M. S. Pindzola, Phys. Rev. A **63**, 012712 (2000)

[45] G. Gwinner and 16 co-authors, Phys. Rev. Lett. **84**, 4822 (2000)

[46] T. Bartsch, S. Schippers, A. Müller, C. Brandau, G. Gwinner, A. A. Saghiri, M. Beutelspacher, M. Grieser, D. Schwalm, A. Wolf, H. Danared, and G. H. Dunn, Phys. Rev. Lett. **82**, 3779 (1999)

[47] T. Bartsch, S. Schippers, M. Beutelspacher, S. Böhm, M. Grieser, G. Gwinner, A. A. Saghiri, G. Saathoff, R. Schuch, D. Schwalm, A. Wolf, and A. Müller, J. Phys. B. **33**, L453 (2000)

[48] A. R. Young, L. D. Gardner, D. W. Savin, G. P. Lafyatis, A. Chutjian, S. Bliman, and J. L. Kohl, Phys. Rev. A **49**, 3577 (1994); D. W. Savin, L. D. Gardner, D. B. Reisenfeld, A. R. Young, and J. L. Kohl, Phys. Rev. A **53**, 280 (1996)

[49] D. B. Reisenfeld, J. C. Raymond, A. R. Young, and J. L. Kohl, Astrophys. J. Lett. **389**, L37 (1992); D. B. Reisenfeld, Astrophys. J. **398**, 386 (1992)

[50] T. Bartsch and 12 co-authors, Phys. Rev. Lett. **79,** 2244 (1997)

[51] K. J. H. Phillips, *Guide to the Sun* Cambridge, (1992)

[52] M. C. Chidichimo, V. Zeman, J. A. Tully, and K. A. Berrington, Astron. Astrophys. Suppl. **Ser. 137**, 175 (1999); K. Butler, Phys. Scripta **T65**, 63 (1996)

[53] D. W. Savin and 15 co-authors, Astrophys. J. Suppl. **Ser. 123,** 687 (1999)

[54] M. F. Gu, S. M. Kahn, D. W. Savin, P. Beiersdorfer, G. V. Brown, D. A. Liedahl, K. J. Reed, C. P. Bhalla, and S. R. Grabbe, Astrophys. J. **518**, 1002 (1999)

[55] M. Niimura,I. Ćadež, S. J. Smith, and A. Chutjian, Phys. Rev. Lett. **88** 103201 (2002)

[56] E. Landi, M. Landini, K. P. Dere, P. R. Young, and H. E. Mason, Astron Astrophys. **135**, 339 (1999)

[57] K. R. Sembach, B. D. Savage, and M. Hurwitz, Astrophys. J. **524**, 98 (1999)

[58] C. W. Mauche and J. C. Raymond, Astrophys. J. **541**, 24 (2000)

[59] A. W. Fullerton and 12 co-authors, Astrophys. J. Lett. **538**, L46 (2000)

[60] L. Spitzer, *Physical Process in the Interstellar Medium* Wiley, New York, (1978) Ch. 3; B. D. Savage and K. R. Sembach, Astrophys. J. **379**, 245 (1991)

[61] E. Träbert, A. G. Calamai, J. D. Gillaspy, G. Gwinner, X. Tordoir, and A. Wolf, Phys. Rev. A **62**, 022507 (2000)

[62] E. Träbert, P. Beiersdorfer, S. B. Utter, G. V. Brown, H. Chen, C. L. Harris, P. A. Neill, D. W. Savin, and A. J. Smith, *Astrophys. J.* **541**, 506 (2000); E. Träbert, Phys. Scripta **61**, 257 (2000)

[63] D. A. Church, Phys. Rep. **228**, 253 (1993); D. A. Church, D. P. Moehs, and M. Idrees Bhatti, Int. J. Mass Spectrom. **192**, 149 (1999); D. P. Moehs, D. A. Church, M. I. Bhatti, and W. F. Perger, Phys. Rev. Lett. **85**, 38 (2000)

[64] P. A. Neill, E. Träbert, P. Beiersdorfer, G. V. Brown, C. L. Harris, S. B. Utter, and K. L. Wong, Phys. Scripta, **62**, 141 (2000)

[65] J. Doerfoert, E. Träbert, and A. Wolf, Hyperfine Interactions **99**, 155 (1996)

[66] A. J. Smith, P. Beiersdorfer, K. J. Reed, A. L. Osterheld, V. Decaux, K. Widman, and M. H. Chen, Phys. Rev. A **62**, 012704 (2000)

[67] D. P. Moehs and D. A. Church, Astrophys. J. Lett. **516**, L111 (1999)

[68] K. M. Aggarwal, A. Hibbert, and F. P. Keenan, Astrophys. J. Suppl. **Ser. 108**, 393 (1997)

[69] S. S. Tayal, J. Phys. B **32**, 5311 (1999); ibid., Atom. Data Nucl. Data Tables **67**, 331 (1997)

[70] N. C. Deb and S. S. Tayal, Atom. Data Nucl. Data Tables **69**, 161 (1998)

[71] S. N. Nahar, Atom. Data Nucl. Data Tables **72**, 129 (1999)

[72] A. K. Pradhan, Phys. Script. **T83**, 69 (1999)

[73] N. C. Deb, K. M. Aggarwal, and A. Z. Msezane, Astrophys. J. Suppl. **Ser. 121**, 265 (1999)

[74] K. M. Aggarwal, P. H. Norrington, K. L. Bell, F. P. Keenan, G. J. Pert, and S. J. Rose, Atom. Data Nucl. Data Tables **74**, 157 (2000)

[75] M. van Noort, T. Lanz, H. J. G. L. M. Lamers, R. L. Kurucz, R. Ferlet, G. Hébrard, and A. Vidal-Madjar, Astron. Astrophys. **334**, 633 (1998); R. L. Kurucz, Phys. Scripta **T47**, 110 (1993)

[76] J-U. Ness, R. Mewe, J. H. M. M. Schmitt, A. J. J. Raassen, D. Porquet, J. S. Kaastra, R. L. J. van der Meer, V. Burwitz, and P. Oredehl, Astron. Astrophys. **367**, 282 (2001)

[77] S. M. Kahn, M. A. Leutenegger, J. Cottam, G. Rauw, J-M. Vreux, A. J. F. den Boggende, R. Mewe, and M. Güdel, Astron. Astrophys. **365**, L312 (2001)

[78] A. K. Pradhan, D. W. Norcross, and D. G. Hummer, Astrophys. J. **246**, 1031 (1981)

[79] C. M. Lisse and 11 co-authors, Science **274**, 205 (1996)

[80] C. M. Lisse, D. J. Christian, K. Dennerl, K. J. Meech, R. Petre, H. A. Weaver, and S. J. Wolk, Science **292**, 1343 (2001)

[81] M. J. Mumma, V. A. Krasnopolsky, and M. J. Abbott, Astrophys. J. **491**, L125 (1997)

[82] V. A. Krasnopolsky, M. J. Mumma, and M. J. Abbott, Icarus **146**, 152 (2000); V. A. Krasnopolsky, J. Geophys. Res. **103** 2069 (1998)

[83] V. A. Krasnopolsky, M. J. Mumma, M. J. Abbott, B. C. Flynn, K. J. Meech, D. K. Yeomans, P. D. Feldman, and C. B. Cosmovici, Science **277**, 1488 (1997)

[84] R. M. Häberli, T. I. Gombosi, D. L. De Zeeuw, M. R. Combi, and K. G. Powell, Science **276**, 939 (1997)

[85] A. Owens and 9 co-authors, Nucl. Phys. B **69/1-3**, 735 (1998)

[86] K. Dennerl, J. Englhauser, and J. Trumper, Science **277**, 1625 (1997)

[87] T. E. Cravens, Adv. Space Res. **26**, 1443 (2000)

[88] B. J. Wargelin and J. J. Drake, Astrophys. J. **546**, L57 (2001)

[89] T. E. Cravens, Astrophys. J. **532**, L153 (2000)

[90] T. G. Northrop, C. M. Lisse, M. J. Mumma, and M. D. Desch, Icarus **127**, 246 (1997)

[91] V. A. Krasnopolsky, J. Geophys. Res. **103**, 2069 (1998)

[92] M. Torney, R. Bingham, J. M. Dawson, B. J. Kellet, V. D. Shapiro, and H. Summers, Phys. Scr. **T98**, 168 (2002)

[93] T. E. Cravens, Geophys. Res. Lett. **24**, 105 (1997)

[94] T. I. Gombosi, D. De Zeeuw, R. M. Häberli, and K. G. Powell, J. Geophys. Res. **101**, 15 233 (1996)

[95] C. M. Lisse, *et al.*, Earth, Moon and Planets **77**, 283 (1999)

[96] V. Kharchenko and A. Dalgarno, J. Geophys. Res. **105**, 18 351 (2000)

[97] N. A. Schwadron and T. E. Cravens, Astrophys. J. **544**, 558 (2000)

[98] R. Wegmann, H. U. Schmidt, C. M. Lisse, K. Dennerl, and J. Englhauser, Planet. Space Sci. **46**, 603 (1998)

[99] H. Ryufuku, K. Sasaki, and T. Watanabe, Phys. Rev. A **21**, 745 (1980)

[100] J. B. Greenwood, I. D.Williams, S. J. Smith, and A. Chutjian, Phys. Rev. A **63**, 062707 (2001)

[101] J. B. Greenwood, I. D.Williams, S. J. Smith, and A. Chutjian, Astrophys. J. Lett. **533**, L175 (2000)

[102] J. B. Greenwood, I. D. Williams, S. J. Smith, and A. Chutjian, p.157 in *Applications of Accelerators in Research and Industry* ed. J. L. Duggan and I. L. Morgan, CP576, AIP, New York, (2001)

[103] J. B. Greenwood, I. D. Williams, S. J. Smith, and A. Chutjian, Phys.Scripta **T92**, 150 (2001).

[104] See, for example, H. Watanabe, D. Crosby, F. J. Currell, T. Fukami, D. Kato, S. Ohtani, J. D. Silver, and C. Yamada, Phys. Rev. A **63**, 042513 (2001)

Chapter 4

HIGHLY CHARGED ION COLLISION PROCESSES IN HIGH TEMPERATURE FUSION PLASMAS

H. Tawara

Laboratory for Applications of Atomic Processes, Obata-Obora 31-33, Tajimi 507-0817, Japan
and
Department of Physics, Kansas State University, Manhattan, KS 66506-2406, USA

Abstract Following an introduction to high temperature nuclear fusion plasma tokamak devices, basic collision processes of highly charged ions relevant to plasmas and their importance to fusion plasma research are described. It is pointed out that the precise knowledge of collision processes results in better understanding of plasma features and their control which can result in achieving sustainable and stable nuclear fusion reactions with high temperature plasmas and thus providing clean energy to human beings in the future.

Keywords: high temperature plasmas, fusion, collision processes

1. Plasma Nuclear Fusion

1.1 Nuclear Fusion Reactions

To find new ways to efficiently produce our future energy with minimum contamination of our environment, the most promising method, among various proposed schemes, seems to be based upon nuclear fusion reactions. Nuclear fusion reactions which can be used to produce sufficient energy are the following two reactions, called D-T (process (4.1)) and D-D (processes 4.2 and 4.3):

$$^{2}\mathrm{D} + {}^{3}\mathrm{T} \rightarrow {}^{4}\mathrm{He} + \mathrm{n} + 17.59MeV. \tag{4.1}$$

$$^{2}\mathrm{D} + {}^{2}\mathrm{D} \rightarrow {}^{3}\mathrm{He} + \mathrm{n} + 3.27MeV \tag{4.2}$$

F.J. Currell (ed.), The Physics of Multiply and Highly Charged Ions, Vol. 1, 103-147.

$$\rightarrow {}^{3}\mathrm{T} + {}^{1}\mathrm{H} + 4.03 MeV. \tag{4.3}$$

Here we should note that the energy emitted in these nuclear fusion reactions is roughly five to six orders of magnitude larger than that in chemical reactions observed in oil- or coal-burning processes.

It is also important to note that, in these nuclear fusion reactions, most of the produced energy is carried by neutrons, which results in intense radio-activation and also radiation damage of wall materials of the vacuum chamber, except for reaction (4.3) where two charged particles of the reaction products share the produced energy. The reaction cross-sections are shown in fig. 4.1(a) as a

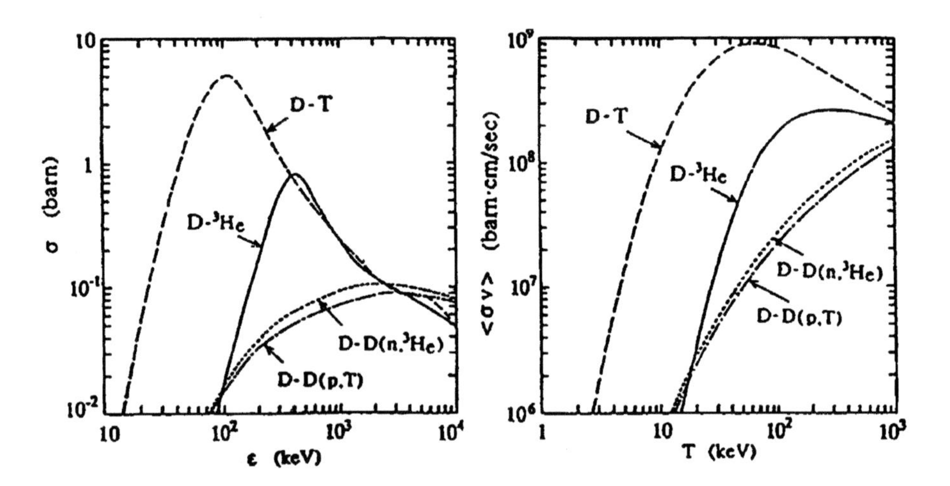

Figure 4.1. Cross-section of some nuclear fusion processes (a) and their fusion reaction rates averaged over the Maxwellian distributions (b). Data for another candidate for plasma nuclear fusion, ${}^{2}\mathrm{D} + {}^{3}\mathrm{He} \rightarrow {}^{4}\mathrm{He} + {}^{1}\mathrm{H} + 18.35$ MeV is also shown. Note that 10 keV is equal to 1.16×10^{8} K and 1 barn is equal to $10^{-24} cm^{2}$.

function of the kinetic energy of particles. Reaction (4.1) has a large cross-sections at relatively low energy. This is why presently the fusion programs are working extensively on the D-T reaction scheme, even though radioactive tritium (^{3}T) is required and thus the whole system becomes more complicated. Reactions (4.2) and (4.3) have roughly similar reaction cross-sections which are smaller than those for reaction (4.1) at low energies (temperatures).

In order to overcome the Coulomb barrier between two nuclei and to initiate nuclear fusion reactions, these particles have to be accelerated or heated up to proper energies or temperatures. The particles in plasmas are usually not mono-energetic but have some distributions of their energy (temperature). In most cases, their energy or temperature distributions can be assumed to be

Maxwellian. The averaged reaction rates for the nuclear reactions are shown in fig. 4.1(b).

The most important point in nuclear fusion reaction schemes, from the view point of energy production, is the fact that the main fuel used in these nuclear fusion reactions, namely deuterium (D), in principle is nearly limitless in sea water which contains roughly 0.015%, with the rest being the ordinary light hydrogen isotope (H). The other main fuel, tritium (T), is radioactive with the lifetime of about 12 years and does not exist in nature. Therefore, it has to be produced artificially through other nuclear reactions such as process (4.3) or more efficient nuclear reactions.

1.2 Comparison of Nuclear Fusion Plasmas with Other Plasmas

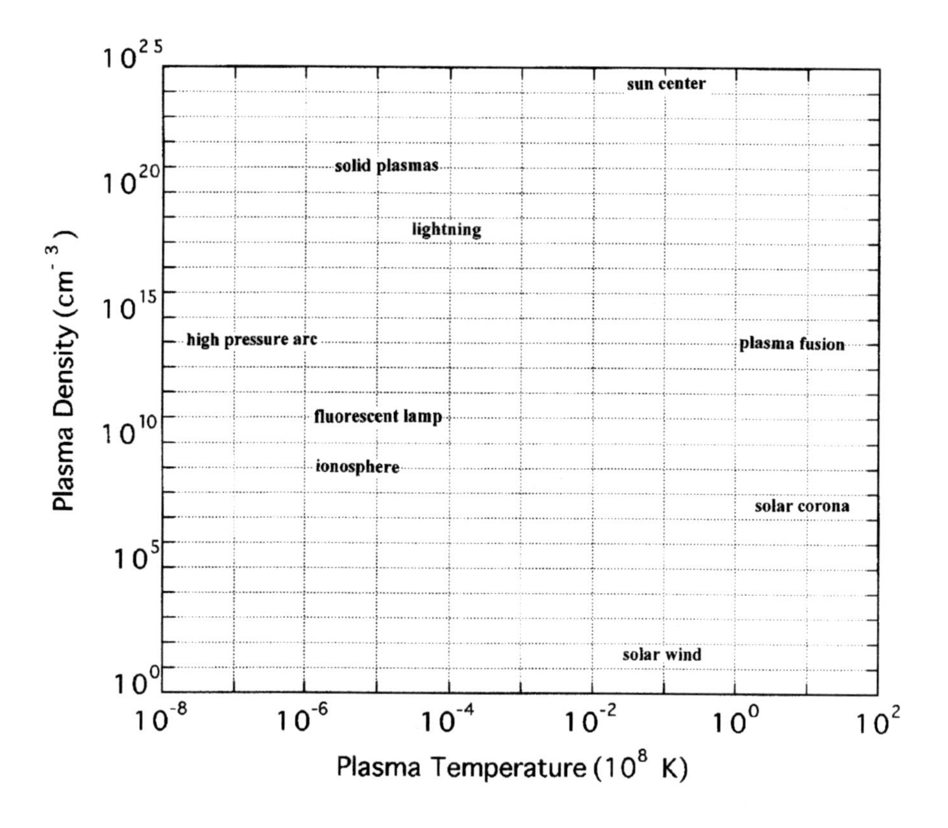

Figure 4.2. Comparison of various plasmas.

In our environment, a number of different kinds of plasmas are present. In fact, nuclear fusion plasmas are expected to be very similar to those in the sun, although there is a big difference between these two types of plasma: the plasma in the sun is confined by its huge gravity. In contrast, laboratory plasmas, including fusion plasmas, are produced in vacuum containers and have to be confined with strong electro-magnetic fields in order to avoid their direct contact with any surface material of the vacuum chamber. In fig. 4.2 typical features of plasmas around us are compared. There are many different plasmas, which differ by many orders of magnitude in their density and temperature. One of the most familiar is the plasma in fluorescent lamps which is typically of low temperature and density.

1.3 Lawson Criterion for Plasma Nuclear Fusion

As the plasma temperature increases through various heating techniques, nuclear fusion reactions occur and, at the same time, there are various energy loss mechanisms such as radiation losses (mainly due to Bremsstrahlung by "hot" plasma electrons) from high temperature plasmas. Thus it is expected that the best plasma conditions exist where the plasma temperatures are not too high. To overcome these problems and to achieve the energy production through high temperature plasma fusion reactions, a criterion, often called the *Lawson criterion*, inxxLawson criterion has to be fulfilled among the density (n), temperature (T) and confinement time (τ) of plasmas.

Typical D-T fusion plasmas need to fulfill the following criterion [1]: $n\tau > 10^{14}\mathrm{s/cm}^3$ at $T > 10$ keV. This means that the plasma temperature should exceed at least 10 keV with the confinement time of 1 second at the plasma density of $10^{14}/\mathrm{cm}^3$. To achieve this goal, many apparatus have been and also are being developed with some limited success so far.

2. Tokamak

To achieve this goal, the two main schemes based upon very different concepts are proposed: *magnetic confinement fusion* (MCF) and *inertia confinement fusion* (ICF). So far a number of different types of plasma confinement apparatus and devices have been proposed and tested. Among them, *Tokamak* type magnetic confinement system seems to be the most successful in many ways, although no energy production system, even as a prototype, has been realised yet. Its principle is based upon the secondary plasma current induced through the primary current in a transformer, as is shown in fig. 4.3. The distinctive feature here is the fact that the combination of the magnetic field induced by the secondary plasma current together with the applied toroidal magnetic field results in a helical field which automatically confines the secondary plasma current itself, producing a stable plasma in a doughnut-shape vacuum vessel.

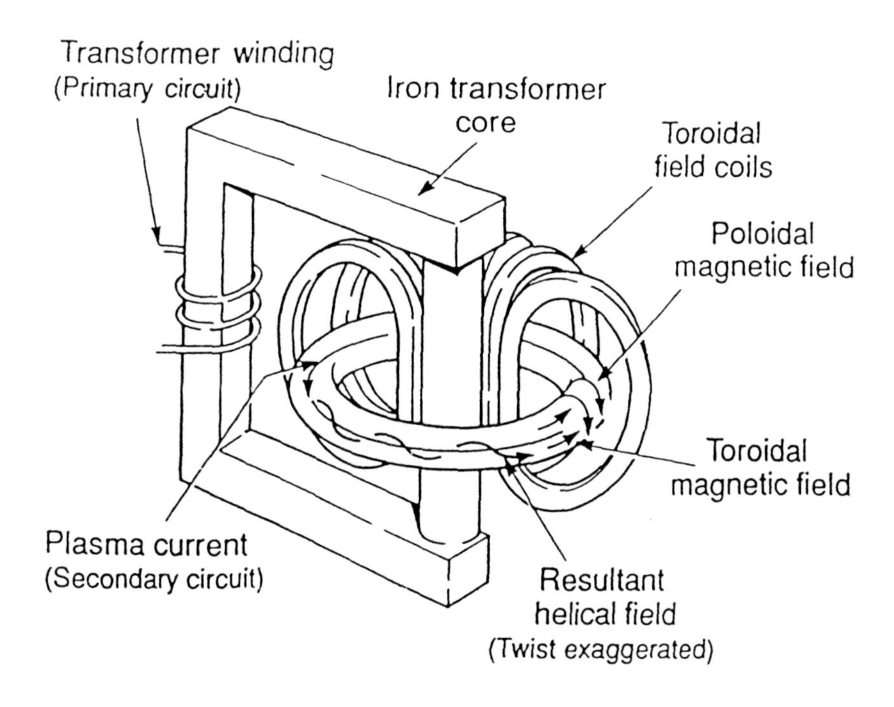

Figure 4.3. Principle of a tokamak. Note that the helical magnetic field is formed through the secondary current (plasma) combined with toroidal magnetic field to self-stabilise the doughnut-shaped plasma.

Table 4.1. Important parameters for the ITER plasma device. Note that a slightly smaller-scale ITER has been redesigned recently. Taken from [2].

Parameter	Value
major plasma radius	8.14 m
minor plasma radius	2.80 m
plasma volume	$\sim$ 2000 m^3
plasma surface	$\sim$ 1200 m^2
nominal plasma current	21 MA
toroidal magnetic field	5.68 T (at R = 8.14 m)
electron density	10^{16} $/m^3$
fusion power	1.5 GW
burning duration	$\geq$ 1000 s
auxiliary heating power	100 MW

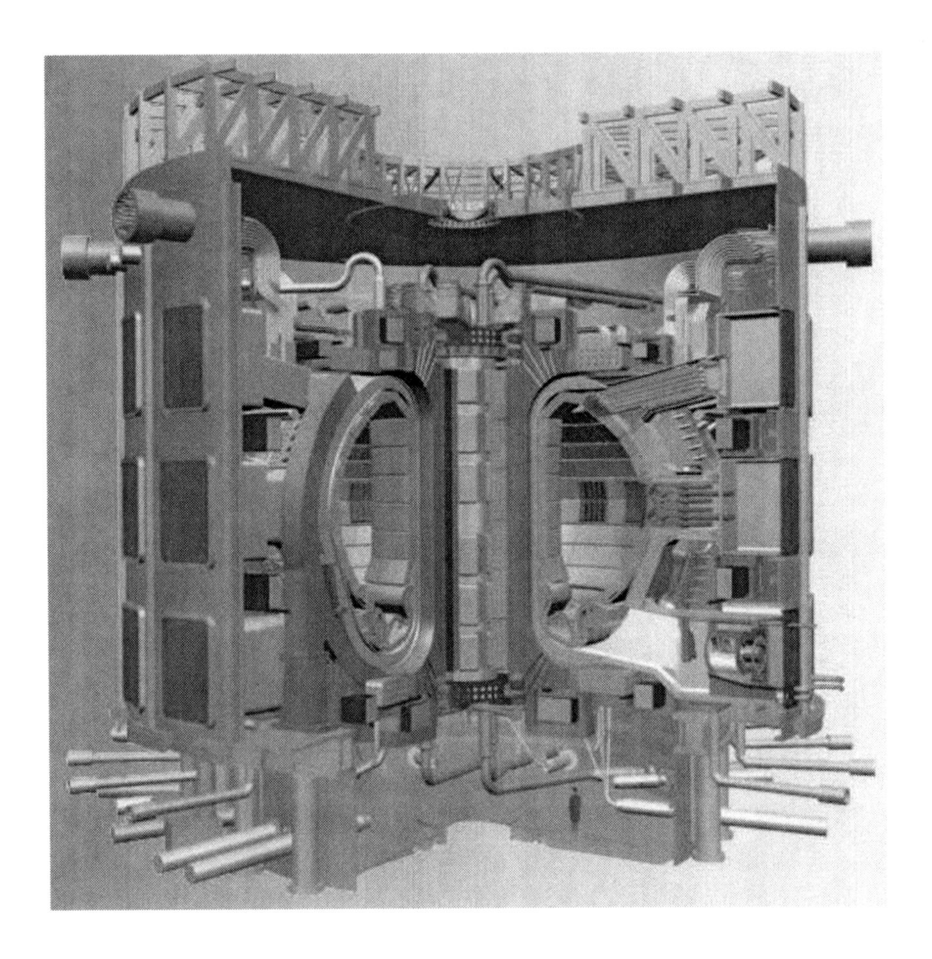

Figure 4.4. Structure of ITER. Compare it with the size of a human at the bottom.

The most advanced of this type is the International Tokamak Experimental Reactor (ITER) [2] which is an extraordinarily big international collaboration project costing more 10 billion dollars and presently under intensive development. The gross structure of ITER is shown in fig. 4.4. Its typical main parameters also are shown in Table 4.1. In fig. 4.5 a cutout view of the doughnut-shape plasmas is shown. Fig. 4.6 shows some detailed plasma features of tokamaks from the viewpoint of atomic processes.

Roughly speaking, tokamak plasmas can be divided into roughly four different categories:

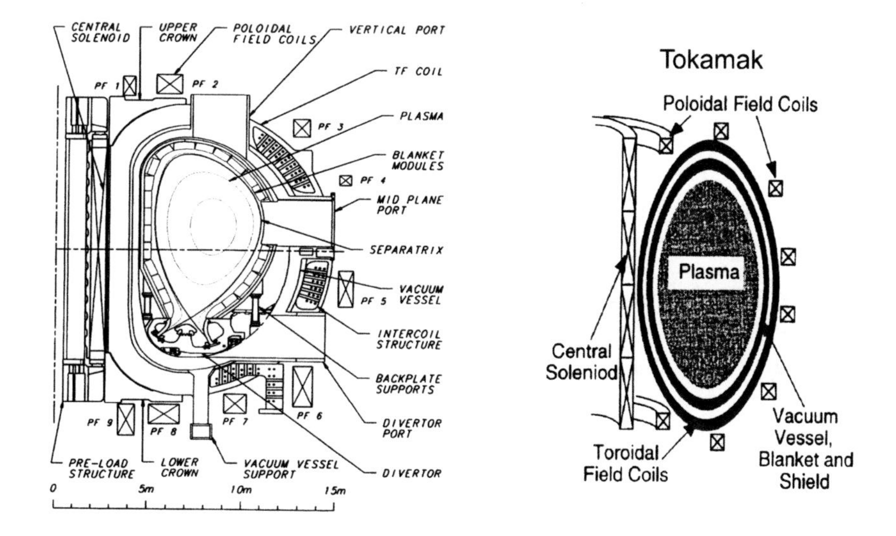

Figure 4.5. A section of ITER plasma showing the main structures and plasma.

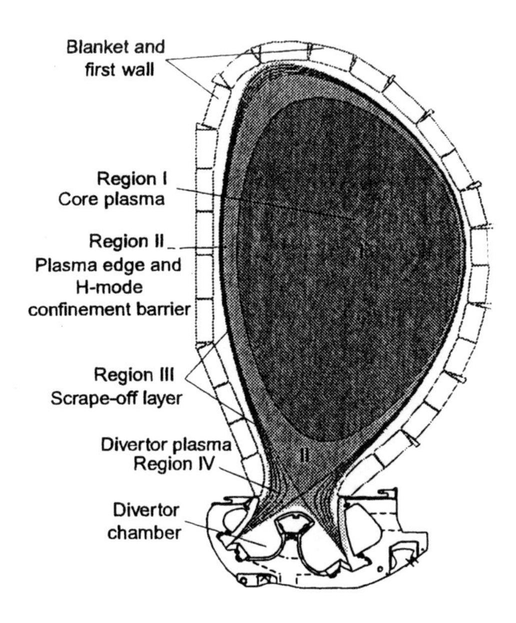

Figure 4.6. Various plasma regions in a tokamak. Note that typically four different types of plasma are present.

1. **Core plasmas** where nuclear fusion reactions occur: The temperatures in the core plasmas have to exceed 10 keV and their densities should be more than $10^{14}/\text{cm}^3$.

2. **Edge (pedestal) plasmas** which are more or less similar to the core plasmas in nature: The plasma temperatures, on average, are only slightly less than those in the core plasmas but can be used to control the core plasmas themselves.

3. **Scrape-off plasmas** which separate high temperature core plasmas from direct contact with the walls of the vacuum vessel over a very thin and narrow (less than 1 cm) region: The plasmas are colder than the core plasmas 1) and edge plasmas 2) and the plasma temperatures are 100-500 eV and their densities are $10^{13} - 10^{14}/\text{cm}^3$. The external poloidal magnetic field is provided to form a separatrix (specially shaped magnetic fields) which isolates plasmas 2) and 3).

4. **Diverter plasmas** which are cold and nearly neutralised but have high density plasma energies and are finally pumped out to the outside of the device as neutral gases: The plasma temperatures are as low as 10 - 100 eV and their densities are higher than those in 1), 2) and 3) and can be $10^{15} - 10^{16}/\text{cm}^3$.

The present chapter concentrates on issues related to highly charged ion, which are important in high temperature core fusion plasmas, although there are other important atomic physics issues in all plasma regions. It should be remembered that there are a number of the interesting but still unsettled atomic and molecular issues involving diverter plasmas [3].

3. Diagnostics and Modelling of Plasmas

General features and relevant atomic physics issues in tokamak plasmas have been discussed in detail [4]. Also the detailed description of the advanced diagnostics techniques recently has been given in connection of the extensive development of the ITER program [2].

3.1 Atomic Data Requirements in Fusion Plasma Research

There are two different research fields of plasmas where atomic data are particularly a requisite: 1) plasma diagnostics and 2) plasma modelling.

Generally, accurate atomic data for some specific atomic transition and collision processes are required in plasma diagnostics. On the other hand, to get an understanding of the overall features of plasmas through modelling, a series of atomic and molecular collision data is indispensable over a wide range of parameters such as plasma temperature or collision energy for the collision partners involved, although lower accuracy can be tolerated (uncertainties of a factor of about two are tolerable) and they should be systematic. Such data should be covered for all the charged states of particle species.

3.2 Plasma Modelling

To understand and model the global features of high temperature plasmas in tokamaks, such as how ions in different charge states behave, how the energy losses are distributed etc., first of all it is necessary to know the temporal and spatial distributions of the electron density and electron temperature. A typical

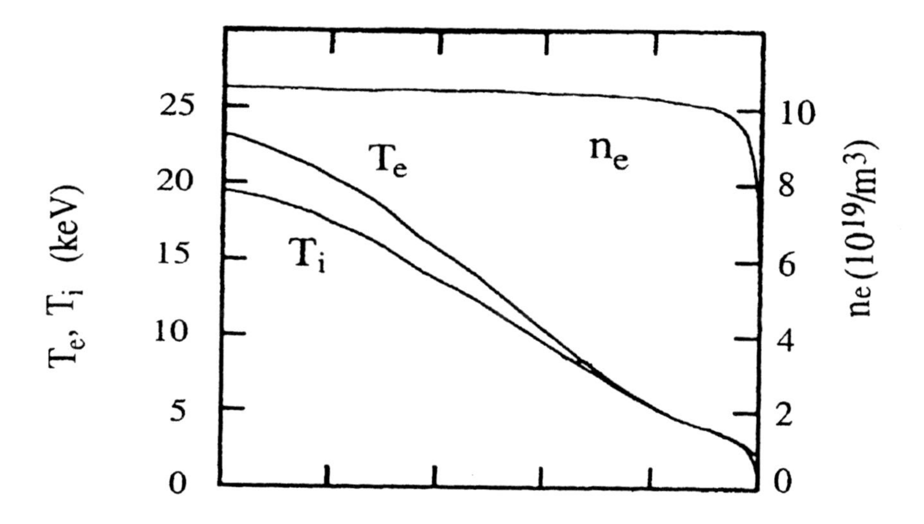

Figure 4.7. Typical distributions of the electron temperatures (T_e), the electron density (n_e) and the ion temperature (T_i) in ITER plasmas. Note that $r/a = 0$ and 1 correspond to the centre and the outer edge of plasmas, respectively.

example of the electron temperature distributions along the radial direction in ITER tokamak plasmas is shown in fig. 4.7. It is clearly noted that the electron temperature (T_e) is maximum near the centre region of the minor radius and decreases toward the edge plasma region and, finally, goes down quickly in the scrape-off plasma region, meanwhile the electron density (n_e) is roughly constant over the whole plasma region, except for in the narrow outer scrape-off region. The ion temperature (T_i) can be assumed to be nearly equal to that of electrons ($T_i \sim T_e$). Their detailed distributions may change, depending on various operating parameters or even on the operation modes.

Assuming that collisions of ions with neutral particles do not play any significant role as the core plasmas are highly ionized, the variations of ions (n_q)

with charge q inside plasmas having the electron velocity v_e (or temperature T_e) and the electron density n_e can be represented as follows [5, 6]:

$$dn_q/dt = [(S_{q-1,q}n_{q-1} + \alpha_{q+1,q}n_{q+1}) - (S_{q,q+1} + \alpha_{q,q-1})\, n_q]\, n_e - \nabla \cdot \mathbf{\Gamma}_q \tag{4.4}$$

where $S_{q-1,q} = \langle v_e \sigma_{q-1,q} \rangle$ is the ionization coefficient for ions changing the charge from $q-1$ to q, $\alpha_{q+1,q} = \langle v_e \sigma_{q+1,q} \rangle$ is the recombination coefficient for ions changing the charge from $q+1$ to q, and $\sigma_{q,q'}$ is the cross-section for charge changing from q to q' in collisions.

The first and second terms in the right-hand-side of eq. (4.4) correspond to the enhancement of ions with charge q due to the ionization from charge $q-1$ and the recombination from charge $q+1$, respectively. The third and fourth terms are due to the reduction of ions with charge q to other neighboring states $q+1$ and $q-1$ due to ionization and recombination, respectively. In some cases, the electron capture in collisions with neutral particles should also be included. The last term is due to the loss or influx of ions with charge q through drift motions parallel to the confining magnetic field (In high temperature plasmas, an additional ion term due to convection also has to be taken into account).

Another important point is the fact that these coefficients, namely the cross-sections, depend strongly on the internal electronic excited states of the collision partners. Indeed, the excitation process, in particular involving the metastable state ions, become important and the following excitation coefficients have to be included in eq. (4.4) above:

$$X_{q,q} =< v_e \sigma_{q,q} \left(nl, n'l'\right) > . \tag{4.5}$$

These are the excitation coefficient from the electronic state (nl) to $(n'l')$ of ion without changing its charge q. Here n and l represent the principal and the azimuthal quantum numbers of the electronic state, respectively. This excitation process becomes important in high density plasmas where multiple collisions play a role. In such conditions, the effective ionization coefficients are significantly enhanced (sometimes by several orders of magnitude). Thus, instead of a simple corona model, a so-called collisional-radiative model becomes a powerful tool to analyze and reproduce the observed plasma features.

It should be noted that these coefficients usually depend not only on the electron temperature (T_e) and electron density (n_e) but also on the neutral particle density (n_0), particularly near the diverter plasma region. As seen in eq. (4.4), a series of coefficients is required in calculating the ion charge distributions in plasmas. In such modelling, systematic coefficients, namely collision cross-sections, are important for all the charge states and over the whole temperature energy range [7]. That is why convenient "scaling" formulas for various data become very important (see section 4.6 later).

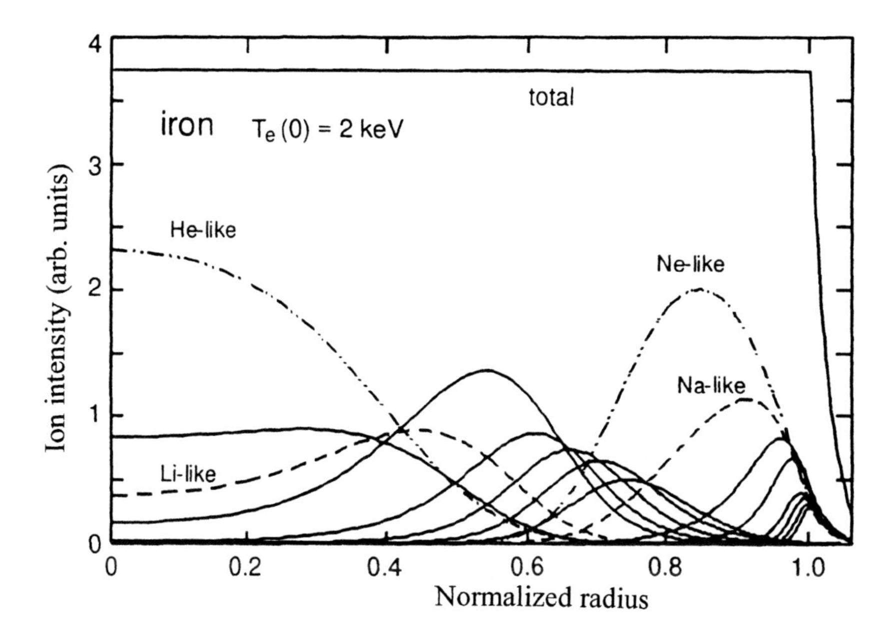

Figure 4.8. Calculated distributions of Fe^{q+} ions along the plasma radius at the electron temperature of 2 keV. Note that the position at radius 1.0 corresponds roughly to the outer edge of the plasma.

A typical calculated example of Fe^{q+} ion distributions in different charge states (q) under a steady (equilibrium) state in tokamak plasmas is shown in fig. 4.8 [8]. Once entering into plasmas, Fe atoms are immediately ionized and the ion charge increases as they move inward the core plasmas. It is clearly seen that, near the edge or outside region of plasmas, ions with low charge state are abundant and more highly ionized ions become dominant near the center region. In this particular example, practically no bare $\left(Fe^{26+}\right)$ as well as H-like Fe^{25+} ions are produced because the plasma temperature is not sufficiently high to ionize the last two electrons in the K-shell of Fe ions whose binding energy is $\sim$ 9 keV). It is also noticed that total ion density is roughly constant inside the core plasma region, except for those in the scrape-off region where parallel loss of ions (here the last term in eq. (4.4)) plays a key role in determining the ion distributions.

The radiation from impurity ions, which is proportional to the square of the atomic number of ions or the effective Z of plasmas, becomes an important energy loss from high temperature core plasmas, which may result in plasma cooling. Therefore, there are serious limitations on the amount of impurity ions allowable, in particular of heavy Z elements (roughly 0.1 - 0.01 % for Ti, Fe, Ni, Mo, W which are the most common materials composing the vacuum chamber and diverter [9]).

3.3 Diagnostics

The plasma diagnostics and precise knowledge of various parameters of plasmas are a requisite for new developments of physics concepts and successful running of the fusion plasma devices:

1 understanding plasma performance and behaviour.

2 operation and control to get the best performance.

3 protection of devices from serious damage which may occur during operation.

For these purposes, a number of plasma parameters including information on their spatial and temporal distributions have to be known with high accuracies and at high resolution (typically mm in the scrape-off region and 10 cm for the core plasmas over the time span of 10 ms in large ITER).

A number of plasma diagnostic techniques have been well developed in many applications over the years. However, some commonly used methods in ordinary plasmas cannot be applied to high temperature, high density tokamak plasmas. Thus new advanced techniques are under intensive development to meet requirements for tokamak plasmas [2, 10]. Some of these techniques, interesting from an atomic physics point of view, are discussed below.

The electron density (n_e) and electron temperature (T_e) have to be known accurately. Most importantly the density (n_i) and temperature (T_i) of hydrogen ions also have to be known to check if they are sufficiently high to achieve nuclear fusion reactions (eq. (4.1) - (4.3)). Generally speaking, the ion temperature is close to the electron temperature but may be slightly different in some cases.

There are two quite different classes of plasma-diagnostics: passive and active.

3.3.1 Passive Diagnostics. Passive diagnostics is simply to directly look at plasmas. For example, a tiny fraction of the main plasma constituents (protons, H^+, or alpha particles if nuclear fusion reactions occur) collide with other particles in plasmas such as impurity ions, B^+, and are neutralized after capturing an electron into the (nl) state (4.6) and finally emit photons to become stabilised:

$$H^+ + B^{q+} \rightarrow H^{0*}(nl) \tag{4.6}$$

$$\rightarrow H^0 + h\upsilon. \tag{4.7}$$

Here * indicates the excited state. The neutral hydrogen particles (H^0) formed in process (4.6) escape out of the confinement magnetic field. After electron-stripping through process (4.8), by passing through a gas or thin foil target at the outside of the plasma confinement region, their kinetic energy (or plasma temperature) can be analyzed precisely with an electro-static or magnetic spectrometer:

$$H^0 + B^+ \rightarrow H^+ + B^0. \tag{4.8}$$

Another way is to carefully observe the energy distributions of photons $(h\upsilon)$ which are emitted after the electron is captured into the (nl) state (4.7). The photon energy distributions are Doppler-broadened due to the ion velocities. Thus the broadened width of the observed photon spectrum gives information on the kinetic energy of particles (protons in this case).

The passive diagnostics techniques are relatively simple but the effectiveness is limited in getting accurate information on the spatial and temporal distributions of plasma parameters and their use for controlling plasmas.

3.3.2 Active Diagnostics. In active diagnostics methods, neutral particles or tiny pieces of solid (so-called pellets) are injected at special locations and at particular times and thus more accurate information on the spatial and temporal distributions and variations of parameters can be obtained. Ions confined in plasmas collide with the injected particles (for example energetic hydrogen atoms, H^0) and capture an electron through the following process (4.9):

$$A^{q+} + H^0 \rightarrow A^{(q-1)+*}(nl) + H^+ \tag{4.9}$$

$$\rightarrow A^{(q-1)+} + h\upsilon. \tag{4.10}$$

Here A^{q+} represents any charged particle including protons (H^+) in plasmas.

After the electron is captured into highly excited (nl) state of the ion by process (4.9), it decays to the stable ground state by emitting a photon. The energy of the emitted photon depends on the ions and their charge and thus can be distributed over the infrared, ultra-violet, vacuum-ultra-violet, and even X-ray energy ranges. By observing the Doppler-broadening of the energy spectrum of the emitted photons, accurate information of plasma ion temperatures can be determined.

For bare carbon ions, which are the prominent impurity ions in most laboratory plasmas, an electron is captured predominantly into highly excited ($n_0 \sim 4$) states (see section 4.4) and thus the emitted photons can be in the soft X-ray or X-ray region. Due to technical reasons, visible photons are much easier to handle and their signals can be transported over very long distances (via fibre optics over more than 100 m). The cross-sections for capture into higher n-states ($n > n_0$) resulting in the visible photon emissions decrease drastically (roughly as n^{-3}). Indeed the electron capture cross-sections for visible line emission ($n = 9 \rightarrow n = 8$ transition) for C^{6+} ions, for example, are about three orders of magnitude smaller than those for the most dominant $n_0 = 4$ channel [11]. Only a few reliable calculations of cross-sections for such high n-state capture are available, as a large number of the states have to be taken into account.

This technique (often called charge exchange spectroscopy - CXS - by plasma analysts) has been established and is commonly used and indeed is one of the most powerful and reliable techniques for plasma diagnostics [12]. The CXS technique can be applied to get information on a number of the important plasma parameters. A typical photon spectrum observed with CXS is shown in fig. 4.9 where photons are produced when 150 keV He^0 beam is injected horizontally into a tokamak He plasma. The observed He II lines are produced in the following electron capture process:

$$\mathrm{He}^{2+} + \mathrm{He}^0(\text{injected}) \rightarrow \mathrm{He}^{+*}(nl) + \mathrm{He}^+ \tag{4.11}$$

$$\rightarrow \mathrm{He}^+ + h\upsilon \tag{4.12}$$

and all are found to be Doppler-broadened at the center of 4685 Å ($n = 4 \rightarrow n = 3$), depending on the velocity (temperature) of He^{2+} ions in the plasma. Analysis of the observed spectrum indicates the following four different components of alpha particles (He^{2+}) in the particular plasmas [13]:

1 a sharp peak due to collisions with slow ($\sim$ 0.08 keV) alpha particles at the scrap-off plasma region.

2 the relatively broad peak due to collisions with lukewarm ($\sim$ 0.3 keV) alpha particles near the edge plasma region.

3 the third peak due to collisions with "hot" ($\sim$ 12.4 keV) alpha particles in the central core plasma region.

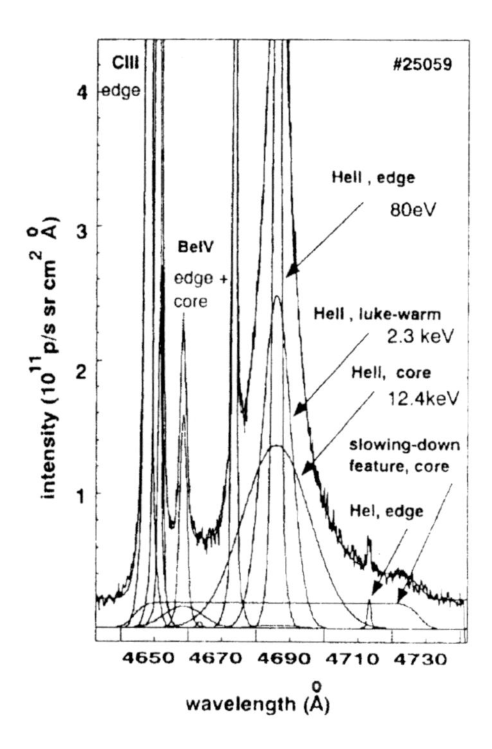

Figure 4.9. Typical photon spectrum observed in a tokamak He plasma while a neutral He^0 beam is injected. Note that the observed spectrum consists of a number of the overlapping peaks with different Doppler-broadening (see the text for details).

4 the fourth broadest peak due to collisions with the injected (150 keV) alpha particles "trapped" in the core plasma region.

The He I line due to collision excitation of the injected neutral He^0 particles, strongly Doppler-shifted, is also observed at higher photon energy. Be II and IV and C III lines are due to collision of these impurity ions in the core and edge plasma region with the injected He^0 particles.

3.4 Important Parameters of Plasmas

Electron density and temperature: A simple Langmuir probe cannot be used in high temperature, high density plasma regions such as those found in a tokamak. There are some advanced techniques to get information on these parameters, based upon interactions of (laser) radiation with plasmas [2]. The reflected radiation of injected laser beams is phase-shifted due to the plasma reflective index. The most reliable method in determining the electron density of high temperature tokamak plasmas is to use the Thomson scattering technique where the scattered radiation spectrum is broadened and gives accurate information on the electron temperature. On the other hand, absolute radiation intensity provides information on the electron density. Typical distributions of these electron parameters already are shown in fig. 4.7. Sometimes the temporal fluctuations of these parameters also have to be known to get stable and controlled plasmas.

Other parameters: There are a number of the important parameters that control and govern the whole plasma behaviour. Some of them can be obtained using methods based upon atomic collision processes.

3.4.1 Impurity Ions: Identification, Density and Temperature. CXS technique is the most reliable for getting such information, as described above. Passive diagnostics can be used with reasonable accuracies, too.

3.4.2 Plasma Rotation. The magnetically confined plasmas inside tokamaks tend to rotate along the toroidal magnetic field lines. Information related to such plasma rotation is important to know the overall stabilities of plasmas. CXS also can be applied to determine the rotation velocity as well as its direction by measuring the Doppler-shift of the peak position of particular photons from the constituent ions produced through process (4.7) or (4.10).

3.4.3 Electrostatic Field Inside Plasmas (Plasma Potential). Any non-uniform local electric field may cause some plasma instabilities. It is not easy to get such information without perturbing the plasma under study. One non-perturbative method is based upon the precise energy measurements of relatively high energy (typ. a few MeV), heavy ions injected from the outside of plasma which are further ionised in collisions with plasmas. If the injected ion with charge q is ionised to q' at a point with the electro-static potential ϕ, the kinetic energy of the ion with charge q' coming out of plasma changes by $(q - q')\,\phi$ which is of the order of a few 100 eV. The small energy variation to be measured requires precise stabilisation and control of the injected particle energy with an energy spread of less than 100 eV. The energy loss of the injected ions in collisions with plasma constituents also has to be known and corrected for, in order to get accurate information of the potential inside the plasma under study.

3.4.4 Magnetic Field Inside Plasmas. The magnetic field inside a plasma is formed by the magnetic coils combined with the large plasma current and plays a key role in determining and maintaining the stability. The emitted photons from particles under the influence of strong plasma magnetic fields are polarised due to the Zeeman effect. By looking at the intensity variation of the components, the magnetic field can be determined.

4. General Collision Processes in Plasmas

In high temperature core plasmas, particularly in the centre, there are plenty of highly charged ions. The fractions and charge distributions strongly depend on the plasma temperature. In fact, from the observed X-ray or other photon spectra the ion charge distributions can be determined. These distributions in turn provide important information on the plasma ion temperatures. Thus it is important to have accurate data of collisions involving highly charged ions.

Collisions involving highly charged ions in plasmas can be categorised into the following four groups:

1 Collisions with electrons.

2 Collisions with atoms (and molecules).

3 Collisions with ions.

4 Collisions with photons.

Under the plasma conditions found in the present-day magnetic fusion devices (temperature and density), the first three catagories of collision, 1) - 3), are dominant. 4), which is generally of a minor importance, becomes significant only in high-power laser produced plasmas such as ICF (see section 4.2).

4.1 Collisions with Electrons

There are a number of collision processes involving highly charged ions colliding with plasma electrons. Indeed they play a critical role in producing highly charged ions in plasmas and also in forming stable plasmas and finally in controlling the radiation losses from high temperature plasmas.

4.1.1 Excitation of Highly Charged Ions by Electron Impact. Though a number of extensive and systematic theoretical investigations have been reported on the excitation processes such as the 'Opacity Project' and the 'Iron Project' [14, 15] so far, experimental measurements of the excitation cross-sections are very limited because of technical difficulties involved in experiments such as extreme low densities of target ions (in particular, of highly charged ions) which is equivalent to the pressure of the order of 10^{-10} torr and low collection efficiencies of the photons produced.

There are some classes of excitation processes involving highly charged ions colliding with electrons. The first process is *direct (potential) excitation* (DE) described as follows:

$$\mathrm{e} + \mathrm{A}^{q+} \rightarrow \mathrm{e} + \mathrm{A}^{q+*}(nl) \tag{4.13}$$

$$\rightarrow \mathrm{A}^{q+} + h\upsilon. \tag{4.14}$$

After excitation, the ion is stabilised by emitting a photon with a specific energy (4.14) or photons through cascades, from which the excited level (nl) is identified. Through observation of photon energy and absolute intensities for process (4.14), the excitation cross-sections can be determined after correcting for the branching and cascade effects, if they are present. Also careful energy loss measurements of the incident electrons in process (4.13) can provide cross-sections.

In the second process called *resonance excitation* (RE), the incident electron, after exciting one of the inner-shell electrons and losing its initial energy, is resonantly captured to form the doubly excited (intermediate) state (4.15) which decays mostly by emitting an electron through autoionisation (4.16):

$$\mathrm{e} + \mathrm{A}^{q+} \rightarrow \mathrm{A}^{(q-1)+**}(nln'l') \tag{4.15}$$

$$\rightarrow \mathrm{A}^{q+*}(n''l'') + \mathrm{e} \tag{4.16}$$

$$\rightarrow \mathrm{A}^{q+} + h\upsilon. \tag{4.17}$$

A number of such resonance states exist and the cross-sections show a series of (experimentally unresolved) resonances which, on average, increase the effective excitation cross-sections. However, the second process is significantly influenced by the following third process through another decay channel (4.20), namely the intermediate state formed (4.18) can also decay through photon emission (note that the intermediate state (4.18) is the same as (4.15)):

$$\mathrm{e} + \mathrm{A}^{q+} \rightarrow \mathrm{A}^{(q-1)+**}(nln'l') \tag{4.18}$$

$$\rightarrow \mathrm{A}^{(q-1)+*}(n''l'') + h\upsilon \tag{4.19}$$

$$\rightarrow \mathrm{A}^{(q-1)+} + h\upsilon' \tag{4.20}$$

which is called *dielectronic recombination* (DR) (see discussion later on). Clearly the third process reduces the excitation cross-sections due to resonance excitation (4.15). Thus, after forming the intermediate state through recombining with an electron, the ion stabilises through two different decay channels, either (4.16) or (4.19), whose probabilities are often cited as autoionisation and radiative rates.

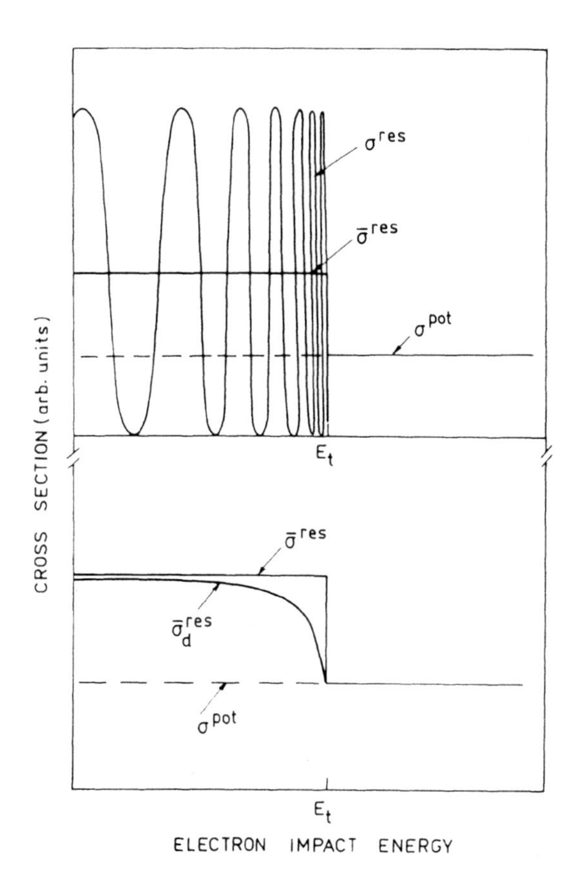

Figure 4.10. Schematic behavior of excitation cross-sections of highly charged ions in electron impact. Here σ^{pot} represents the direct excitation, σ^{res} the resonant excitation, ${\sigma_d}^{res}$ dielectronic recombination, and E_t the threshold energy for excitation processes, respectively.

Typical features of excitation cross-sections of ions under electron impact are shown in fig. 4.10. This figure shows how three different processes (DE, RE and DR) contribute and change as a function of the electron impact energy. Electron impact excitation is also closely related to other processes such as ionisation and recombination, described in detail later on.

4.1.2 Ionisation of Highly Charged Ions by Electron Impact. One of the relatively well established processes is the *"direct" single ionisation* process (DSI) resulting in increase of the ion charge by unity from q to $(q+1)$:

$$e + A^{q+} \to e + A^{(q+1)+} + e. \tag{4.21}$$

Here one of the electrons in the outer-most shell is ionised. In many cases this process is dominant over multi-electron ionisation processes (see (4.22)). For relatively light ions and highly charged, heavy ions with a few electrons, both theoretical and experimental understanding has been obtained and quite accurate data are available [16]. When a comparison between theory and experiment is made, it should be noted that sometimes significant fractions of the metastable state ions may be included in the primary ion beam which is produced in an ion source. This effect is easily noticed from the observed cross-sections, which show up before the expected threshold energy of the electron ionisation.

For many-electron ions, the situation is not so good. *"Direct" multiple ionisation* (DMI) involves more than one electron in the same or nearby shells:

$$e + A^{q+} \to e + A^{(q+n)+} + ne \quad (n > 1)\,. \tag{4.22}$$

Thus the incident ion increases its charge by n from q to $(q+n)$. As n increases, probabilities decrease quickly. In such multiple electron ionisation collisions, strong correlation among electrons can play a role.

Furthermore, if the electron impact energy is sufficiently high, the contribution from *inner-shell electron ionisation* becomes noticeable and results in a significant enhancement of the fraction of higher charge state ions. Inner-shell electron ionisation is also well understood and the resulting enhancement or distributions of highly charged ions can be known through model calculations [17].

Probably the most interesting and also important phenomena involving ionisation of many-electron ions relate to *inner-shell excitation, followed by autoionisation (EA),* which is described as follows:

$$e + A^{q+} \;\to\; e + A^{q+**}\left(nln'l'\right) \tag{4.23}$$

$$\to\; A^{(q+1)+*}\left(n''l''\right) + e \tag{4.24}$$

$$\to\; A^{(q+1)+} + h\upsilon. \tag{4.25}$$

In this process, generally called *indirect ionisation,* where one of the inner-shell electrons is excited, the ion forms the doubly excited state (4.23) which in turn decays via auto-ionisation by emitting one of these electrons in the excited states (4.24) before stabilising to the ground state (4.25). Cross-sections for this process are sometimes an order of magnitude larger than those for direct ionisation (4.21), although they strongly depend on the electronic configurations of the ions under consideration.

Another important “indirect" ionisation process involving highly charged ion + electron collisions is *resonant-excitation-double autoionisation (REDA)*:

$$\mathrm{e} + \mathrm{A}^{q+} \quad \rightarrow \quad \mathrm{A}^{(q-1)+**}\left(nln'l'\right) \tag{4.26}$$

$$\rightarrow \quad \mathrm{A}^{(q+1)+} + 2\mathrm{e} \tag{4.27}$$

where firstly the incident electron excites one of the inner-shell electrons, thus losing its initial kinetic energy and then is resonantly captured into a Rydberg state of the ion (4.26) (the same as (4.15)) and subsequently this intermediate state is auto-ionised through simultaneous double auto-ionisation, emitting two electrons (4.27). Thus, the ionisation cross-sections are expected to show resonance-like behaviour.

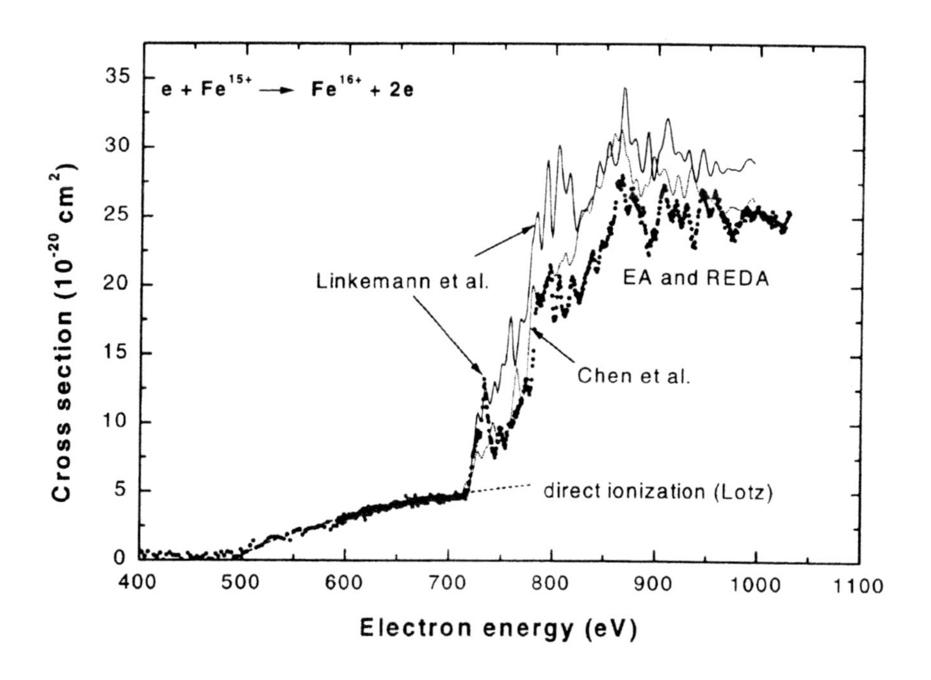

Figure 4.11. Single-electron ionisation cross-sections for Fe^{15+} by electron impact showing the contribution from different ionisation processes.

The most important features among these three ionisation processes can be ionization seen in fig. 4.11 where the direct single ionisation for Fe^{15+} ions (process (4.21)) is clearly a minor part but the indirect ionisation processes (4.23

to 4.25) and (4.26 to 4.27) are dominant [18]. It is clear that the calculations reproduce reasonably well the observed cross-sections [19, 20].

Some recent systematic investigations have been reported for EA and REDA [18, 21]. Furthermore, a comprehensive compilation of data for ionisation cross-sections involving various ions has been given [16].

4.1.3 Dielectronic and Radiative Recombination.

In electron + highly charged ion collisions, an electron is recombined with or is captured into an ion through the following three different processes, all accompanied by photon emission:

The first process, called *radiative recombination* (RR), occurs mainly in low energy electron collisions:

$$\mathrm{e} + \mathrm{A}^{q+} \rightarrow \mathrm{A}^{(q-1)+} + hv \tag{4.28}$$

where the incident electron is simply captured into an empty excited state of the ion (4.28), meanwhile the excess energy is emitted as a photon while the primary ion charge decreases by one. As the electron energy increases, the RR cross-sections decreases steadily above the threshold energy where they show a step-wise increase.

The second recombination process, *dielectronic recombination* (DR), has already been described above in connection with the excitation process [22]. Generally DR becomes important at collision energies just below the excitation threshold. Here the incoming electron, while exciting one of the inner-shell electrons in an ion and losing its energy, is captured into a relatively highly excited state of the ion and, thus forms the intermediate doubly excited state. Then while one of the electrons in the ion goes down to a lower orbital, a photon due to the transition between the excited state and the ground (or sometimes less excited) state is emitted and finally the ion becomes stabilised. Thus, it is clear that DR occurs resonantly at particular electron energies and its cross-section shows some resonance structures. This process forms the same intermediate doubly excited state as REDA mentioned above (4.26).

Due to significant developments of experimental techniques in observing DR using high energy ion storage rings [23] or low energy ion traps such as EBIS or EBIT [24, 25], recently reliable experimental high resolution DR cross-section data has been obtained. To understand the detailed features observed, impressive progresses in theories have also been reported and some theories, for example, the independent-processes and isolated resonance (IPIR) approximation, have been found to reproduce excellently the observed results [26].

Fig. 4.12 shows a typical observed behaviour of DR cross-sections for Li-like Cu^{26+} $\left(1s^2 2s\right)$ as a function of the electron energy [27]. Note that there are two series of DR ($2p_{3/2}nl$ and $2p_{1/2}nl$) resonances superimposed on a nonresonant

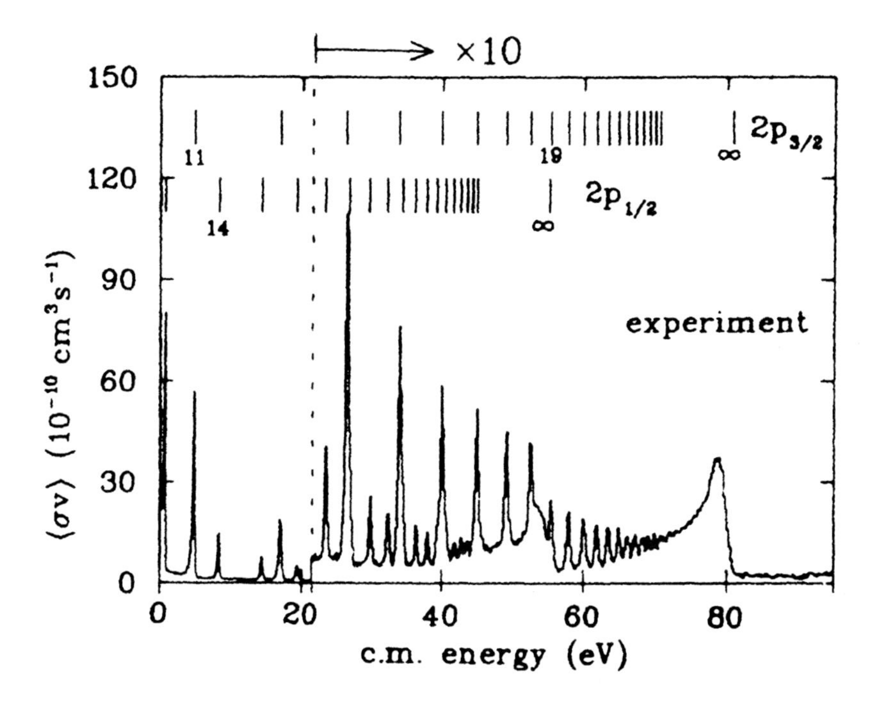

Figure 4.12. Observed DR cross-sections, averaged over the electron velocity, $\langle \sigma v \rangle$, for Li - like Cu^{26+} $(1s^2 2s)$ ions. Note that two series of peaks are seen.

RR continuum, mentioned above (4.28). The most intense observed peaks corresponds to $2p_{3/2}3l$ resonances and a series of lines are clearly seen, meanwhile weak peaks in-between correspond to $2p_{1/2}nl$ (fine-structure splitting).

The dielectronic recombination process is one of the dominant energy loss mechanisms from high temperature fusion plasmas (if impurity ion species exist) and also plays a key role in diagnostics of plasmas through observation of the emitted photons, mostly in the X-ray region.

4.1.4 Three-body Recombination Involving a Third Particle.

Another important recombination process, called *three-body recombination* (TBR), also occurs through the (high density) presence of a third particle, either an electron or a heavy particle M:

$$\mathrm{e} + \mathrm{A}^{q+} + \mathrm{M} \rightarrow \mathrm{A}^{(q-1)+} + \sum \mathrm{M} \qquad (4.29)$$

Here $\sum$ denotes the summation over all the possible states. This process results in the loss of highly charged ions. However, this becomes important only in very high density regions, such as the diverter region in tokamak plasmas where neutral atom/molecule densities are high or in high power laser-produced plasmas or ICF mentioned above.

4.2 Collisions with Neutral Particles (Atoms or Molecules)

Collisions between a highly charged ion and a heavy particle (including another ion) open more resulting channels, compared with electron collisions, as both collision partners can be ionised or excited, in addition to the electron capture (sometimes called electron transfer).

4.2.1 Electron Capture. Plenty of highly charged ions are present in high temperature core plasma regions where the plasma is almost completely ionised and practically no neutral particles are present. Although not many ion-atom collisions occur over the core plasma region itself, they play a key role, particularly in plasma diagnostics.

Even in such high temperature plasmas, collisions of highly charged ions with neutral particle species are limited to relatively low (kinetic) energy, less than 100 keV. In such a low energy region, except for proton collisions, the *electron capture* process is dominant over other processes such as ionisation [28]. In particular the electron capture cross-sections for highly charged ions are quite large and sometimes as high as $10^{-14} - 10^{-13} cm^2$.

Indeed, with its close relationship to applications in fusion plasma research, one of the most investigated collision processes between highly charged ions and neutral particles is the electron capture process where a projectile ion captures one or more electrons from a target (atom or ion). Mechanisms of electron capture and, thus, its probabilities depend strongly on the collision velocity between the collision partners, relative to that of electrons concerned. At low collision energies, the two collision partners form a short-lived quasi-molecular ion during collisions where electrons bound to the outer-shells in the atom before the collision are confined and governed under the influence of the combined two-centre Coulomb field. When they separate from each other after the collision, the electron originally in the neutral atom can move resonantly into the highly charged ion if proper energy levels are available in the ion, matching with the initial binding energy of the electron in the target.

In the core plasma region, such electron capture occurs between highly charged (impurity) ions (as intrinsically the core plasmas consist of nearly 100% hydrogen ions) and a very small amount of the non-ionised hydrogen atoms or of those intentionally injected to heat or diagnose plasmas. The (single) electron capture between an ion with charge q and neutral species can be represented as follows:

$$\mathrm{A}^{q+} + \mathrm{B} \rightarrow \mathrm{A}^{(q-1)+*}(nl) + \mathrm{B}^{+} \tag{4.30}$$

$$\rightarrow \mathrm{A}^{(q-1)+*}(n'l') + h\upsilon \tag{4.31}$$

$$\rightarrow \cdots \rightarrow \mathrm{A}^{(q-1)+} + h\upsilon'. \tag{4.32}$$

In such a process, the electron is captured into relatively highly excited Rydberg state (nl) of the ion (4.30) and then the ion emits photons (4.31), sometimes several photons are emitted with different energies before complete stabilisation to the ground state (4.32). Note that the direct transition to the ground state is less likely to occur but a cascade series can be dominant [29–31].

So far a number of experimental studies have confirmed the theoretical models and the calculated results on electron capture processes. Indeed, it has been found that, in the electron capture processes at low collision energies, the classical picture can reproduce the observed results quite nicely [32, 33].

Generally, single electron capture is dominant over multiple electron capture. In some cases, such as in many-electron atom/molecule target collisions, double electron capture becomes significant [34, 35]. As these two electrons tend to be captured into highly excited states $(nl, n'l')$ (4.33), thus forming autoionising states, one of the electrons can be emitted (ionised) into the continuum (4.34) and finally the ion is stabilised by emitting photons (4.35) and thus the incident ion charge changes only by unity, just like single electron capture (processes 4.30 to 4.32):

$$\mathrm{A}^{q+} + \mathrm{B} \rightarrow \mathrm{A}^{(q-2)+}(nl, n'l') + \mathrm{B}^{2+} \tag{4.33}$$

$$\rightarrow \mathrm{A}^{(q-1)+}(n''l'') + \mathrm{e} \tag{4.34}$$

$$\rightarrow \cdots \rightarrow \mathrm{A}^{(q-1)+} + h\upsilon. \tag{4.35}$$

Some detailed discussion on the double electron capture processes involving formation of the auto-ionising states and their decays can be found in a review article [35]. It is expected that, as the number of the captured electrons increases, the autoionisation process becomes dominant and then more than one-electron is emitted. On the other hand, as the collision energy increases, the electron capture processes become weaker because only overlapping wave functions allow an electron to change its position from target to projectile ion. At relatively high collision energy, a number of the interesting phenomena such as ionisation or stripping or excitation occur in highly charged ion + neutral particle collisions [28].

4.2.2 Excitation. In collisions between heavy particles, both projectile ion and target atom can be excited. Generally as the excitation energy of the highly charged ion is higher than that of the target neutral atom, the target excitation is more likely to occur in a high temperature core plasma region:

$$\mathrm{A}^{q+} + \mathrm{B} \rightarrow \sum \mathrm{A}^{q+} + \mathrm{B}^{*}(nl) \tag{4.36}$$

$$\rightarrow \mathrm{B} + hv. \tag{4.37}$$

Thus the excitation cross-sections can be determined through observation of emitted photons (4.37). However, as determination of the absolute efficiencies of photon detection systems is quite difficult, experimental measurements of the absolute cross-sections are available only for a limited number of targets such $1^1\mathrm{S} \rightarrow n^1\mathrm{P}\,(n = 2-5)$ in H and $1^1\mathrm{S} \rightarrow n^1\mathrm{L}\,(n = 2-4, \mathrm{L} = \mathrm{S}, \mathrm{P}, \mathrm{D})$ in He targets.

In the excitation process between heavy particle collisions, the following parameter conveniently defines the collision dynamics [36]:

$$\xi = v_i / \left(qE_{ex}\right)^{1/2}. \tag{4.38}$$

Here v_i and q are the velocity and the charge of the projectile ion, respectively, and E_{ex} the transition (excitation) energy. When $\xi \gg 1$, the perturbation approach can be applied, meanwhile $\xi \ll 1$ the adiabatic treatment can work. In $\xi \sim 1$ region, more rigorous treatments such as the close-coupling method are necessary.

4.2.3 Ionisation (Stripping). In highly charged ion + heavy particle collisions, the ionisation process below (4.39) becomes important only at higher energies, where the electron capture probabilities, process (4.30), decrease rapidly. Yet, the ionisation process, particularly for highly charged ions, play only a very minor role in high temperature (core) plasmas. It seems to be sufficient to take into account only the ionisation of neutral atoms (4.40).

$$\mathrm{A}^{q+} + \mathrm{B} \rightarrow \mathrm{A}^{(q+1)+} + \sum \mathrm{B} \tag{4.39}$$

$$\rightarrow \quad \sum \mathrm{A}^{q+} + \mathrm{B}^{r+} + re \tag{4.40}$$

$$\rightarrow \quad \mathrm{A}^{(q-1)+*}(nl) + \mathrm{B}^{+}. \tag{4.41}$$

$\sum$ denotes summation over all the possible states. The electron capture process (4.41) is the same as in (4.30). Typical behaviour of these cross-sections for O^{7+} + He collisions is shown in fig. 4.13 as a function of the collision energy. This figure clearly shows that at low collision energies electron capture (4.41) is dominant and its cross-sections decrease very rapidly above the velocity of 1 a.u. ($\simeq$ 25 keV/u) and the ionisation of target atoms (4.40) becomes dominant, meanwhile the ionisation (or stripping) of the incident ion (4.39) also shows similar increase but its cross-sections are still too small due to the large binding energy of the projectile ion (870 eV for the K-shell electron in O^{7+} ion, compared to 24.6 eV for an electron in He). The behaviour of these three processes varies as the ion charge (q) changes: for lower charge ions, processes (4.40) and (4.41) decrease but process (4.39) increases as the binding energy of the ions decreases [28, 37].

Similar electron capture between positive and negative ions, sometimes called recombination (or neutralisation), for example between highly charged ion and negative hydrogen ion (H^-), occurs as follows:

$$\mathrm{H}^- + \mathrm{A}^{q+} \rightarrow \mathrm{H}^0 + \mathrm{A}^{(q-1)+*}(nl)\,. \tag{4.42}$$

This is expected to be significant only at relatively low collision energy up to a few keV as the binding energy of negative ions is very small (0.75 eV for H^-) and the cross-sections decrease quickly when the collision energy increases.

Another interesting aspect of this process involving negative ions is *electron stripping* (or ionisation) from negative hydrogen ions

$$\mathrm{H}^- + \mathrm{A}^{q+} \rightarrow \mathrm{H}^0 + \mathrm{e} + \sum \mathrm{A}^{q+} \tag{4.43}$$

which will be discussed from the view point of plasma applications later on (see section 4.5).

5. Collision Processes in Various Plasmas in Tokamaks

Depending on the plasma temperatures (see the four categories of tokamak plasmas in section 4.2), the relevant atomic processes vary. Outside the core plasma region, the plasma temperatures are low and a lot of neutral particles, not only atomic species but also molecules, are present. Thus, collisions between neutral particles and ions with relatively low charge and sometimes even chemical reactions become important.

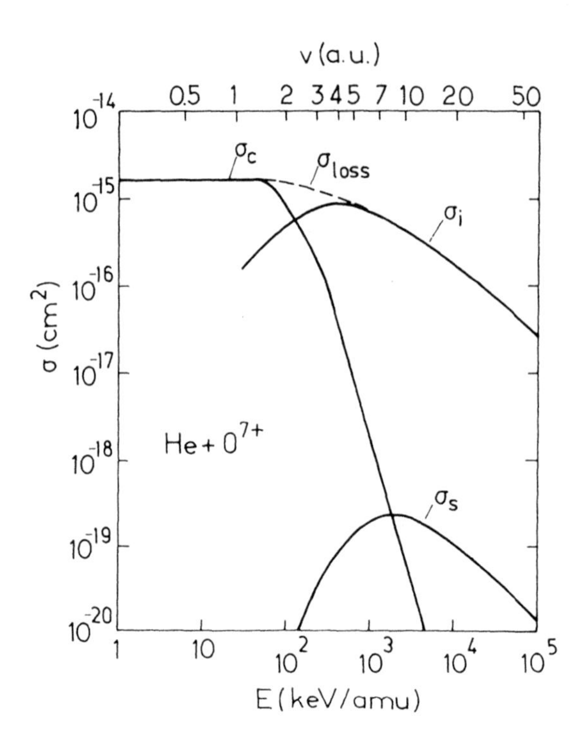

Figure 4.13. Typical variation of cross-sections for projectile ion ionisation (4.39), target ionisation (4.40) and electron capture (4.41) in O^{7+} + He collisions as a function of the collision energy. Here σ_c indicates the electron capture cross-section, σ_i the ionisation cross-section of He atom, σ_s the stripping cross-section of 1s electron in O^{7+} ion, respectively.

Here, we concentrate on important issues involving atomic (and molecular) collisions in the core and edge plasmas as we are interested in highly charged ions. In the high temperature core and edge plasmas, most of particles including impurities, either sputtered from the walls or injected for plasma diagnostics/monitoring, are highly ionised and some of them are fully ionised. Most of these highly charged ions are concentrated near the plasma centre region but some can escape from the confinement and move toward the outside of the confinement region due to drift or diffusion. In the scrape-off plasma region, most of the particles may be in relatively low charge states due to the low temperature. However, some of them which have drifted out of the core plasmas, for example due to imperfect magnetic field alignment or a unstable magnetic field, are still highly charged.

One of the interesting collision processes involving highly charged ions in the edge plasma region is the collisions of such highly charged ions with the excited

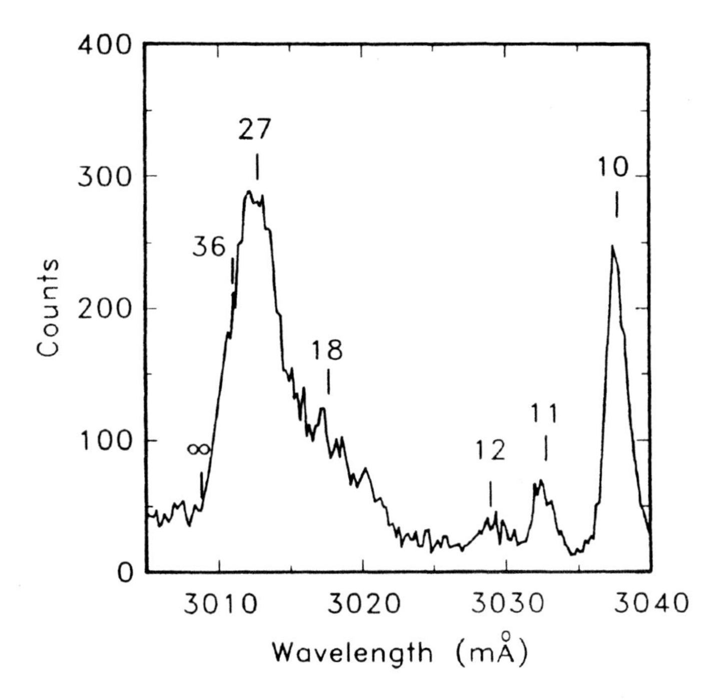

Figure 4.14. Typical observed K-X-ray spectrum originated from Ar^{17+} ions colliding with ground and excited hydrogen atoms in tokamak plasmas.

neutral species such as hydrogen atoms in the excited states, $H^*(n)$, which are most likely to be produced in collisions of plasmas with the inner-surfaces of vacuum chamber walls. Then the following collisions occur:

$$A^{q+} + H^*(n) \rightarrow A^{(q-1)+*}(n'l') + H^+ \quad (4.44)$$

$$\rightarrow A^{(q-1)+*}(n''l'') + hv \quad (4.45)$$

$$\rightarrow \cdots \quad (4.46)$$

$$\rightarrow A^{(q-1)+} + hv' \quad (4.47)$$

where an electron from the excited hydrogen atom is captured into highly excited states of the ion (4.44) which becomes stabilised through photon emission (4.45 and 4.47), similar to (4.30) to (4.32). This process itself is very interesting but has not been investigated much experimentally or theoretically so far. In fact due to the small binding energy of the excited state atomic hydrogen, $H^* (n)$, the cross-sections of the electron capture process (4.44) are expected to be quite large (see (4.54) later), compared with those for the ground state hydrogen atom [38, 39]. There is no basic collision experiment to quantitatively confirm this suggestion, as any experiment involving such excited atomic hydrogen atoms is very difficult. In particular, it is difficult to know the absolute density of the $H^* (n)$. Only a single observation of X-rays originating from collisions with Ar^{17+} ion in a tokamak has been reported, as shown in fig. 4.14 [40]. The observed K X-ray spectrum clearly indicates, in addition to their collisions with the ground state hydrogen atoms ($H\,(n = 1)$), there is a significant contribution of the excited state hydrogen atoms, although their peaks from different final n'-states are not well resolved. In fact X-ray peaks to the right in the figure are due to transitions from different $n' = 10 - 12$ originated from Ar^{17+} colliding with ground state hydrogen atoms. Meanwhile the broad peak at the left is due to overlapping peaks due collisions with excited state ($n = 2 - 4$) hydrogen atoms. Note that collisions with hydrogen atoms in the initial $n = 2, 3$ and 4 states result in the predominant electron captured states $n_0' = 18$, 27, and 36, respectively, which are much higher than $n_0 = 8$ for ground state atomic hydrogen (see (4.53) in section 4.6.2).

It is also important to know the distributions of the final $(n'l')$ states after collisions in order to estimate the densities of these excited state atoms through photon yield measurements. This is also an important and basic atomic research issue, but no reliable data are available for such collision processes so far, except for a few calculations [38, 39].

5.1 Atomic Processes in Plasma Heating: Plasma Charge Converter (neutraliser)

Confined plasmas have to be heated to initiate nuclear fusion reactions and sustain them. Among various techniques for plasma heating, the neutral beam injection (NBI) method is the most promising and efficient power injection method for heating plasmas. The most commonly used technique for NBI is to use high energy (100 keV - 1 MeV) neutral hydrogen atoms (H^0). These atoms are derived by charge-conversion from hydrogen ions before injection into the magnetic field. If an energy of a few 10 keV is sufficient for the neutral H^0 atoms, they can be produced by passing protons through a gas, through electron capture (see 4.30 to 4.32):

$$H^+ + B \rightarrow H^0 + B^+. \tag{4.48}$$

Here B represents any neutral gas atom/molecule. This process has been investigated in detail and the relevant data are well established [41, 42]. The cross-sections for this process become increasingly small when the neutral beam energy necessary for NBI injection increases. Above 100 keV, the conversion efficiencies through electron capture into protons ($H^+ \rightarrow H^0$) decrease to only a few $\%$ or less and most of the primary protons are simply wasted without conversion.

It is easily noticed however from the known cross-sections [41] that at higher energies the (single) electron stripping from negative hydrogen ions (H^-) (see (4.43) above) is much more efficient to get neutral hydrogen atoms for NBI:

$$H^- + B \rightarrow H^0 + e + \sum B. \tag{4.49}$$

In fact this is a process of ionisation or stripping of an electron from H^- ions. Based upon the data available, the highest conversion efficiencies into atomic hydrogen through process (4.49) are estimated to be about 55 - 60 $\%$ at a few 100 keV, which is necessary for large fusion research devices. This number seems to be reasonably good from a scientific point of view but still too small from a practical engineering point of view. For example, if 100 MW NBI power is required for plasma heating as expected in ITER NBI [2], roughly 40 MW of the injected negative hydrogen ions are lost without conversion to neutral H atoms even under the optimum conditions and this can not be tolerable in engineering fusion reactors.

This conversion efficiency can be enhanced to much higher values through another atomic process, using a highly ionised plasma target instead of neutral gas target, based upon the following atomic process:

$$H^- + B^{q+} \rightarrow H^0 + e + \sum B^{q+} \tag{4.50}$$

$$\rightarrow H^+ + 2e + \sum B^{q+}. \tag{4.51}$$

Here processes (4.50) and (4.51) result in one- and two-electron stripping, respectively. Collisions of ions with other ions are interesting from a basic as well as a technical point of view. Unfortunately as precise measurements of the cross-sections for such processes (4.50) and (4.51) are far more complicated and, thus, require advanced techniques, compared with those for neutral gas atoms or molecules, only a limited amount of reliable data is available so far. Few collision experiments for these species have been performed using well-advanced techniques such as crossed-ion beams and related techniques.

The effectiveness of plasma neutralisers through (4.50) for NBI easily is understood simply by looking at the data available. The electron stripping cross-sections of H^- ions in collisions with neutral H_2 gas ($5 \times 10^{-16} cm^2$ at 50 keV) is roughly one order of magnitude smaller than those for a H^+ ion target ($5 \times 10^{-15} cm^2$ at 50 keV). The limited data show that the cross-sections increase roughly with $q^{1.3}$ and thus those for Ar^{8+} become on the order of $10^{-13} cm^2$ [43].

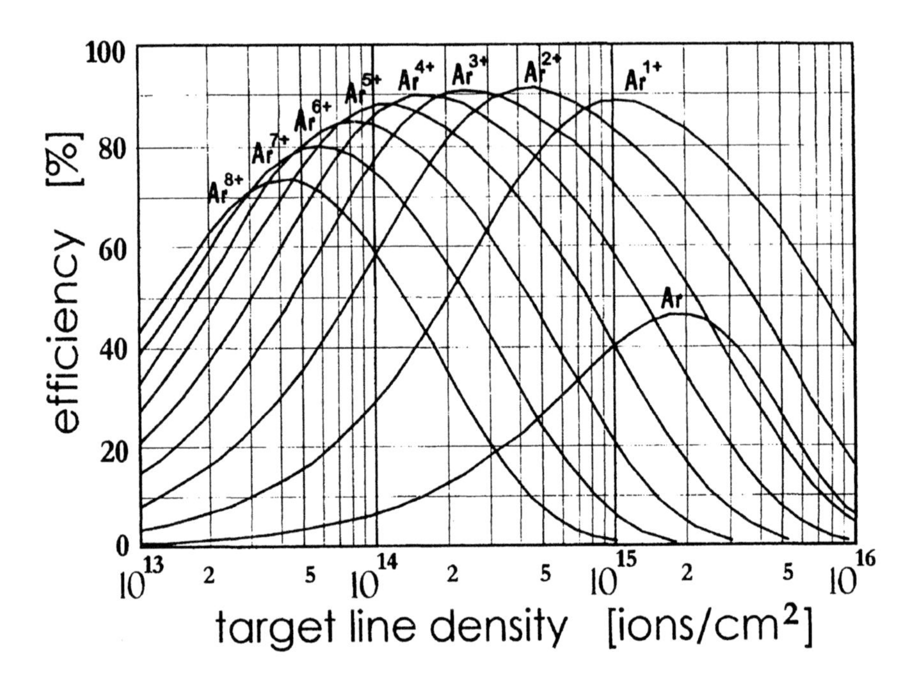

Figure 4.15. Variation of neutral hydrogen (H^0) fraction produced in electron stripping from negative hydrogen ions (H^-) passing Ar^{q+} ion plasma targets for different values of q. Note that the conversion efficiency drastically increases when the target charge increases from $q = 0$ (neutral atom) to $q = 1$ (singly charged ion). Then, as q increases, the conversion efficiency stays nearly constant, meanwhile importantly the optimum target densities decrease. Further increase of q results in the loss of the conversion efficiency to neutral hydrogen. This suggests that completely ionised low Z plasmas are preferable in applications to NBI.

The optimum conversion efficiencies estimated using these limited data can be as high as 90% which shows much better power conversion, compared with those in neutral gas systems which presently are commonly used. As shown in fig. 4.15, the conversion efficiencies are strongly dependent upon the degree of ionisation or fractions of the ionised (charged) component of the plasma target. It is interesting to notice that at higher ion (plasma) charge the conversion efficiencies to H atoms decrease [44]. This is easily understood because, due to the increased charge, the cross-sections for double-electron stripping (4.51) increase and thus the fraction of neutral H atoms decreases.

Though the plasma charge converter seems to be a very powerful tool, collision mechanisms in the energy range of interest, such as for ITER ($\sim$ 1 MeV), are not yet well understood. For example, one of the big differences between plasma and neutral gas targets is the fact that the effective charge of plasma targets is higher than that of neutral atoms due to less screening by the surrounding electrons. Another important point is that in the plasma target it is possible to get much higher densities over short distances, compared with those of gas target. As neutral particles are less efficient in conversion from H^- to H^0, plasmas including highly charged ions, particularly completely ionised plasmas, should be the best for this purpose.

Therefore, one of the most challenging issues involving plasma charge converters (neutralisers) is, in addition to collision investigations from the point of view of basic atomic physics, how to get "completely ionised" plasmas with sufficient densities. A plasma ion density of $10^{14}/cm^2$ has been achieved with high power micro-wave discharge techniques, although with relatively low ionisation efficiencies. However, this is still not sufficient to act as a plasma charge converter [45]. The detailed understanding of physics and features of such plasmas as a collision target are still far from complete at present [46]. Hence, another challenge is to get "completely ionised" plasmas.

6. Scaling Properties of Atomic Processes and their Cross-Sections

The most dominant ion/atom species in fusion plasmas is hydrogen (H) and its isotopes (D, T). Data pertaining to collision processes involving these atoms or their ions have been accumulated over many years and are thought to be reasonably well understood [41, 42]. Yet there are still a number of the unpursued collision processes relevant to the understanding of hydrogen plasmas and their behaviour. In order to understand the overall features of plasmas and to analyse the observed results, we need to have a large number of atomic data sets involving various ion species.

The collision processes are known to strongly depend on a number of parameters of the collision partners including the excitation energy (E_{ex}), ionisation energy (I_b), ionic charge (q), internal energy (E_i), energy defect (ΔE) (internal energy difference before and after collisions), nuclear charge (Z) as well as collision energy (E) (or collision velocity (v) or temperature (T)) or sometimes their combined forms which are often more convenient for applications (for example, see (4.38)).

There are too many sets of ion species with different charge states and also with different excited states relevant to fusion plasma. To support such data requirements, in addition to numerical databases, a number of theoretical approaches, either crude or precise, have been developed and are easily available at a number of the web sites, for example www-amdis.iaea.org/GENIE at the International Atomic Energy Agency (IAEA). Still it is impossible to theoretically or experimentally investigate all these processes in detail. Therefore it is important if we can find some "scaling laws" which can be used to get the unknown atomic parameters such as cross-sections based upon the known atomic parameters, either through simple extrapolation or interpolation. In some cases, reasonably reliable scaling formulas having some of these parameters can be found simply based upon the data available combined with a theoretical background.

6.1 Ionisation by Electron Impact

There are exceptionally good scaling laws for the *direct ionisation* of ions and atoms as long as there are few electrons in the target ions. One of the most convenient formulas was proposed by Lotz [47] many years ago and is still conveniently in use in a number of cases. His empirical formula for single electron ionisation is:

$$\sigma_i^1 = \sum_j a_j \mathrm{N}_j ln\left(u\right)/uI_b^2 \left[1 - b_j exp\left(-c_j\left(u-1\right)\right)\right] \qquad (4.52)$$

where u represents ratio of electron collision energy (E) to the ionisation threshold energy (I_b), $u = E/I_b$, and N_j the number of electrons in a particular j^{th} shell, a_j, b_j, and c_j the constants for the jth shell given in his tables. $\sum_j$ represents the summation over all the possible electron shells. It has been found that this formula is quite reliable at least for highly charged, few-electron ions, though it is well known that it deviates significantly from the observed cross-sections for many-electron ions, particularly near the threshold region, where the indirect ionisation contributes significantly to total ionisation cross-sections (see [48, 49] and section 4.4).

6.2 Electron Capture Involving Highly Charged Ions

In *electron capture* processes at low collision energies ($v_i < 1$ a.u.) relevant to fusion plasmas, the electron from the outer-most orbital of the target is captured predominantly into a highly excited (Rydberg) state (see processes 4.30 to 4.32).

The predominant principal quantum state, n_0, for single electron capture can be estimated from the following empirical formula:

$$n_0 = q^{0.75}/\left(I_b/13.6\right)^{0.5} \qquad (4.53)$$

where I_b represents the ionisation energy of target species in units of eV. For example, $n_0 = 9$ for $Ar^{17+} + O_2$ collisions. This empirical formula is based upon the classical picture of the electron capture process and is found to reproduce the experimental data over a wide range of parameters involved [50, 51].

A number of the interesting common features in electron capture involving highly charged ions at low collision energies ($v_i < 1$ a.u.) can be summarised as follows:

1. The predominant electron-capturing n_o state is determined largely by the electronic charge (q) of ions (but not by the nuclear charge (Z) of ions) and also by the target atom (see (4.53)).
2. The electron is more likely to be captured into high l-states, except for at very low energies ($\sim$ eV/u) [39].
3. The l-distributions can be significantly different even for the same ionic charge with ions of different nuclear charge (Z) [28].

Total electron capture cross-sections for highly charged ions $(q \geq 5 - 6)$ at low collision energies are roughly independent of the collision energy, and can be estimated reasonably well (within 30 - 50 %) with some empirically obtained formula based upon the classical picture. According to this model, total cross-sections are simply proportional to the ion charge and the inverse square of the ionisation energy of the target:

$$\sigma_t = 2.6 \times q/I_b^2 \left(10^{-13} cm^2\right) \tag{4.54}$$

where q is the charge of the initial charge of the primary ion and I_b the first ionisation energy of the target (in units of eV). This and similar empirical formulas have been confirmed to reproduce quite nicely the observed data for ions with $q \geq 6$ colliding with rare gas atoms over a wide range of parameters (q, I_b) [50, 51].

As the collision energy increases ($v_i > 1$ a.u.), the most probable electron capture state, n_0, becomes small and also the n-distributions become relatively broad and tend to disperse among a number of different n-states. Finally, at high energies, $n = 1$ becomes dominant as those for high n-states decrease as n^{-3}. The overlapping of the wave-functions between collision partners, which is necessary for electron capture (transfer), decreases and, thus, the electron capture cross-sections decrease very rapidly ($\propto v_i^{-12}$ at asymptotically high energies) [28].

Therefore, as the collision energy increases, an electron in the most loosely bound outer-shell ceases to play a role, and instead for a "fast-moving" projectile ion at higher energies, capture into the projectile from one of the inner-shell

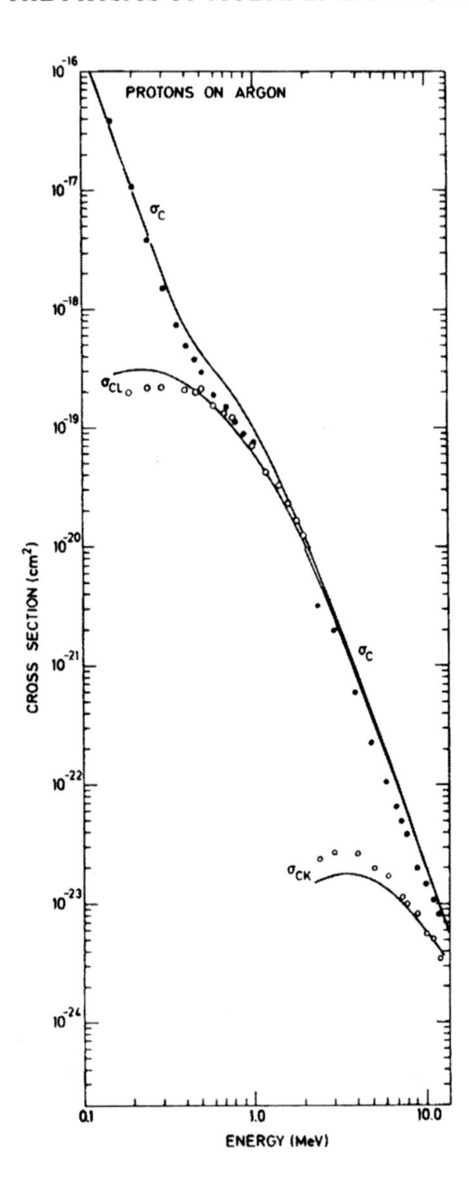

Figure 4.16. Electron capture cross-sections in $H^+ + Ar$ collisions as a function of the collision energy. Note that the contribution from different shells changes drastically from the M-shell electron to the L-shell (σ_{cL}) and then to the K-shell (σ_{cK}) as the collision energy increases. σ_c represents total electron capture cross-sections which are dominated by the M-shell electron capture at low collision energies. Note that the cross-sections vary by seven orders of magnitude when the proton energy changes by two orders of magnitude.

electrons of the target is more likely. This feature is clearly seen in fig. 4.16 for H^+ + Ar collisions [52]. At the intermediate collision energy region, the contribution of the electron transfer from the M-shell of Ar atoms becomes diminishingly small and is overcome by that from the L-shell (note a shoulder at around 0.6 MeV where the L-shell contribution becomes dominant) which in turn is dominated by capture into the K-shell at still higher energies (the maximum collision energy shown in the figure is still not sufficiently high for K-shell capture to dominate).

Total electron capture cross-sections at relatively high energies can be obtained through some scaling formulas [53]. On the other hand, the exact (nl) distributions should be known for many applications. At present there are no general formulae to estimate the (nl) distributions available at low-intermediate collision energy. Only recently an empirical formula for the (nl) distributions in electron capture processes has been proposed based upon the calculated (but not observed) cross-sections obtained with the classical trajectory Monte Carlo (CTMC) method [38, 39].

6.3 Excitation by Ion Impact

Based upon the limited experimental data, the following empirical scaling formulae have been found for the dipole-allowed and -forbidden excitation of target atoms by highly charged ions. For both cases, general formulas with the reduced energy (ξ), defined in (4.38), and the reduced cross-section are given by a single universal curve as follows [36]:

$$g\sigma^{ex} = \mathrm{F}\left(\xi^2\right) \tag{4.55}$$

where the factor g has different forms for different types of transition

$$g = \mathrm{E}_{ex}^2/qf \tag{4.56}$$

for dipole-allowed transitions and

$$g = n_0^2 n_f^3 \mathrm{E}_{ex}^4/q \tag{4.57}$$

for dipole-forbidden transitions, where f is the oscillator strength for the transition and n_0 and n_f represent the principal quantum numbers of the initial and final states in excitation. A typical example of the universal plots is shown in fig. 4.17 where the reduced cross-section, $g\sigma^{ex}$, is plotted against the reduced energy of $25\xi^2$.

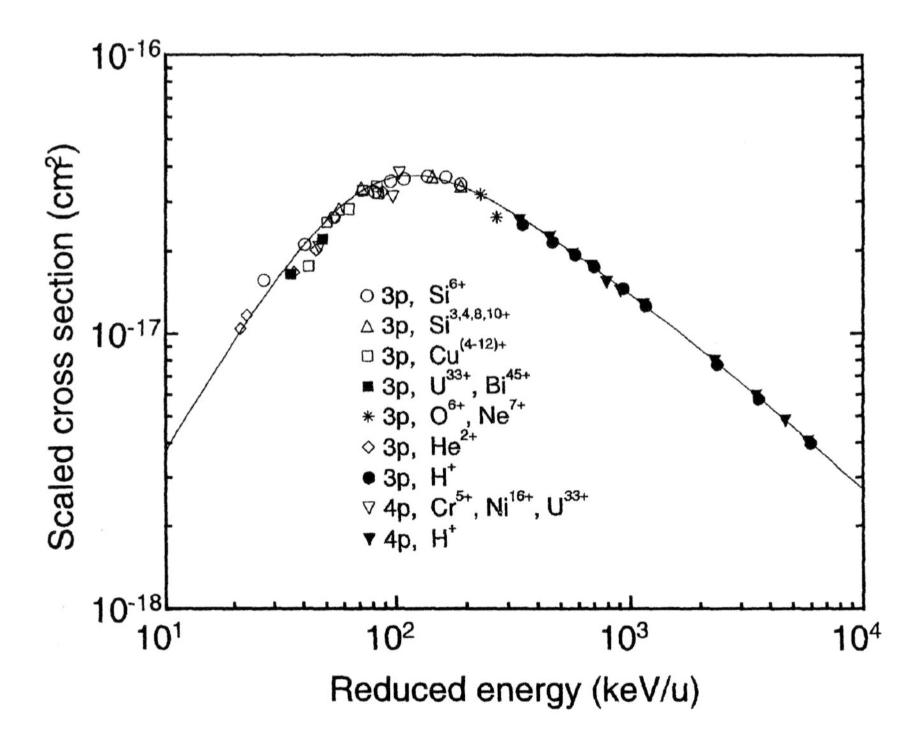

Figure 4.17. Scaled universal cross-section curves for excitation of He by impact of various highly charged ions as a function of the reduced collision energy. The solid line shows the best fit to the experimental data.

6.4 Ionisation of Target Atoms by Highly Charged Ion Impact

At relatively high collision energies, ionisation processes of neutral atoms by highly charged ions (4.40) become important; meanwhile the electron capture probabilities (4.41) decrease drastically.

Generally speaking, ionisation by highly charged ion impact has a similar dependence to that for electron impact. In fact, at high collision velocities, the ionisation cross-sections of neutral atoms (He) are found to be practically the same for both electron and proton impact at the same velocities, as shown in fig. 4.18 where the data for He ionisation are compared [54]. It is clear that the ionisation cross-sections of He atoms above 250 eV (about 10 times the ionisation energy) for electron impact and for 500 keV proton impact having equal velocity converge onto a single curve. It is also noted that the cross-sections by proton impact become maximum near the threshold energy region where the

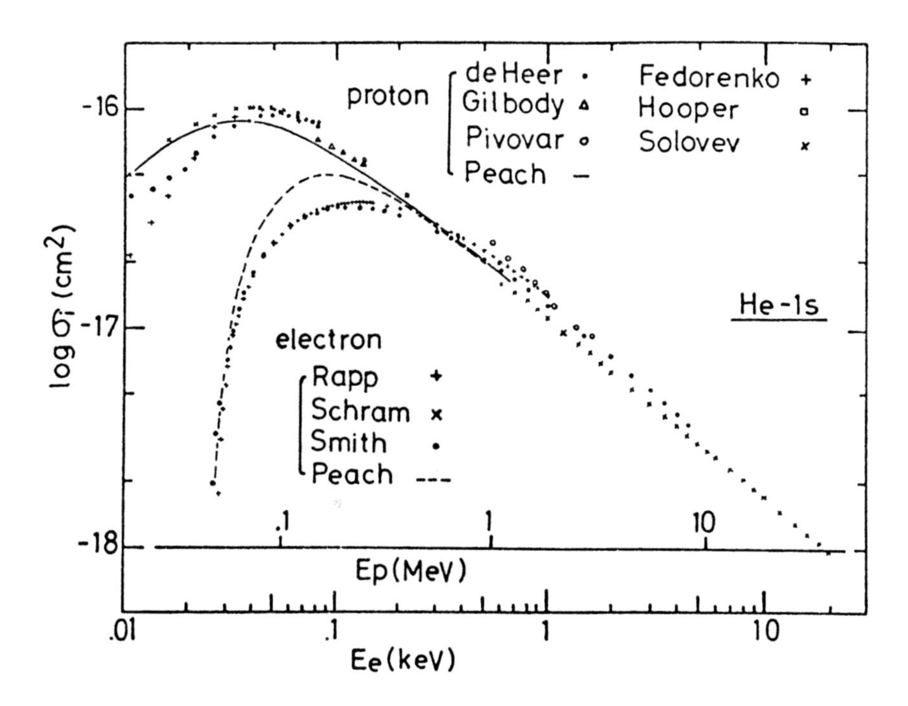

Figure 4.18. Ionisation cross-sections of He under proton and electron impact scaled to the same velocity. Note that, at sufficiently high velocities, the ionisation cross-sections are practically the same for both particle impacts.

momentum transfer to target electrons is maximum. Similar situations have been found for the inner-shell electron ionisation as well [54].

By taking into account the projectile ion charge (q), the ionisation energy of the target (I_b) and also the collision energy (E), the ionisation cross-sections (σ_i) with proper scaling parameters for both energy and cross-section are given in the following simple universal formula:

$$\sigma_i/qI_b^2 = f\left(E/qI_b\right). \tag{4.58}$$

This form is very similar to (4.52) for electron impact. This feature can be understood from the fact that the distant (large impact parameter) collisions are dominant for outer-shell electron ionisation.

6.5 Common Features in Highly Charged Ion Collisions with Electrons and neutral atoms

As mentioned already, experiments in highly charged ion + electron collisions are technically very difficult. On the other hand, those for highly charged ion + neutral atom collisions at the equivalent relative velocities seem to be straightforward if proper accelerators are available.

Fortunately there are some common features among collisions involving electrons and neutral atoms with highly charged ions. The electron capture from neutral atoms, for example, is equivalent to the recombination in electron collisions in some ways. The difference is the fact that the electrons in neutral atoms are not mono-energetic but have a relatively broad energy distributions described as "Compton profiles" which depend on the electronic state and its binding energy. To compare both processes, the folding of the Compton profile is necessary for neutral atoms.

Recently, new techniques have been developed and established to study and simulate electron-highly charged ion collisions using high energy (MeV/u), highly charged ions colliding with neutral atoms [55]. The theoretical background for this is based upon the fact that the double differential cross-section in ion-atom collisions (σ^{ia}) is related to the single differential cross-section in electron-ion collisions (σ^{ei}) through the following formula:

$$\frac{d^2\sigma^{ia}(E,\theta)}{d\Omega dE} = \left[\frac{d\sigma^{ei}(E,\theta)}{d\Omega}\right] \times \left[\frac{J(P_z)}{(v_p + P_z)}\right] \tag{4.59}$$

where v_p is the projectile velocity, $J(P_z)$ the Compton profile of the electron momentum in target, and P_z the momentum component of the target electron projected to the projectile ion direction. Thus electron-ion collision processes can be pursued through ion-atom collision experiments.

Similarly, dielectronic recombination (DR) also has been investigated through the *resonant transfer excitation* process (RTE) in high energy, heavy ion collisions. Some detailed discussion on similarity of ion collisions with electron and heavy particle has been given [56, 57].

7. Concluding Remarks

After a short introduction to one of the most hopeful magnetically confined "tokamak" plasma devices, ITER, based upon nuclear fusion using high temperature plasmas, we have described various plasmas and their features in tokamaks which follow some detailed discussion on the relevant atomic collision processes there. It is shown that the atomic collisions involving highly charged ions and their data play a critically important role and the detailed knowledge of atomic physics is essential in modelling and diagnosing high temperature plasmas and in achieving the nuclear fusion reactions sustained and, furthermore, in providing stable energy in the future. Finally, it is pointed out that new atomic collision physics issues and topics are continuously arising through close collaboration with development of fusion plasmas.

Acknowledgments

In part, the present work was supported by the Chemical Sciences, Geosciences and Biosciences Division, Office of Basic Energy Science, Office of Science, U.S. Department of Energy. The author would like to thank Prof. P. Richard and Dr. J.D. Gillaspy for their kind support, Dr. M. Shimada for providing recent information of ITER and also Prof. V.P. Shevelko for his help.

The author would like to acknowledge the publishers of the following articles for their permission to use the figures in the text :

Fig. 4.4 : taken from the ITER website, http://www.iter.org/ITERpublic/ITER/contents1.html

Fig. 4.5 : taken from Nuclear Fusion vol. 39, no. 12 (special issue on ITER,1999) p. 2140, with permission from International Atomic Energy Agency. Copyright in these materials is vested in the International Atomic Energy Agency, Vienna, Austria, from which permission for republication must be obtained.

Fig. 4.6 : taken from Nuclear Fusion vol. 39, no. 12 (special issue on ITER, 1999) p. 2144, with permission from International Atomic Energy Agency. Copyright in these materials is vested in the International Atomic Energy Agency, Vienna, Austria, from which permission for republication must be obtained.

Fig. 4.8 : taken from Proceedings of International Conference on Atomic and Molecular Data and Their Application, vol. 2 (2000) p. 133, K. Behringer, with permission from American Institute of Physics.

Fig. 4.9 : taken from Plasma Physics and Controlled Fusion, vol. 35 (1993) p. 799, M.G. von Hellermann et al. : Observation of alpha particle slowing-down spectra in JET helium beam fuelling and heating experiments, with permission from Institute of Physics Publishing.

Fig. 4.10 : taken from Physics of Highly Charged Ions (R.K. Janev, L.P. Presnyakov and V.P. Shevelko, 1985) p. 60, with permission from Springer Verlag.

Fig. 4.11 : taken from Atomic Physics with Heavy Ions (ed. H.F. Beyer and V.P. Shevelko,1999) p. 286, with permission from Springer Verlag.

Fig. 4.13 : taken from Physics of Highly Charged Ions (R.K. Janev, L.P. Presnyakov and V.P. Shevelko, 1985) p. 346, with permission from Springer Verlag.

Fig. 4.14 : taken from Physical Review Letters, vol. 56 (1986) p. 50, J.E. Rice, E.S. Marmar, J.L. Terry, E. Kaellne and J. Kaellne, Observation of charge-transfer population of high-n levels in Ar16+ from neutral hydrogen in the ground and excited states in a tokamak plasma, with permission from American Physical Society (copyright 2002).

Fig. 4.15 : taken from Nuclear Instruments and Methods in Physics Research B, vol. 99 (1995) p. 98, F. Melchert, M. Beneer, S.K. Kruedener and E. Salzborn, Neutralization of H- beams in "plasma neutralizers", with permission from Elsevier Science.

Fig. 4.17 : taken from Atomic Physics with Heavy Ions (ed. H.F. Beyer and V.P. Shevelko, 1999) p. 300, with permission from Springer Verlag.

Fig. 4.18 : taken from Physics Letters, vol.59A (1976) p. 199, H. Tawara, The equivalence of the innershell ionization by proton and electron impact at sufficiently high velocities, with permission from Elsevier Science.

References

[1] F.F. Chen, Introduction to Plasma Physics (Plenum Press, New York, 1977)

[2] Nuclear Fusion Vol. **39**, no. 12 (1999) (Special issue: This whole issue is dedicated to recent development report of ITER)

[3] R.K. Janev (ed.), Atomic and Molecular Processes in Magnetic Fusion Edge Plasmas (Plenum, 1995)

[4] R.K. Janev, Comm. At. Mol. Phys. **26**, 83 (1991)

[5] D.J. Campbell, Phys. Plasmas **8**, 2041 (2001)

[6] D.E. Post, Atomic and Molecular Data and Their Applications, AIP Conference Proceedings Vol. **434** (ed. P.J. Mohr and W.L. Wiese, 1998) p. 233

[7] H.P. Summers, Adv. At. Mol. Phys. **33**, 275 (1984)

[8] K. Behringer, Atomic and Molecular Data and Their Applications, AIP Conference Proceedings Vol. **543** (ed. K.A. Berrington and K.L. Bell, 2000) p. 129

[9] D.E. Post, J. Nucl. Mat. **220/222**, 143 (1995)

[10] H. R. Griem, Principles of Plasma Spectroscopy (Cambridge Monographs on Plasma Physics Vol. 2) (Cambridge University Press, 1997)

[11] H. Tawara and W. Fritsch, Phys. Scripta **T28**, 58 (1989)

[12] R. Hoekstra, Comm. At. Mol. Phys. **30**, 361 (1995)

[13] M.G. von Hellermann, W.G.F. Core, J. Frieling, L.D. Horton, R.W.T. Köning, M. Mandl, and H.P. Summers, Plasma Phys. Control. Fusion **35**, 799 (1993)

[14] M.J. Seaton, Y. Yu, D. Mihalas, and A.K. Pradhan, Mon. Notice Royal Astro. Soc. **266**, 805 (1994)

[15] K. Butler, Atomic and Molecular Data and Their Applications, AIP Conference Proceedings Vol. **434** (ed. P.J. Mohr and W.L. Wiese, 1998) p. 23

[16] H. Tawara and M. Kato, NIFS-DATA-51 (National Institute for Fusion Science, Toki, Japan, 1999)

[17] T. Mukoyama, T. Tonuma, A. Yagishita, H. Shibata, T. Koizumi, T. Matsuo, K. Shima, and H. Tawara, J. Phys. B **20**, 4453 (1987)

[18] A. Müller, Atomic Physics with Heavy Ions (eds. H.F. Beyer and V.P. Shevelko, Springer, Berlin, 1999) p. 271

[19] D.C. Griffin and M.S. Pindzola, J. Phys. B **21**, 3253 (1988)

[20] D.L. Moore and K.J. Reed, Adv. At. Mol. Opt. Phys. **34**, 301 (1994)

[21] M. Khouilid, S. Cherkani-Hassan, S. Rachafi, H. Teng, and P. Defrance, J. Phys. B **34**, 1727 (2001)

[22] W.G. Graham, W. Fritsch, Y. Hahn, and J.A. Tanis (eds.), Recombination of Atomic Ions NATO ASI series B Vol. **296** (Plenum, 1992)

[23] A. Wolf, Atomic Physics with Heavy Ions (eds. H.F. Beyer and V.P. Shevelko, Springer, Berlin, 1999) p. 3

[24] D.H.G. Schneider, J. Steiger, T. Schenkel, and J.R. Crespo Lopez-Urrutia, Atomic Physics with Heavy Ions (eds. H.F. Beyer and V.P. Shevelko, Springer, Berlin, 1999) p. 30

[25] J. Gillaspy (ed.), Trapping Highly Charged Ions: Fundamentals and Applications (Nova Science Publishers, New York, 1999)

[26] M.S. Pindzola, D.C. Griffin, and N.R. Badnell, Atomic, Molecular and Optical Physics Handbook (ed. G.W. Drake, American Institute of Physics, 1996) p. 630

[27] G. Kilgus, D. Habs, D. Schwalm, A. Wolf, N.R. Badnell, and A. Müller, Phys. Rev. A **46**, 5370 (1992)

[28] R.K. Janev, L.P. Presnyakov, and V.P. Shevelko, Physics of Highly Charged Ions (Springer, Berlin, 1985)

[29] H. Tawara, P. Richard, U.I. Safronova, and P.C. Stancil, Phys. Rev. A **64**, 042712 (2001)

[30] J.B. Greenwood, I.D. Williams, S.J. Smith, and A. Chutjian, Astrophys. J. **533**, L175 (2000)

[31] P. Beierdorfer, R.E. Olson, G.V. Brown, H. Chen, C.L. Harris, P.A. Neill, L. Schweinkhard, S.B. Utter, and K. Widmann, Phys. Rev. Letters **85**, 5090 (2000)

[32] H. Ryufuku, K. Sasaki, and T. Watanabe, Phys. Rev. A **21**, 745 (1980)

[33] A. Niehaus, J. Phys. B **19**, 2925 (1986)

[34] V.P. Shevelko and H. Tawara, Atomic Multielectron Processes (Springer, Berlin, 1998)

[35] M. Barat and P. Roncin, J. Phys. B **25**, 2205 (1992)

[36] R.K. Janev, Atomic Physics with Heavy Ions (eds. H.F. Beyer and V.P. Shevelko, Springer, Berlin, 1999) p. 291

[37] V.P. Shevelko, D. Böhne, B. Franzke, and Th. Stöhlker, Atomic Physics with Heavy Ions (eds. H.F. Beyer and V.P. Shevelko, Springer, Berlin, 1999) p. 305

[38] K.R. Cornelius, K. Wojtkowski, and R.E. Olson, J. Phys. B **33**, 2017 (2000)

[39] J.A. Perez, R.E. Olson, and P. Beierdorfer, J. Phys. B **34**, 3063 (2001)

[40] J.E. Rice, E.S. Marmar, J.L. Terry, E. Källne, and J. Källne, Phys. Rev. Letters **56**, 50 (1986)

[41] H. Tawara and A. Russek, Rev. Mod. Phys. **45**, 178 (1973)

[42] R.K. Janev, W.D. Langer, K. Evans, and D.E. Post, Elementary Processes in Hydrogen-Helium Plasmas (Springer, 1987)

[43] F. Melchert, M. Beneer, S. Kruedener, and E. Salzborn, Nucl. Instr. Meth. B **99**, 98 (1995)

[44] F. Melchert, W. Debus, M. Lieher, R.E. Olson, and E. Salzborn, Europhys. Letters. **9**, 443 (1989)

[45] G.D. Alton, R.A. Sparrow, and R.E. Olson, Phys. Rev. A **45**, 5957 (1992)

[46] H. Tawara, Heavy Ion Physics **1**, 649 (Budapest, 1995)

[47] W. Lotz, Z. Phys. **206**, 205 (1967) ; **216**, 241 (1968)

[48] H. Tawara, Photon and Electron Interactions with Atoms, Molecules and Ions, Landolt- Börnstein New Series Vol. **I/17** A (ed. Y. Itikawa, Springer, 2000) p.2/56

[49] H. Tawara, Photon and Electron Interactions with Atoms, Molecules and Ions, Landolt-Börnstein New Series vol. **I/17** B (ed. Y. Itikawa, Springer, 2001) p.3/211

[50] M. Kimura, N. Nakamura, H. Watanabe, I. Yamada, A. Danjo, K. Hosaka, A. Matsumoto, S. Ohtani, H.A. Sakaue, M. Sakurai, H. Tawara, and M. Yoshino, J. Phys. B **28**, L643 (1995)

[51] N. Selberg, C. Biedermann, and H. Cederquist, Phys. Rev. A **54**, 4127 (1996)

[52] M. Rodbro, E. Hosdal-Pedersen, C.L. Cocke, and J.R. Macdonald, Phys. Rev. A **19**, 1936 (1979)

[53] V.S. Nikolaev, Sov. Phys.-Uspekhi **8**, 269 (1965)

[54] H. Tawara, Phys. Letters **59A**, 199 (1976)

[55] P. Richard, C.P. Bhalla, S. Hagmann, and P. Zavodsky, Phys. Scripta **T80**, 87 (1999)

[56] D. Brandt, Phys. Rev. A **27**, 1314 (1983)

[57] Y. Hahn, Recombination of Atomic Ions NATO ASI series B Vol. **296** (eds. W.G. Graham, W. Fritsch, Y. Hahn, and J.A. Tanis (Plenum Press, 1992) p. 11

Chapter 5

RADIOBIOLOGICAL EFFECTS OF HIGHLY CHARGED IONS

Their relevance for tumor therapy and radioprotection in space

G. Kraft

Biophysik, Gesellschaft fur Schwerionenforschung
Planckstr. 1, 64291 Darmstadt, Germany
and
Technische Universitat Darmstadt
G.Kraft@gsi.de

Abstract The radiobiological effects of highly-charged-ion beams are of interest for tumortherapy and for radioprotection in space. In tumor therapy, high-energy protons and carbon ions exhibit an inverse dose profile, i.e. an increase of energy deposition with penetration depth. This allows a greater tumor dose for protons and carbon ions than for photons. In addition, for the heavier carbon ions, this increase in dose is potentiated by a greater relative biological efficiency (RBE). On the other hand, the greater RBE of particles is the concern of space-radioprotection because the radiation burden of the cosmic galactic radiation consists of heavy charged particles from protons to iron ions. In this chapter, the physical and biological basis of particle radiotherapy and its present status will be presented. For space radioprotection, the particle spectrum will be given and the risk of cancer induction and long-term genetic mutation will be discussed. In contrast to the cell inactivation problem for tumor therapy, where physics-based models have been developed, the genetic changes are more complex in their mechanisms and only rough estimations can be given for the time being.

Keywords: linear energy transfer, ion beam therapy, relative biological efficiency, radioprotection in space, particle tracks, delta electron emission

F.J. Currell (ed.), The Physics of Multiply and Highly Charged Ions, Vol. 1, 149-196.

1. Introduction

Radiobiology is mostly concerned with the influence of sparsely ionising radiation like x- or gamma-rays and electrons. In the last 50 years, however, the radiobiological action of heavy charged particles like protons or heavier ions has been studied with increasing intensity. This is for two reasons: first, heavy charged particles represent the best tool for an external radiotherapy of inoperable tumors. This is due to the favourable depth dose distribution where the dose increases with penetration depth and because of the small lateral and longitudinal scattering that allows irradiation of deep-seated target volumes with optimum precision. In addition, for particles heavier than protons, i.e. in the region of carbon, the biological efficiency increases at the end of the beam's range while it is low in the entrance channel, thus allowing a better inactivation of otherwise very radio-resistant cells of deep-seated tumors.

The second important field for particle radiobiology is the application in radiation protection in space research. In space outside the magnetic shielding of the earth, high-energy protons from the sun and heavier particles up to iron from interstellar sources pose a genetic and carcinogenic risk for man and can also influence and destroy semiconductor devices like computers [1]. Because of the very high energies of these particles, shielding becomes difficult and extremely expensive. Therefore, the action of these particles should be known precisely in order to minimize the necessary shielding. In the following paragraphs, both problems will be dealt with in some detail but therapy will be paid more attention because of its greater relevance for our life.

2. Physics Relevant for Particle Radiobiology

2.1 Microscopic Structure of an Ion Track

The knowledge of a particle beam's microscopic structure is essential for the understanding of the mechanism of cell inactivation and genetic effects [2]. The macroscopic dose of a particle beam, which is the main parameter in radiobiology, is given by the number of particles traversing the mass unit and the dose deposited by each particle, called linear energy transfer - LET. According to the Bethe-Bloch-formula the energy is transferred to the target electrons that are emitted as delta electrons (High-energy delta electrons can form individual short tracks, called delta rays.) Measurements of delta-electron spectra revealed that more than three quarters of the dissipated energy is used for the ionisation process and only 10 to 20% are left for target atom excitation. Therefore, it is the action of the liberated electrons that - together with the primary ionisation – determines the biological action of the ions [3].

There are several processes that contribute to the features and the spectrum of the emitted electrons. For a high-energy transfer, the binding energy of

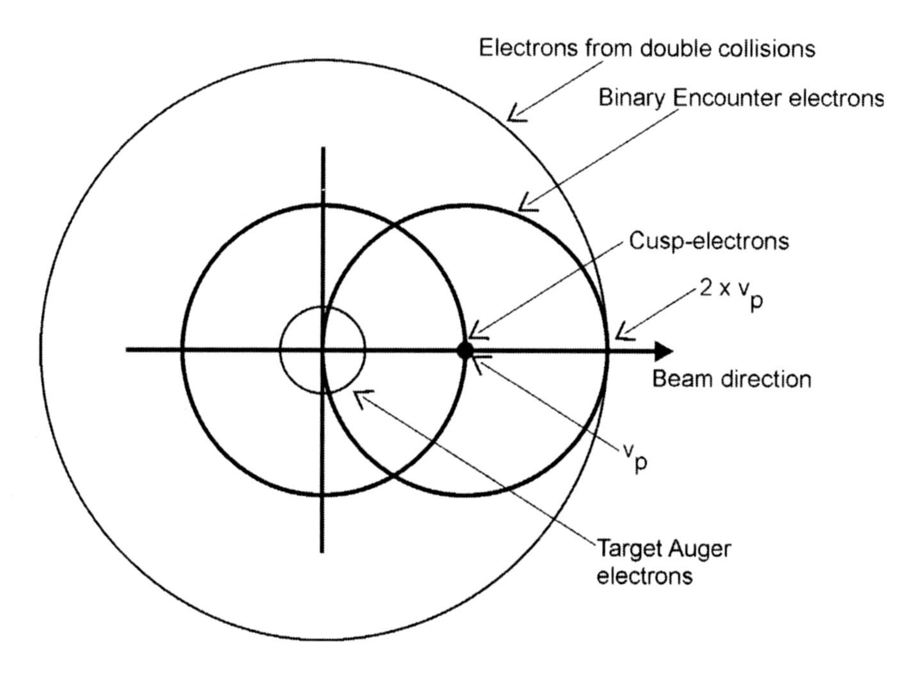

Figure 5.1. Velocity coordinates of the electron emission in ion-atom collisions.

the electrons can be neglected and electrons are treated as a free electron gas undergoing binary collisions. These electrons are emitted in radial symmetry around the centre of mass system that coincides with the projectile system because of the large mass excess of projectiles compared to electrons (figs. 5.1 and 5.2)

At lower energy transfers, the binding to the target atom becomes more relevant and the collision has to be calculated as a three–body interaction yielding a large number of electrons emitted with low velocity. At low particle energies, weakly bound electrons are exchanged with the target yielding a high–intensity peak at the velocity of the projectile. These are called Cusp-electrons because of the shape of the peak in the energy spectrum. At higher energies, this peak disappears because no electron can jump into bound states. Finally, Auger electrons from both, target and projectile atoms, can be emitted. The main characteristic of these emission processes is shown in a velocity plot in fig. 5.2 Most of the electrons are emitted in a forward direction.

For the calculation of the radial dose distribution around the particle track, the transport and energy deposition of these electrons has to be followed through the target material. The transport is characterized by elastic and inelastic scattering.

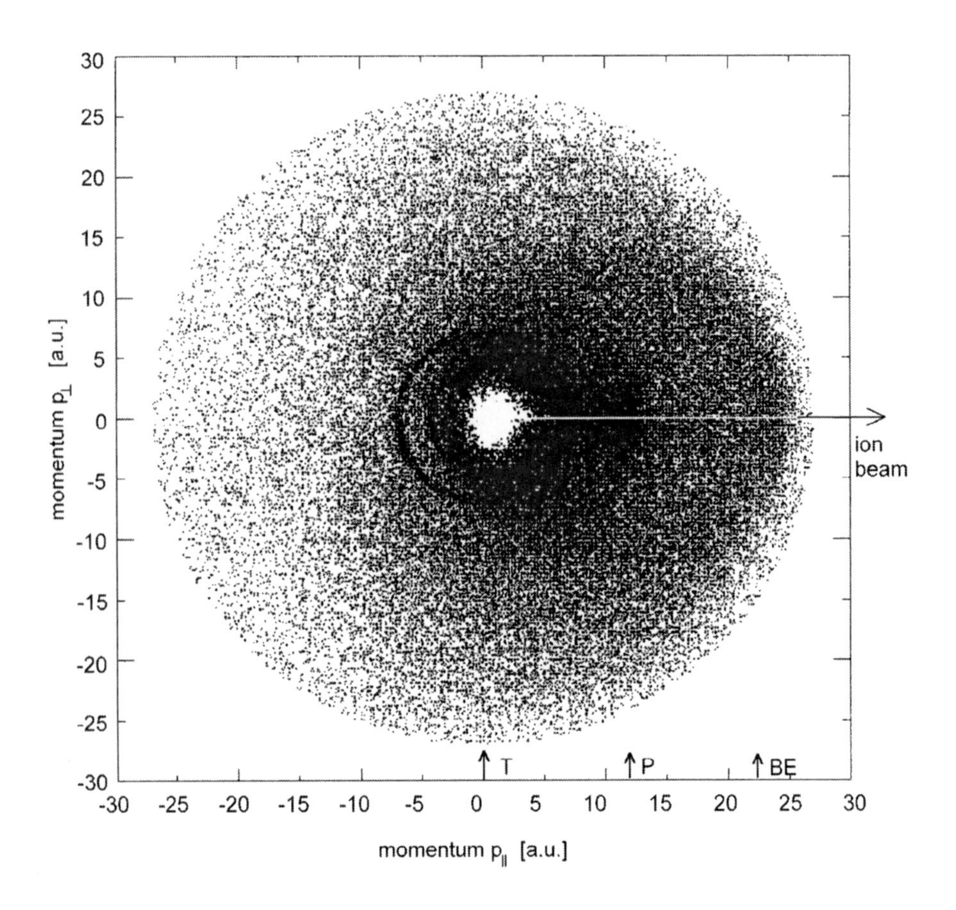

Figure 5.2. Angular distribution of the measured electron emission [4].

While the elastic scattering changes only the direction, the inelastic processes i.e. ionisation and excitation can cause further biological damage.

The ionisation cross section has a maximum around 100 eV kinetic energy of the electrons. This maximum is almost independent of the composition of the stopping material as long as the material consists of organic compounds. There, the mean free path is in the range of 10 nm, which is about the dimension of DNA: high-energy electrons cause multiple ionisation events at the end of their range in a distance that corresponds to the cross section of a DNA molecule.

As it makes no difference whether an electron is created by photon impact or by the impact of a heavy charged particle, there is no difference in the biological action of each individual electron either. There is, however, a big difference in the spatial distribution between electrons being created along a track of a heavy charged particle and the random distribution of electrons being created in photon beams and there is also a difference in ionisation density between particles of different energies and atomic numbers like protons and carbon ions.

For protons, mostly independent electron tracks are produced and the situation at the DNA level is similar to that of the photon-produced electrons: Locally correlated DNA damage can only be produced by increasing the total number of electrons i.e. by increasing the macroscopic dose. For the heavier carbon ions it is obvious that at low energies many electron tracks are produced that cause locally multiply damaged sites (LMDS) within the DNA. It is the reduced reparability of this clustered damage that causes the high relative biological efficiency (RBE) [6] (see 5.3).

2.2 Depth Dose Distribution (Bragg Curves)

The main reason to use heavy charged particles in therapy instead of photons is the inverse dose profile i.e. the increase of energy deposition with penetration depth up to a sharp maximum at the end of the particle range, the Bragg peak, named after William Bragg, who measured an increase of ionisation at the end of the range of alpha particles in air [7].

This increase of dose with penetration depth for a particle beam produced by an accelerator was first reported by R. Wilson who recognized its potential for a medical application in tumor therapy [8].

The favourable depth dose distribution is a direct consequence of the interaction mechanism of heavy charged particles with the penetrated material and is different from that of electromagnetic radiation. Within the energy range used for therapy, heavy charged particles interact predominantly with the target electrons and the interaction strength is directly correlated with the interaction time. At high energies of the projectiles, the interaction time is short and the energy transfer to the target small. When the particles are slowed down and

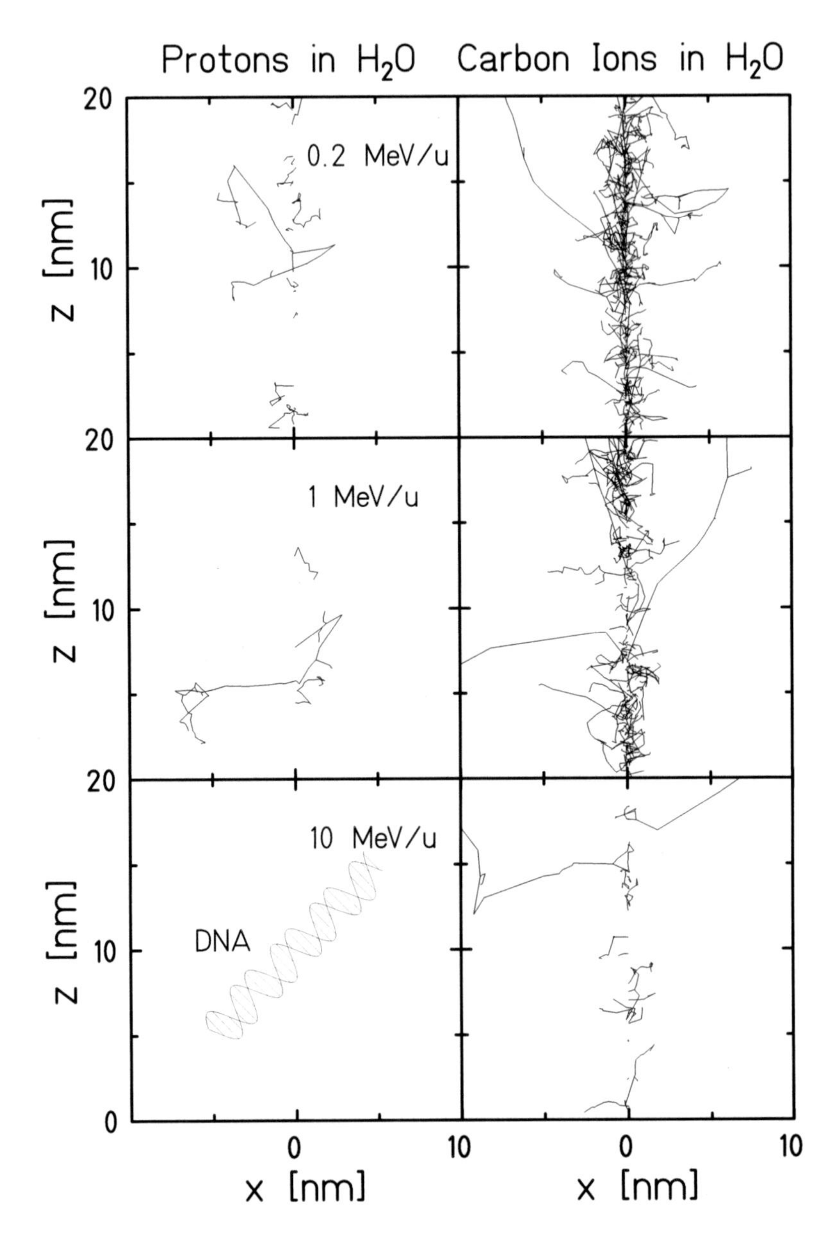

Figure 5.3. The microscopic structure of proton and carbon tracks in water is compared to a schematic representation of a DNA molecule. Protons and carbon ions are compared for the same specific energy before, in and behind the Bragg maximum [5].

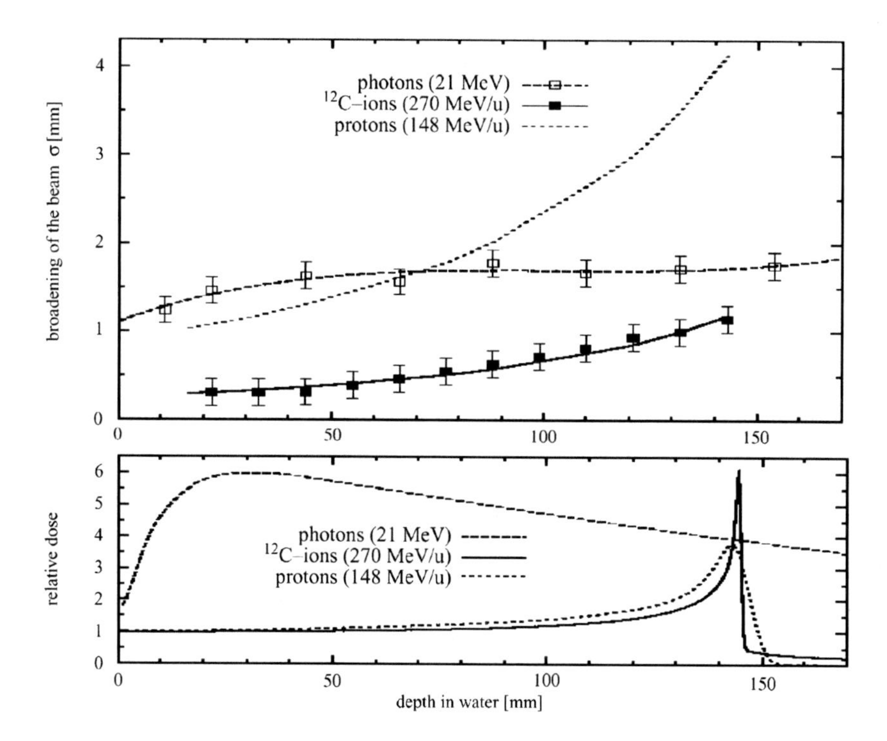

Figure 5.4. Comparison of the lateral scattering of photon, proton and carbon beams as function of the penetration depth (top) and the depth dose correlation (bottom) [9].

close to the end of their range the interaction time becomes larger and the value of the energy transfer is at its maximum.

The energy loss as function of particle energy and atomic number is given in the Bethe-Bloch-formula [10–12]:

$$-\frac{dE}{dx} = \frac{4\pi e^4 Z_{eff}^2 N}{m_e v^2} \ln \frac{2mv^2}{I(1-\beta^2)} + \text{relativistic} \tag{5.1}$$

where m_e is the electron mass, v the projectile velocity, N the density of the electrons of the target material, e the elementary charge and I the mean ionisation potential. Finally, Z_{eff} is the effective charge empirically approximated by Barkas [13]. For high energies, all electrons are stripped off the projectile and the effective charge equals the atomic number. At small energies, electrons are collected from the target and the effective charge decreases, yielding zero when the particles stop. The change of Z_{eff} is the main reason for the sharp decrease of the energy loss at lower energies, which is the essential criterion for the use of heavy particles in therapy.

2.3 Lateral and Longitudinal Scattering

When the energy loss is plotted over the penetration depth a Bragg curve for a single particle results in a dose ratio from plateau to peak of 1 to 2 orders of magnitude: Measured Bragg curves, however, have a much lower dose ratio because of the statistics of the energy loss process. The interaction of the projectile with the electrons is a process of very many collisions and most of the differences in the individual energy transfer are compensated but yield a small straggling of the particle range [14, 15] .

Range-straggling broadens the individual Bragg curve and decreases the peak to plateau ratio. Because this process strongly depends on the atomic number of the projectiles, the Bragg peak is always broader for protons than for carbon ions. As a matter of fact, this is rather irrelevant for therapy because the tumors to be treated are larger than the natural width of the Bragg peak and various methods are now being used to extend the Bragg region over the size of the target volume. Only at the distal side of such extended or smeared-out Bragg peaks (SOBP) the natural decay is visible. Fig. 5.5 compares the extended Bragg peaks of protons and carbon ions at different penetration depths.

The differences between protons and carbon ions in the entrance channel as well as at the distal side are small. More important than the longitudinal straggling for therapy is the lateral scattering. In practical applications, a target in the proximity of a critical structure will not be treated in such a way that the beam stops in front of the critical site because of possible range uncertainties. These tumors are treated in a way that ensures that the beam passes the critical sites. Then, the lateral scattering determines the closest approach possible. Fig. 5.4 shows that for protons the beam broadening is less than the photon value of 2 mm for a penetration depth up to 10 cm in tissue. For carbon ions, the broadening is smaller than for photon beams up to a penetration depth of 20 cm. This is why extremely accurate fields can be produced with carbon beams also for deep-seated tumors while for more superficial tumors e.g. in the eye the accuracy of protons is sufficient.

2.4 Nuclear Fragmentation

A very important feature of particle beams is their nuclear fragmentation. When heavy ions pass through a thick absorber like the human body or a thick shielding of a space craft even small cross sections for nuclear reactions produce a significant amount of lighter reaction products. In radiotherapy the change in biological efficiency between the primary ions like carbon and the lighter secondaries has to be taken into account in treatment planning, as well as their longer range. However, radioactive positron-emitting isotopes are very useful to track the beam path inside the patient. For space research, fragmentation represents a major obstacle: In order to shield against the very numerous low-

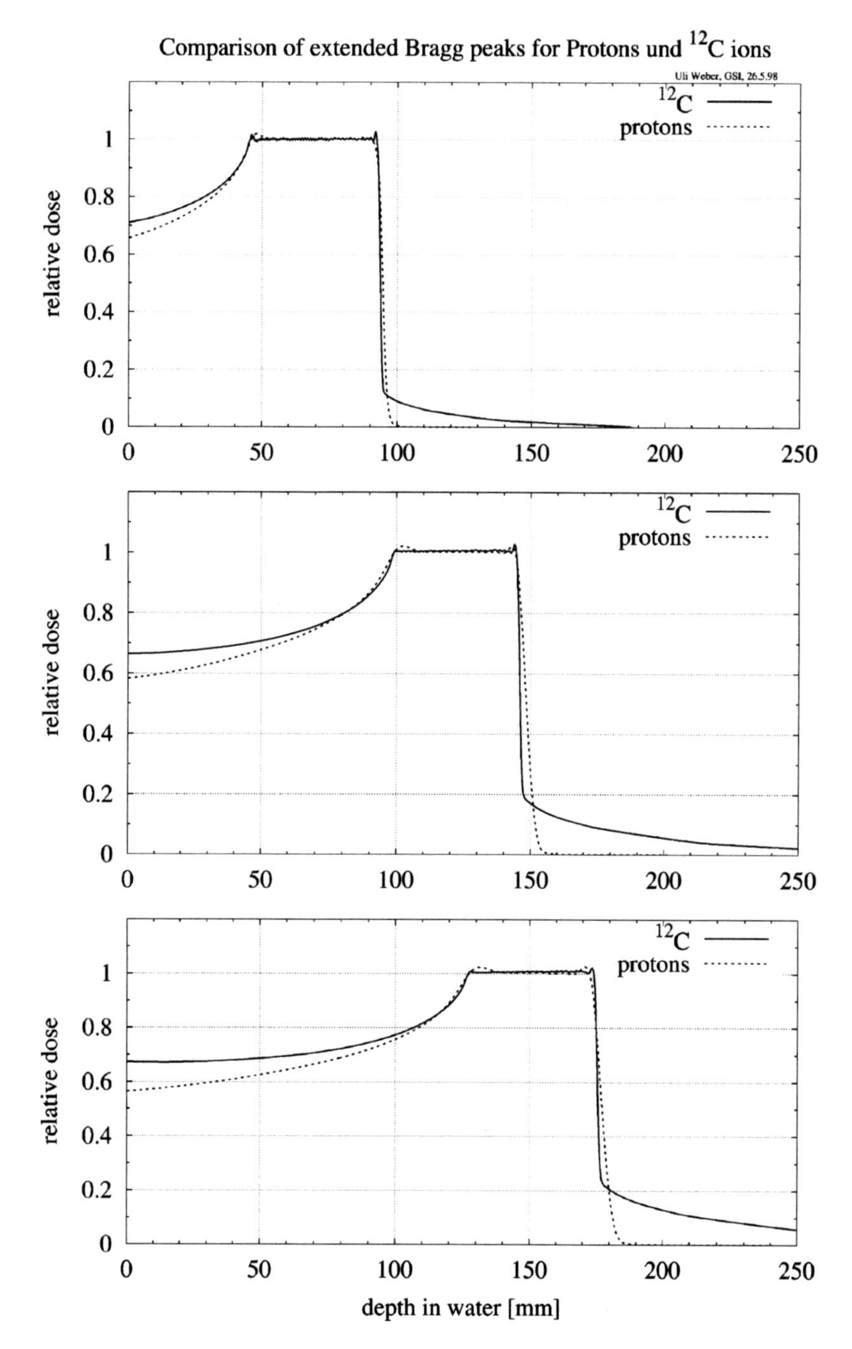

Figure 5.5. Comparison of calculated extended Bragg peaks of protons and carbon ions at different penetration depths.

energy particles of a few hundred AMeV, very efficient energy absorbers are needed, in which the few high-energy particles produce showers of light low-energy particles. It has been calculated that an increase of shielding material does not necessarily mean a decrease in exposure. For high-energy heavy ion collisions, it is the pure geometry that mainly determines the reaction: For the more probable and therefore more frequent distant or glancing collisions, the majority of the projectile and target nuclei are not affected by the nuclear reaction. But in between, a zone of high "temperature" is created that can be deexcited by the evaporation of a few nucleons. Because of the reaction kinematics the lighter fragments have the same velocity as the primary ions at the collision [16, 17]. The range of these fragments is given by the formula

$$R_{fr} = R_{pr} \frac{{Z_{pr}}^2}{M_{pr}} \cdot \frac{M_{fr}}{{Z_{fr}}^2} \tag{5.2}$$

with R being the range, Z the atomic number and M the masses of fragments (subscript *fr)* and projectiles (subscript *pr)*, respectively. According to this formula, the fragments with a lower atomic number have a longer range. Because all lighter fragment nuclei may be produced - from the primary nucleus down to protons - these nuclei form a tail of dose beyond the Bragg maximum of the primary beam.

2.5 Positron Emission Tomography

When the carbon projectiles have lost one or two neutrons yielding ^{10}C or ^{11}C these carbon isotopes have a shorter range and stop before the stable ^{12}C-isotope. The neutron-deficient isotopes are of special interest for therapy because they are positron emitters and their stopping point can be monitored by measuring the coincident emission of the two annihilation quanta of the positron decay. Another positron emitter is ^{15}O, which again can be produced by nuclear reactions of carbon or protons inside the patient's body [18]. To date, only positron emitting isotopes induced by carbon beams have been used for beam verification. Although the pattern obtained by positron-emitting isotopes is not identical with the dose distribution it is possible to monitor the range of the primary beam [19]. This is very useful information because a critical point in any particle treatment is the proper calculation of the particle range. According to the Bethe-Bloch-formula the energy loss and consequently the particle range critically depend on the density of the material traversed. In the human body, there are density differences between fat, bone, and muscles of some thirty percent. A special problem for treatment planning are air-filled holes like those in ear and brain, that have a much lower density (fig 5.6).

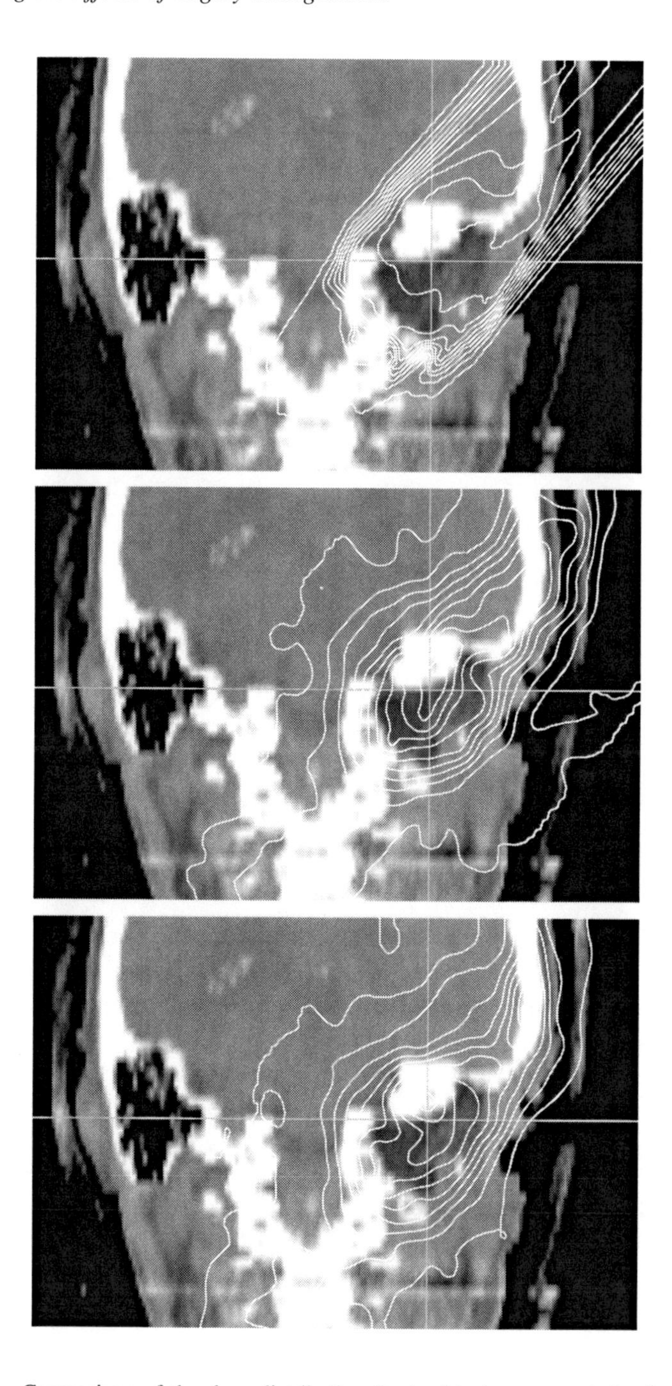

Figure 5.6. Comparison of the dose distribution (top) with the expected distribution of the β^+ decay (middle) and the measured β^+ distribution (bottom). The measurement shows good agreement with the simulation and no range differences [19].

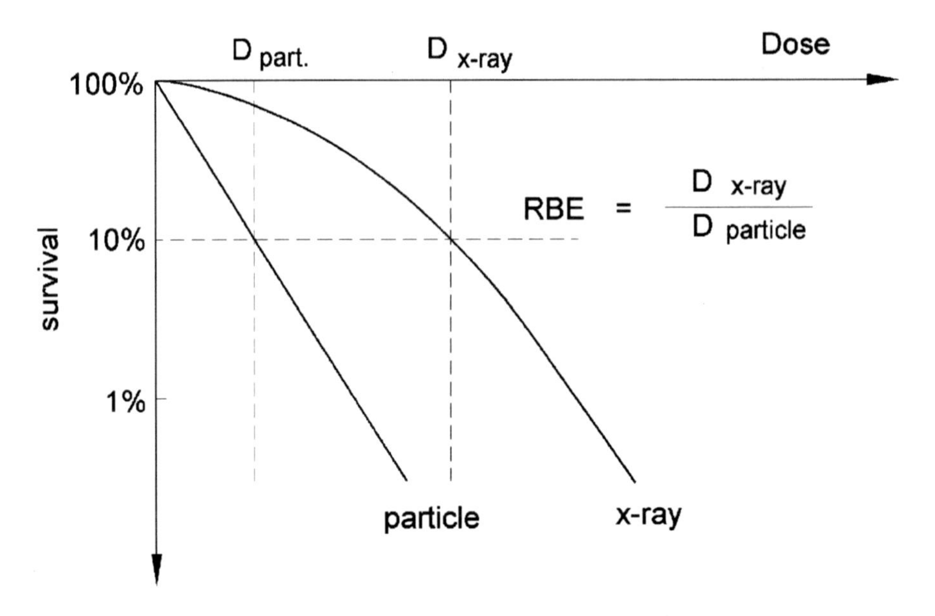

Figure 5.7. Definition of the relative biological effectiveness RBE, illustrated for cell survival curves.

3. The Relative Biological Effectiveness

3.1 Definition of RBE and its Dependence on Dose, Atomic Number and Repair

The problem of radiation therapy is to effectively kill tumor cells while protecting the normal tissue as far as possible. First, this is a problem of dose delivery precision and second, a biological problem to optimize the biological efficiency of the radiation to be used [20].

Experimentally it is found that densely ionising radiation like α-particles or heavier ions generate a greater biological effect than the same dose of X-rays [21]. In fig. 5.7 a schematic dose effect curve for cell inactivation for particles is compared to that of X-rays. The X-ray dose response is characterized by a biphasic behaviour: at low doses it has a large shoulder and at high doses a steep exponential tail. In practice, this graph can be approximated by a linear-quadratic response curve because of the limited - mostly low - dose range.

$$S = S_0 exp - (\alpha D + \beta D^2) \tag{5.3}$$

S/S_0 is the ratio of survival, α and β are fit parameters that describe the response to the dose D. For particles, the quadratic compound decreases with increasing ionisation density and then the "survival curve" is characterized by a pure exponential decay with dose [22].

In order to compare the different radiation qualities the relative biological efficiency is defined as the ratio of X-ray dose to particle dose in order to achieve the same biological effect

$$\mathrm{RBE} = \frac{D_x}{D_{ion}} \tag{5.4}$$

with D_x and D_{ion} being the X-ray and ion doses, respectively. Because of the non-linearity of the X-ray curve, the RBE - according to this definition - is not an invariable value, characterising the biological action of a particle beam in relation to photons. In fact, RBE strongly depends on the effect level and is greatest for lower doses and decreases with higher doses [23]. At very high doses, X-ray and particle response curves are usually parallel. Consequently biologically optimized treatment plans that are only correct for one dose level. If the doses are changed in order to increase or decrease the effect, the biological optimisation has to be redone.

DNA is the main target for cell inactivation by ionising radiation. Therefore, all those dependencies of RBE on the various parameters become at least qualitatively evident from the mechanisms of DNA damage: at low X-ray doses, mainly isolated damage such as single strand breaks, etc. is produced. The cell has a very efficient repair system for this type of damage that occurs very frequently and is not only caused by ionising radiation. Even simultaneous damage at both DNA strands, i.e. double strand breaks, can be repaired by the cell rather quickly with a reduced but still high fidelity. But if the local damage is enhanced by higher local doses more complex DNA damage (clustered damage) is produced which is less reparable. This is visible in the steep decay in the X-ray survival rates at higher doses. There, the increment of biological damage is larger for the same dose than at low X-ray doses (fig 5.7).

Another way to increase the local dose is the use of particles. For particles, high local doses are already produced in the centre of a single ion track (fig 5.5). For light ions like protons, only at the very end of the track i.e. at the last few micrometers, a clustering of DNA damage can be realized. This seems not to be of major importance for therapy. For protons, a small increase of RBE of 10 % is used throughout clinical therapy. Very heavy ions - argon or heavier - are extremely efficient in cell killing but the efficient region also extends into the normal tissue in front of the tumor, causing heavy late damage.

For ions between protons and argon, in the region of carbon, the RBE dependence is very favourable for therapy (fig. 5.8). At high energies, the local ionisation is low. Therefore, individual DNA lesions are produced with a large repair potential and the damage produced in the normal tissue in front of the

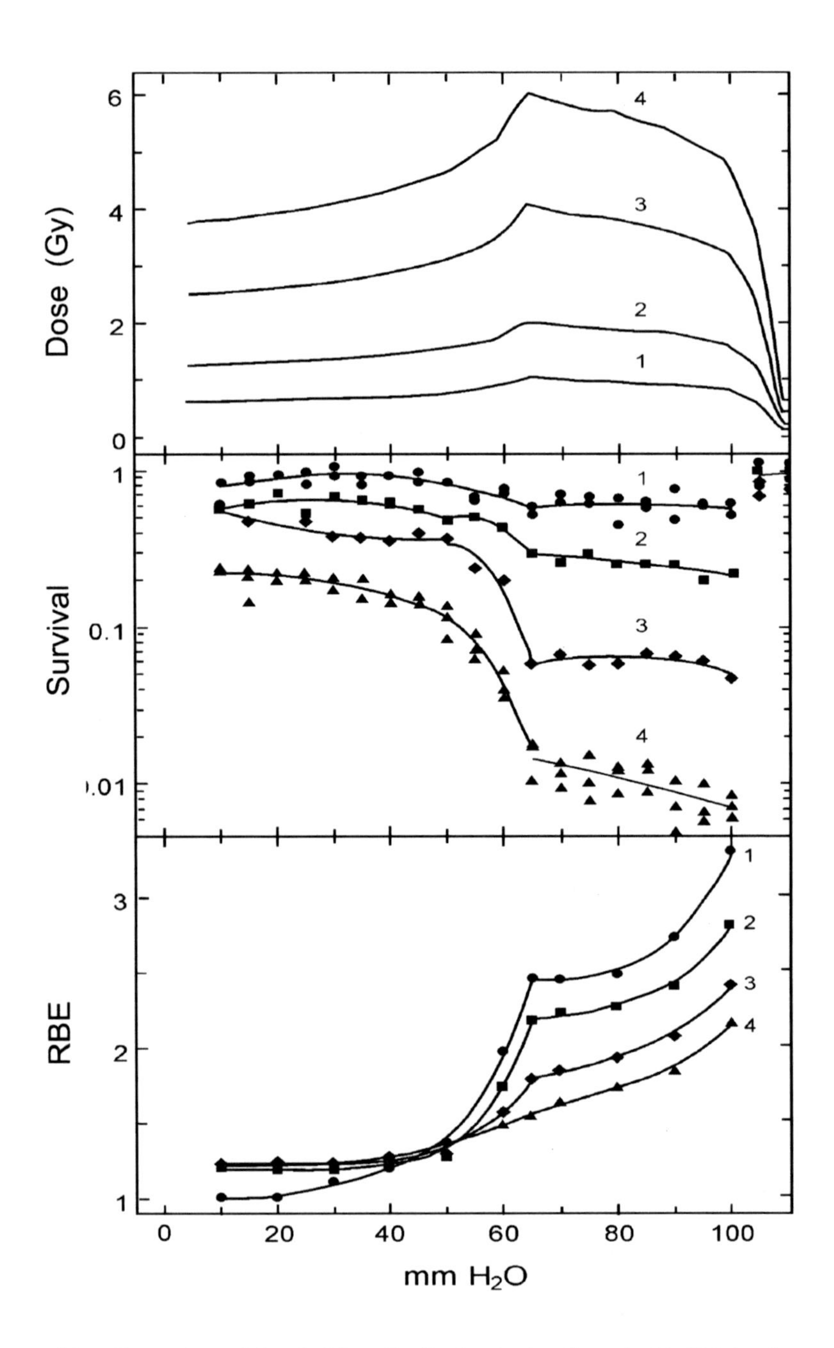

Figure 5.8. Comparison of absorbed dose distribution (top), cell survival (middle) and extracted RBE values (bottom) for the carbon irradiation of a simulated 4 cm tumor 8 cm in depth. All dose values show an increase of RBE with penetration depth [23].

target is to a large extent repairable as it is for X-rays. This yields an RBE close to 1. At the end of the carbon ions' range, in the last centimetres, the local ionisation density reaches a level where a majority of irreparable DNA damage is produced in each single particle track yielding high RBE values and efficient tumor inactivation.

There, RBE values up to three or four are found and the absorbed dose D as given in Gray has to be replaced by the effective dose (D_e) given in Gray-equivalent (Gye)

$$D_e[Gye] = D[Gy] \times RBE \tag{5.5}$$

The comparison of beams of different dose levels reveals the dose dependence of RBE for one specific tissue. But for the practical application these data cannot be transferred directly to treatment planning. Provided RBE is mainly determined by DNA repair, the intrinsic repair capacity of the different cells plays an important role in the RBE effect. It has been shown experimentally that cells having little or no repair capacity exhibit little or no RBE increase when exposed to densely ionising radiation [24]. On the other hand, tumors that are usually very radio-resistant because of their enhanced repair capacity become very sensitive to heavy ion exposure because of the larger RBE effect. It should be mentioned that the RBE dependence on atomic number has also to be taken into account with regard to the dose contributions of the lighter fragments produced by the carbon ions.

In summary, the increase in the relative biological efficiency, RBE, depends on the physical side of the track structure with high local ionisation densities causing clustered lesions. In addition, RBE depends on the intrinsic repair capacity of the tissue and shows the largest effect for x-ray radioresistant tumors.

3.2 Tailoring RBE for Therapy

The RBE determines the weighting factors for the superposition of the various Bragg curves in order to achieve a homogeneous killing effect over the complete target volume. The influence of RBE on the treatment plan very much depends on the method used for beam shaping. Using passive beam shaping systems the RBE dependence has to be in-built in the shape of the range modulators and is fixed [25]. For active beam shaping systems the higher conformity with the target field has the consequence that RBE varies over the complete target volume in three dimensions [26].

Due to a gain of knowledge in time, RBE has been determined in a different way at each of the three heavy ion centres: Berkeley, Chiba and Darmstadt. At Berkeley, *in vitro* data from cell cultures were used for RBE determination. In many experiments, cell survival was determined in pristine and also in smeared-out Bragg peaks of variable depth and width. These measurements were used to design a set of ridge filters for the therapy of the tumor sites in the corresponding depth and extension. However, the implicit assumption is that these *in vitro* data of special human cell lines should be valid for all the human tissue affected by irradiation [27].

At the National Institute for Radiological Science (NIRS) in Chiba, a different strategy for the incorporation of RBE into the ridge filter design was used [28, 29]. Again, the basis were cell experiments using human salivary gland HSG cells. Survival was measured for HSG-cells and fitted to a linear-quadratic dose dependence. For a mixed radiation field, the fit co-efficients were interpolated as function of LET. Thus, a nearly universal profile of the spread-out Bragg peak (SOBP) could be obtained for different carbon energies and different widths of the SOBP. With this universal ridge filter a flat survival response of HSG-cells could be produced independent of carbon energy but again, differences were found for other cell lines. Another important relationship was found in the experiments: RBE in the mid of the SOBP of carbon ions was close to that found in the experiment with neutron irradiation. Consequently, the dose prescription and fractionation at Chiba was adapted to the large clinical experience of neutron therapy [30].

At GSI, the RBE is calculated separately for each element of the treatment volume. In contrast to the other facilities RBE is not deduced from in-vitro data. Instead, known X-ray sensitivities of the same tumor histology are the basis of the calculation. These data can be dose-effect curves for the specific tumor types or fractionation curves from which the X-ray sensitivity can be deduced. The basis for the calculation is the Local Effect Model (LEM) [31] where the RBE is calculated according to the size of the cell nucleus, the radial dose distribution and the X-ray sensitivity curve. In this model, the biological response to particle radiation is the convolution of the induction probability of lethal damage as measured with X-ray irradiation with the different values of the radial dose distribution integrated over the size of a cell nucleus [32].

$$S_{ion} = e^{-N_{lethal}} \tag{5.6}$$

with

$$N_{lethal} = \int\limits_V \frac{\ln S_x(D)}{V} \bullet dV \tag{5.7}$$

being the number of lethal lesions, $S_x(D)$ the X-ray dose effect curve for cell killing, V the nuclear volume in the cell and S_{ion} the calculated survival after ion exposure. The surviving fraction for irradiation with a particular ion of various energies can be calculated if the X-ray effect is known. Then the

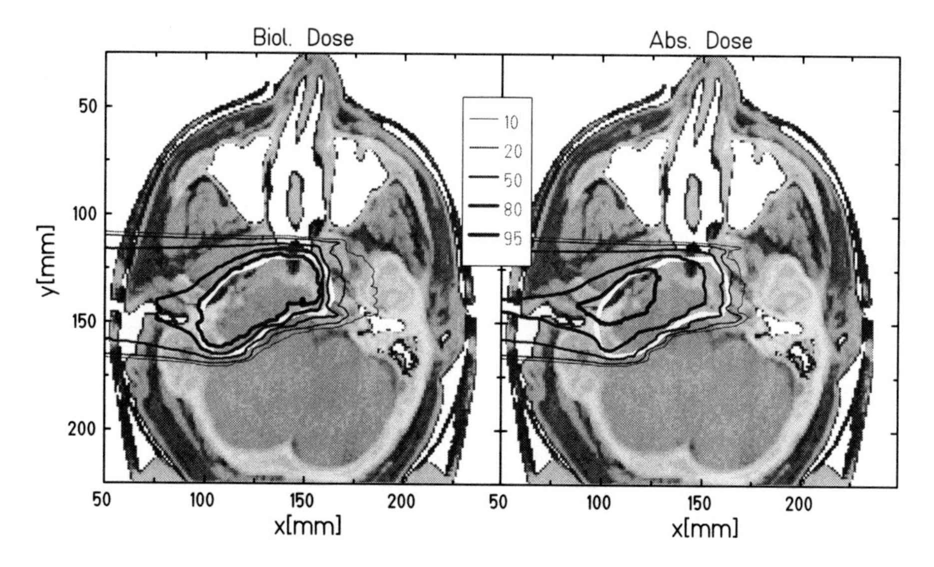

Figure 5.9. Comparison of the absorbed dose with the biologically effective dose for a tumor in the base of the skull. The biological effectiveness increases towards the distal end of the target volume. Therefore, a lower absorbed dose yields the same biological effect.

RBE is extracted for each volume element separately yielding a map of RBE values over the complete irradiation volume.

Figure 5.9 shows the biological and physical dose of a tumor in the base of skull. Beside the differences in the dose contours it is important that the physical dose is 0.5 Gy at its maximum while the biological dose is almost 2 Gye. Thus, RBE values up to 4 are applied to this slowly growing tumor.

The clinical experience with the patients treated at GSI with the biologically optimized dose profiles calculated according to LEM confirms the reliability of the calculations. Up to now, no recurrent tumors have been observed in the treatment fields indicating the application of the correct RBE values [33]. In summary, the RBE of the carbon beam is favourable compared to that of protons and represents an additional advantage in the improvement of the efficiency in tumor cell inactivation. This means at the same time that the very efficient carbon beams have to be applied with the utmost care in order to avoid any damage to the normal tissue.

4. Heavy Particle Therapy

4.1 Motivation for Particle Therapy

The term *cancer* stands for a large variety of diseases that have in common that cells of patient have lost their growth regulation and start to proliferate

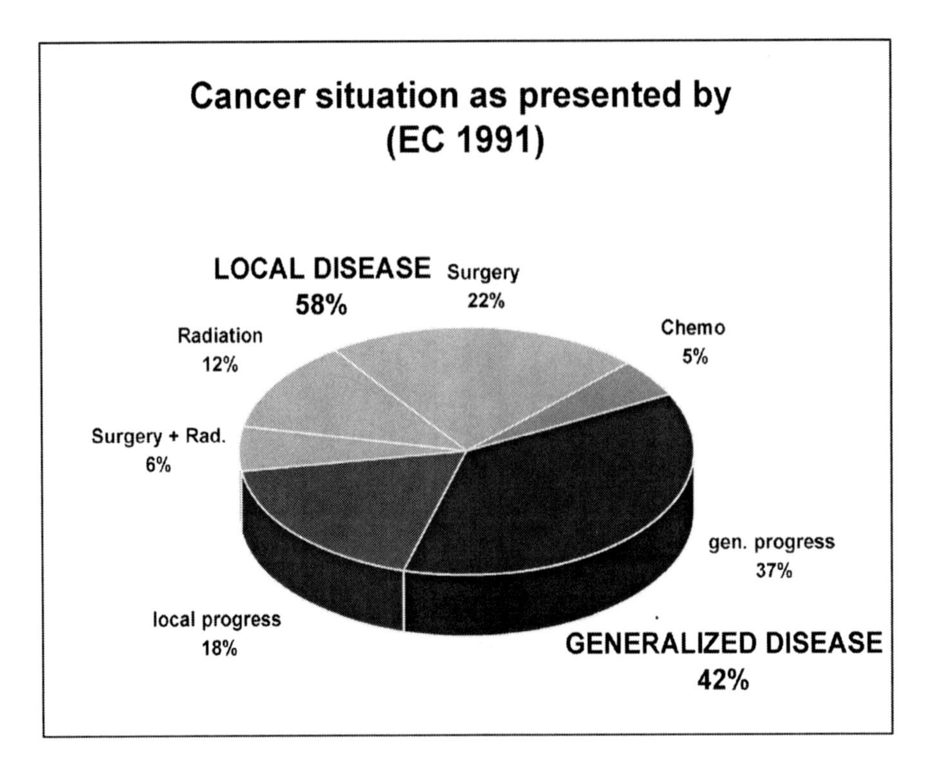

Figure 5.10. Distribution of the approximately 1 million new cancer patients of the European Union per year. Patient diagnosis with local diseases i.e. one single primary tumor has a greater chance for a long-term cure by the various treatments.

in an uncontrolled way, thereby producing a tumor. For benign tumors, this tumor growth stays local resulting in a single but ever enlarging tumor volume. In malign tumors, cells are spread throughout the body, seeding new tumor colonies everywhere, the so-called metastases.

The understanding of origin and mechanism of tumor diseases has made large progress with the discovery of the oncogenes and their behaviour on the microscopic level but up to now, this knowledge is not sufficient to cure tumors. In contrast to a bacterial infection, where foreign cells invade the body from the outside and can be destroyed because they are different from the host's body cells, the growth of a tumor is the growth of oncogenetically deregulated but otherwise normal cells of the respective patient. In addition, oncogenes are also present in the non-tumor cell, i.e. in all normal cells controlling cell growth but their influence on cell growth normally comes to a halt when the organ has reached its genetically determined volume. Not so tumor cells.

Up to now, it has been impossible to destroy a macroscopic tumor with molecular methods although large effort is being made in this direction and great hope is put on this field. At present, the most efficient way to kill a tumor is to remove it by surgery as long as the tumor is small and solid i.e. not metastatic (fig. 5.10). Chemotherapy uses drugs to stop proliferating cells i.e. tumor growth in order to extend the patient's life. Finally, a large fraction of tumors can be eliminated by radiotherapy with ionising radiation. In principle, any tissue can be destroyed by radiation if only the dose is high enough. In practice, the dose that can be given to a tumor is limited by the tolerance of the normal tissue surrounding the tumor and the normal tissue dose is determined by physical parameters, e.g. the depth dose dependence and the lateral scattering of the radiation.

Conventional therapy started with low-energy X-rays. In order to increase the precision of the delivered dose electromagnetic radiation such as ^{60}Co-gamma rays and finally bremsstrahlung photons from linacs were used because of the smaller scattering and the increase of dose for the first three centimetres followed by an exponential decay with depth. An entirely different behaviour can be expected from heavy particles. For heavy ions the dose even increases with depth and lateral scattering can be minimized.

A second strategy for a higher irradiation efficiency in the tumor was to use radiation of different "biological quality". This was attempted in neutron therapy where an almost threefold biological effect can be found for the same absorbed dose. Neutron therapy produced an excellent tumor control because of the elevated relative biological efficiency (RBE). However, these positive results were correlated with severe side effects due to the very poor depth dose profile of neutrons. Thus, the high efficiency for tumors also affects the normal tissue surrounding the tumor.

In contrast, heavy charged particles in the region of carbon produce a high RBE which is comparable to that of neutrons but which is restricted to the end of the particle range. Therefore, tumor cells can be killed with greater efficiency but the normal tissue in front of the tumor is not exposed to high RBE radiation. For these reasons – a very high precision in dose delivery and a greater biological effect but restricted to the tumor – heavy ions promise a much better cure rate.

Heavy particle therapy is predominantly applied to deep-seated tumors where the advantage of the inverse dose profile is most significant. In general, the size of the target volume is much larger than the spot of an unmodified beam as produced by the accelerator. Therefore, the beam has to be enlarged laterally and longitudinally. In the beginning of particle therapy in the sixties and seventies, passive beam shaping devices were developed in which the beam is distributed at each instant over the complete target volume. These mechanical devices had the great advantage of simplicity and of not being sensitive to intensity fluctuations from the accelerator. However passive systems had to be tailored individually for each patient and even then the congruence between the optimum target volume and the actually irradiated volume was not satisfying.

Active systems make use of the possibility to deflect a beam of charged particles by magnets and to change its range by energy variation of the accelerator. With active systems, a most conform beam delivery can be achieved without patient-specific hardware. However active systems require a more sophisticated control system, which only became possible after more powerful computers had become available. Active systems are more sensitive to accelerator fluctuations but offer great flexibility and do not need patient-specific hardware.

4.2 Passive Beam Spreading Systems and Intensity-Modulated Particle Therapy

For a passive shaping of the ion beam, many hardware variations exist that cannot all be described here but are reported in great detail in [25]. For the lateral enlargement, scatter systems as given in fig 5.11 have been used. In order to avoid beam fragmentation in passive devices, magnetic deflection systems like wobblers or scanners have been introduced but these magnetic deflection systems had no feedback to the beam delivery from the accelerator. Homogeneity over the target area was achieved by repeating the deflection pattern so many times in a random way that all fluctuations in the beam intensity were averaged.

In order to shape the beam in longitudinal direction absorbers of variable thickness were introduced. These absorbers consist of regions of different thickness. A part of the beam penetrates a thicker absorber having then a short range in the patient while another part penetrates a thin absorber having a longer range in the patient. This principle has been realized in many ways as for instance in linear or spiral ridge filters or in propellers of different thickness.

Finally, the distal part of the target field can be shaped by compensators or boli in front of the patient in order to spare critical structures behind the target volume. This shaping, however, is done at the expense of an extended high-dose field in front of the target volume (fig. 5.11).

In summary, passive systems allow only a limited adaptation of the irradiated field to the target field and in order to exploit the advantages of heavy charged particles to a full extent active systems are indispensable.

In order to reduce the unavoidable waste of the primary beam when using passive systems and to reduce the fragmentation of heavy-ion beams active lateral scattering systems have been developed mostly at Berkeley [25]. In these systems, magnets that are perpendicular to each other and to the beam axis deflect the ion beam. Two types of beam pattern have been realized - linear and zigzag - raster scanners and circular patterns (wobbler). In the linear system, two fixed triangle frequencies are used, for instance a slow one to move

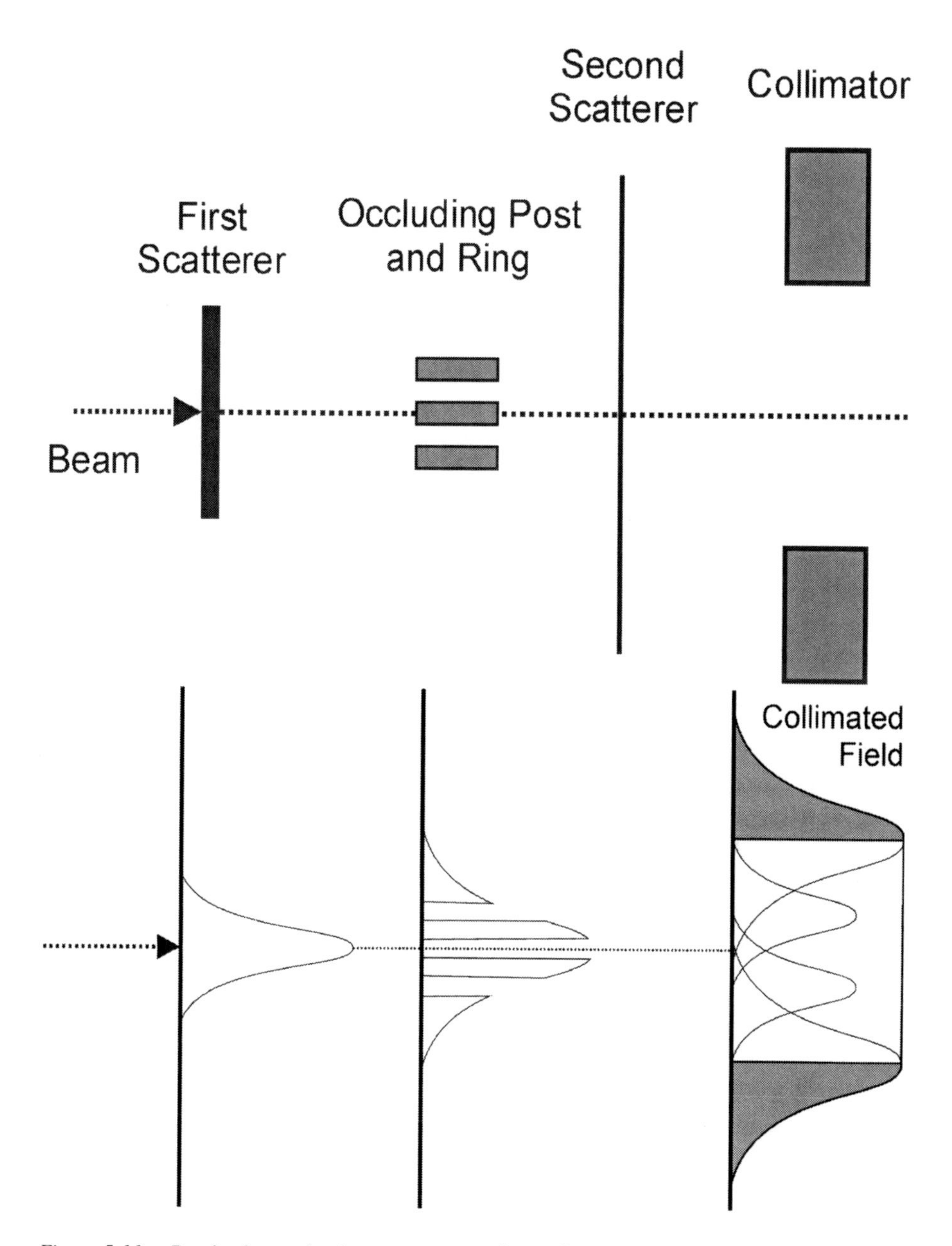

Figure 5.11. Passive beam shaping system to produce a flat dose distribution over a large field. Uniform fields are produced by a first scatterer and an annular ring plus a second scatterer, redrawn according to [25].

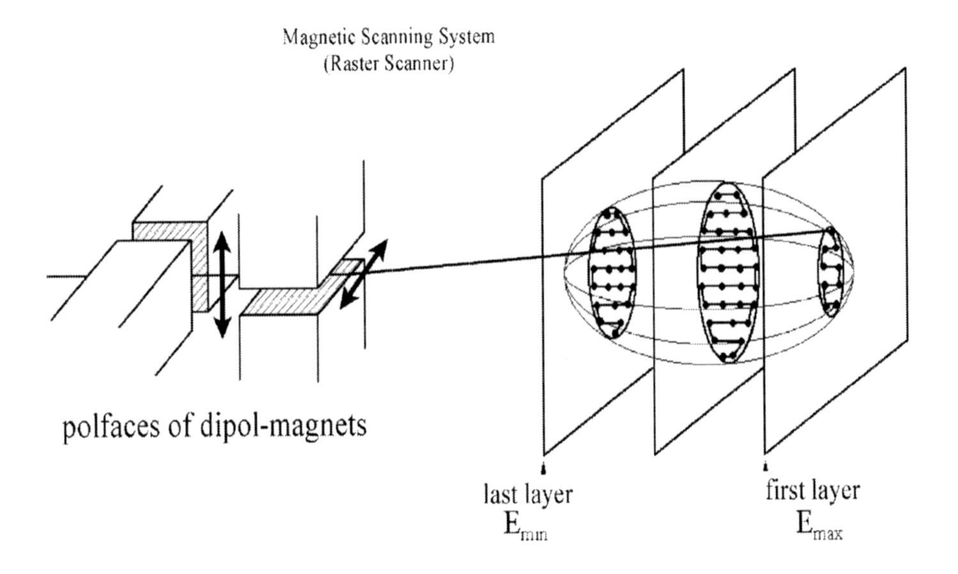

Figure 5.12. Schematic drawing of the intensity-controlled raster scan. The target volume is dissected into layers of equal particle range that are covered by a net of picture points. The beam is guided along the imaginary line of picture points by two fast magnets.

the beam up and down and a fast one to write horizontal lines similar to those in the TV-system. In the circular wobbler system, sinusoidal frequencies of the same amplitude are applied to the magnetic deflection system having a phase shift of 90° between horizontal and vertical deflection magnets. By selection of 3 different amplitudes, 3 different circular distributions that are stacked with the correct intensity are used to achieve a flat field larger than the treatment area. However, there was no feedback between the fluctuation beam intensity and the speed of the deflection. In order to achieve a homogeneous dose distribution the target field has to be covered many times with the hope to average out the inhomogeneities of the accelerator fluctuations.

Intensity-controlled scanning systems for 3 dimensions have been developed by PSI in Villigen for protons [34] and by GSI in Darmstadt for carbon ions in order to achieve the best target-conform irradiation fields possible [35, 36].

The target volume is dissected into layers of equal particle range. Using two deflecting magnets driven by fast power supplies, a "pencil beam" is scanned in a raster-like pattern over the layers, starting with the most distal one. After this layer has been painted the energy of the beam and consequently the range is reduced and the next layer is treated (fig 5.12).

The difficulty arising from this method is obvious: the more proximal layers are already partly covered with dose when the distal layers are being treated.

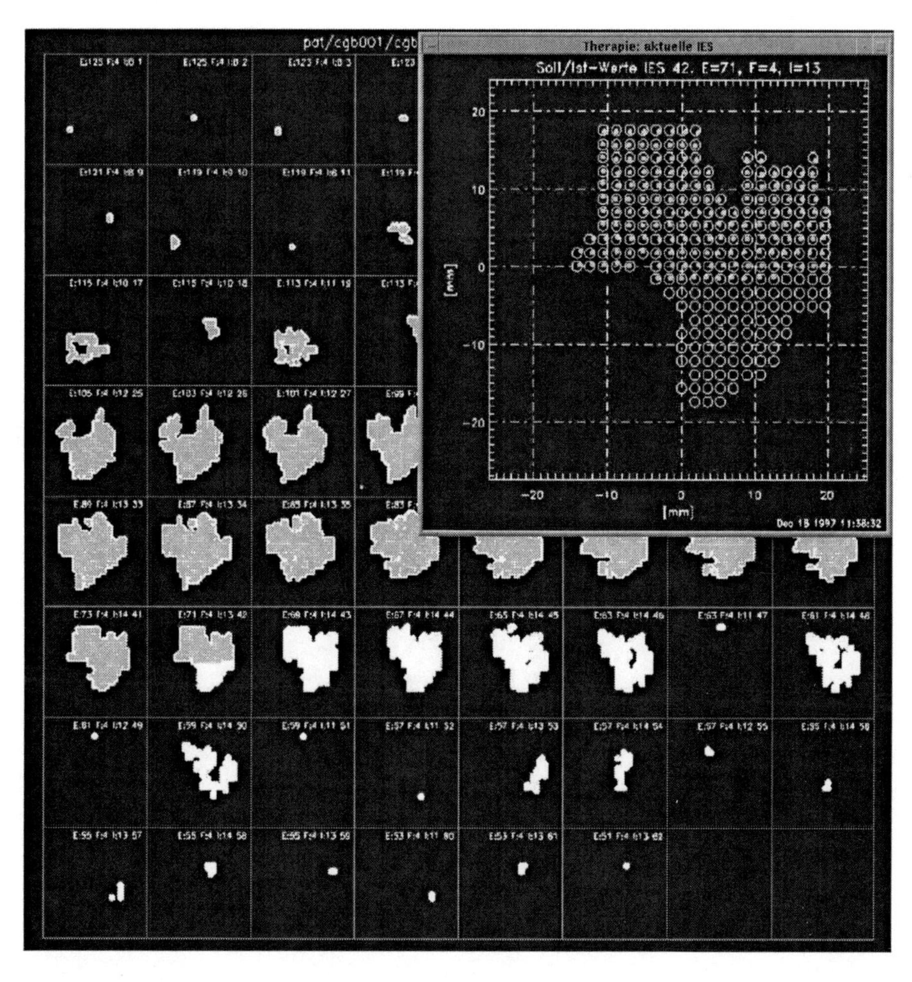

Figure 5.13. Compilation of the different range slices of a treatment volume. In each panel one slice is shown, in the magnified panel the circles represent the calculated centre positions of the beam that are filled with the measured centre of the beam. The beam diameter is larger than the circles and overlaps many positions yielding a homogeneous distribution [37].

Consequently, these layers have to be covered during the continued irradiation with an inhomogeneous particle distribution in order to reach a homogeneous dose distribution or a homogeneous distribution of the biological effect over the total target volume in the end. In order to achieve the desired inhomogeneity the beam path is divided into single picture points (pixels or voxels) over the individual areas for which the individual particle covering has been calculated before [36]. In this intensity-modulated particle therapy (IMPT), up to 30,000 pixels per treatment volume are filled with an individually calculated number of particles. Because of the finite range of particle beams the IMPT has the depth as an additional free parameter compared to similar X-ray techniques, called IMRT.

Although each layer may differ in contour, the shape of a target volume can be "filled" with high precision. The different layers are of great complexity as is shown in fig. 5.13 that depicts the energy slices of a plan of a patient treated at GSI with carbon ions.

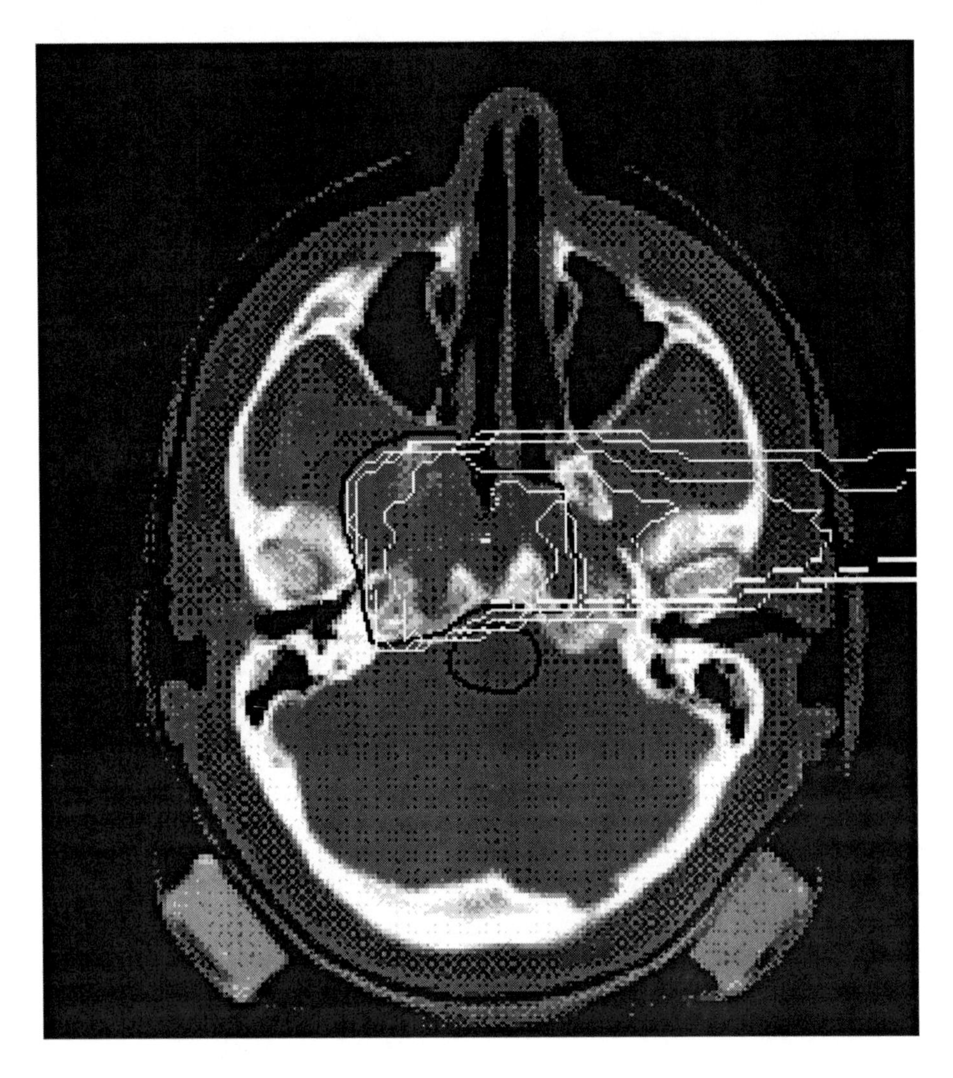

Figure 5.14. Computer tomography section through the head of a patient. The treatment plan is superimposed, showing isodose contours [26].

The small spots of high or low energy ions in the beginning and at the end of scanning are due to the density inhomogeneities of the tissue. Small areas of high density like bones have to be treated with higher energies. This accounts for the small spots of higher energy irradiation. Vice versa, areas of low density have to be filled with low energy particles [26].

In fig. 5.14, a final treatment plan is projected on a CT scan of a patient. The good conformity of target volume and treatment area as well as the steep gradients of the dose in areas adjacent to critical sites are visible.

4.3 Comparison of Beam Spreading Systems

Ion beam therapy with protons and heavier ions started in the late sixties when the stability of accelerators was rather poor and no sophisticated control system was available due to the lack of fast computers. In addition, in conventional therapy, during the transition from X-rays to Co-gamma rays at tolerable dose distributions were achieved, frequently, the tumor dose was not high enough to reach a permanent cure.

Back then, the transition to the very advantageous inversed dose profiles of particles represented major progress. When using passive beam delivery systems the quality of the beam distribution is widely independent of the performance of the accelerator and from intensity fluctuations. For small proton fields, like those for the treatment of eye tumors, the passive beam application still offers the best and cheapest solution.

For larger and deep-seated tumors the comparison becomes more difficult. Conventional therapy, however, has made large progress over the last years. Co-gamma units are widely replaced by electron linacs and sophisticated methods for treatment planning and beam delivery keep up with the meanwhile improved accuracy of diagnostic methods like CT, NMR and PET. Using intensity-modulated radiation therapy (IMRT) dose distributions very close to the target volume can be achieved that are comparable and sometimes even better than the ion beam application using passive absorbers [38]. Nevertheless, active particle beam application is always better in its conformity than any passive beam shaping method and also better than X-ray-IMRT. The scanning methods developed at PSI and GSI represent an optimum 3-dimensional type of "particle IMRT" since the beam intensity can be controlled not only for each direction but also for each depth position separately.

4.4 Treatment Planning

4.4.1 Static Fields. Prior to the development of tomographic methods the available information on the location and the size of a tumor was rather poor and based on a few X-ray images that gave the projection of the tumor volume only but not an exact 3-dimensional image. This poor visual information corresponded in some way to the limited possibilities to deliver the dose using X- or gamma-rays. In tomographic imaging, computer-assisted X-ray tomography

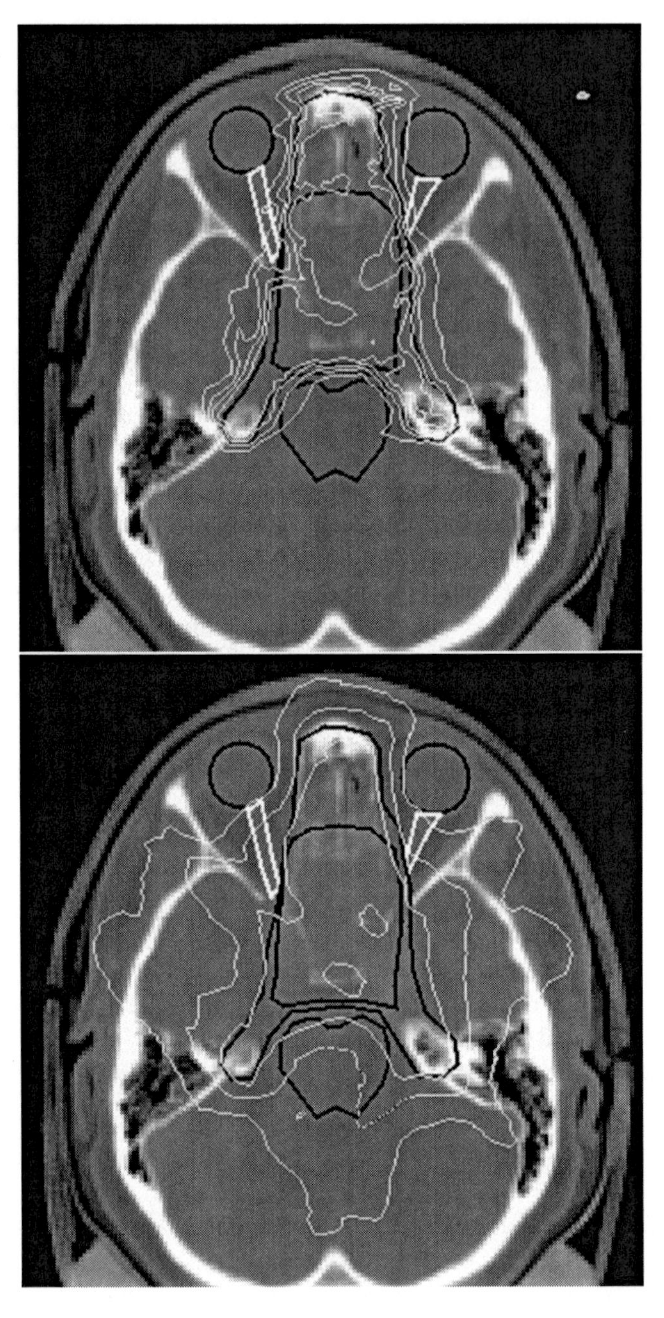

Figure 5.15. Comparison of Treatment Plans. Heavy Ions (above) 2 fields. IMRT (below) (With the courtesy of O.Jäkel, DKFZ Heidelberg)

(CT) came first, followed by positron emission tomography (PET) and magnetic resonance imaging (MRI), now allowing delineation of the contour of the tumor with millimetre resolution and in three dimensions.

In particle therapy, these images are used to deliver the beam in a most conformal way to the tumor volume. Thus, the first task of treatment planning is to identify the target contours that are usually larger than the tumor volume. It is the task of the physicians to graph the border lines in each slice in the tomogramme and to select the best possible entrance channels as well as to mark the organs of risk such as brain stem, optical nerves, eyeballs, etc.. This procedure is independent of the type of radiation to be used. The next step of particle treatment planning very much depends on the system to be used for beam delivery.

For passive systems, the contour of the largest projection in beam's eye view is translated into a collimator to be mounted in front of the patient. A limited depth modulation is obtained by the deployment of ridge filters and compensators or boli that will also be installed in front of the patient. For the design of these compensators, the density inhomogeneities in the penetrated tissue have to be known. In tissue, there are density differences between fat, bone and soft tissue up to 30% or 40%. In addition, "vacuoles" i.e. air-filled gaps or metal implants (as bone substitutes) exhibit even larger density variations. In general, the density can be obtained from the CT image: after calibration, the gray values of the CT can directly be translated into water-equivalent density values. Higher densities are then translated into a longer path in a water-equivalent target and correspondingly, lower densities are translated into shorter water-equivalent ranges. This translation is possible because the dose is defined as energy deposition per mass unit and the compression or the dilatation of the absorber material changes the energy loss proportionally to the density while keeping the dose constant. The translation into water-equivalent densities is the basis for the calculation of compensators for passive delivery systems and for the treatment planning in active beam delivery systems.

4.4.2 Inverse Treatment Planning. In active beam delivery systems, the target volume is dissected into layers of equal particle range and the beam is moved over these layers from pixel to pixel. Treatment planning for heavy ions is just "inverse" to this procedure. It starts from a pixel in the most distal layer and - in a first approximation - calculates the number of stopping particles necessary to produce the required dose. This is done for all pixels of this distal layer. Then, the next and more proximal layer is treated in the same way. In order to determine the correct dose, the fraction of the dose deposited during the irradiation of the deeper layer has to be subtracted. Thus, layer after layer is treated for the planning of the entire volume. Due to the fact that heavy ions have a dose tail beyond the Bragg maximum a few more steps are required before the absorbed (i.e. the physical dose) fits perfectly to the target volume.

If two or more entrance channels are used the planning can be carried out independently, generating a homogeneous field for each entrance channel. However, with regard to the geometry of the tumor, it is frequently more appropriate to plan inhomogeneous fields that will result in a homogeneous field if all entrance channels are used.

For heavy ions like carbon, further complexity arises from the variation of the RBE over the treatment field. In each volume element, the dose originates from a mixture of primary and secondary ions of different energies. Because the RBE depends on atomic number and energy, this mixture has to be known in order to calculate the local RBE values correctly. After the optimisation of the physical dose local RBE values have to be calculated as weighting factors of the dose. Then, the irradiation field has to be re-optimized. As RBE also depends on the dose level, local RBE values have to be calculated iteratively in this second step and in all further steps of optimisation. In one treatment field, 10,000 up to 50,000 individual pixel points are irradiated and have to be optimized beforehand. Consequently, the final biological optimisation is far more time-consuming than the optimisation of the absorbed dose alone.

Because this planning starts from the dose to be delivered in different pixels to the target and calculates the number of particles and their energy to be delivered, this type of planning is called inverse planning or intensity-modulated particle treatment (IMPT). A similar method has recently been developed and is now being used in patient treatment with high-energy photons from electron linac bremsstrahlung. Here, the photon field from the accelerator is first restricted in its contours and the intensity in each point of the field is modulated with a variable collimator (multi-leaf collimator). Using photons with an inverse planning and delivery scheme, the so-called Intensity-Modulated Radiation Therapy (IMRT) a dose distribution can be produced that is very often comparable to that of a proton beam but requires less investment. However, the precision of a carbon beam cannot be reached with IMRT.

4.5 Patient Positioning

The fundamental advantage of particle therapy is the high precision of beam application. Up to now, the beam has been delivered in the frame of a coordinate system fixed with respect to the treatment room. Therefore, the patient has to be immobilized and adjusted within the same coordinate system. For this purpose, a large variety of techniques have been adapted from high-precision photon therapy like stereotactic treatment or IMRT [39]. For the immobilisation of head and neck, thermoplastic masks and bite blocks are used. These devices are manufactured for each patient individually. Bite blocks allow a precision and reproducibility of 1-2 mm. Thermoplastic masks cover the head completely and allow a positioning accuracy better than 1 mm. For a total immobilisation of the body foam moulds and thermoplastic total body masks have been developed that guarantee accuracy almost to the millimetre. In any case, the position of the patient is checked prior to the treatment to an extent that correlates with the precision of the treatment. For passive beam shaping using proton beams this precision may be less than for active beam scanning with carbon ions where a precision better than 1 mm must be guaranteed throughout the entire treatment.

In the thorax, the external immobilisation is counteracted by the internal motions due to heartbeat and breathing. These motions are more or less cyclic and of regular amplitude. Up to now, the treatment of tumors in the lung has been "gated" with the breathing for passive beam shaping and the beam is applied only during a short period of time of relative stability. This procedure is not applicable for active beam scanning because of a drastic elongation of the exposure time as well as the possibility of residual interference patterns caused by small motions producing hot and cold spots. However, the active beam scanning is - in principal - able to follow patient motion if the scanning is significantly faster than the internal motion but it has not been realized to date.

First attempts to adjust the position of the patient's body directly to the irradiation have been made in NAC South Africa, mainly with the intention to correlate the initial position of the patient with the coordinate system of the beam delivery. For this purpose, light marks are fixed on the patient and monitored via a TV system. Because patient positioning is a very time-consuming and expensive part of any precision treatment major efforts for the improvement are being made in this field [40].

4.6 Safety and Control System

It is the purpose of the safety and control system to ensure a safe and precise delivery of the prescribed dose to the target volume. In practice, very different systems for beam delivery are being used. Therefore, it is impossible to describe a general safety and control system for all therapy units [41, 42, 37, 43]. It is, however, common to all systems that they can be divided into an accelerator section and an application section.

Regarding passive beam shaping, these two sections are separated because the quality of the beam as extracted from the accelerator does not influence the quality of the application. As long as energy and atomic numbers are correct, the right beam shape is produced by the absorbers, range shifters, compensators, etc. The safety of the treatment only requires that the exact number of particles as measured in an ionisation chamber enters the system and that all necessary components are in the right place. Of course, the patient has to be positioned correctly, too [25].

For an active beam delivery system, the interplay of scanning system and accelerator is more complex [37, 44]. First, the energy has to be changed stepwise without a change of the zero position of the beam and its diameter. Second, the beam extraction has to be smooth in time. Although intensity fluctuations can be rectified by the feedback of the scanning system, a "grass"-like structure with large intensity spikes makes scanning difficult. In addition to the beam energy, the diameter of the beam in the target has to be changed on request from the scanner in a controlled way. For safety reasons, a monitor system has to be installed in front of the patient that controls these parameters independently. At the GSI scanner unit, the heart of the monitor system consists of two ionisation chambers and two multiwire chambers. The combination of a wirechamber and an ionisation chamber measures the position of the beam and its current intensity. This combination is read out 6,000 times per second and at least four times for each voxel. If more than one measurement per voxel disagrees with the requested number the beam is aborted in the accelerator within less than half a millisecond. In addition to this fast interruption cycle, slow monitor systems control other important parameters such as the vacuum in the beam pipes, gas flow in the monitors or access control and so on. Similar control and safety systems adapted to the respective needs of the beam application system are installed in all therapy units. It is, however, also common to all therapy projects that the required manpower for the system software is usually completely underestimated.

4.7 Gantry

The term "gantry" stands for a beam delivery system that allows one to administer the beam from all directions to the patient in a supine position. For conventional therapy using photon beams with the exponential depth dose distribution, the use of a gantry is mandatory for any treatment of a deep-seated tumor in order to distribute the large integral dose outside the tumor over a large volume.

In photon therapy, the radiation source or the electron linac is moved around the patient. In particle therapy, it is impossible to move the accelerator around the patient. Instead, deflection magnets are mounted on a turnable system. The most frequently discussed systems are given in fig. 5.16

At Loma Linda, three corkscrew gantries are in operation for therapy but at present the systems are not equipped with scanning. An excentric gantry including a voxel scan system has been built at PSI, Villigen, and three gantry systems are installed at the NPTC, Boston.

For carbon ions, no gantry system has been realized yet. Because of the greater ion energies necessary to obtain the same penetration depth and because of the higher magnetic rigidity the design of a carbon gantry cannot be

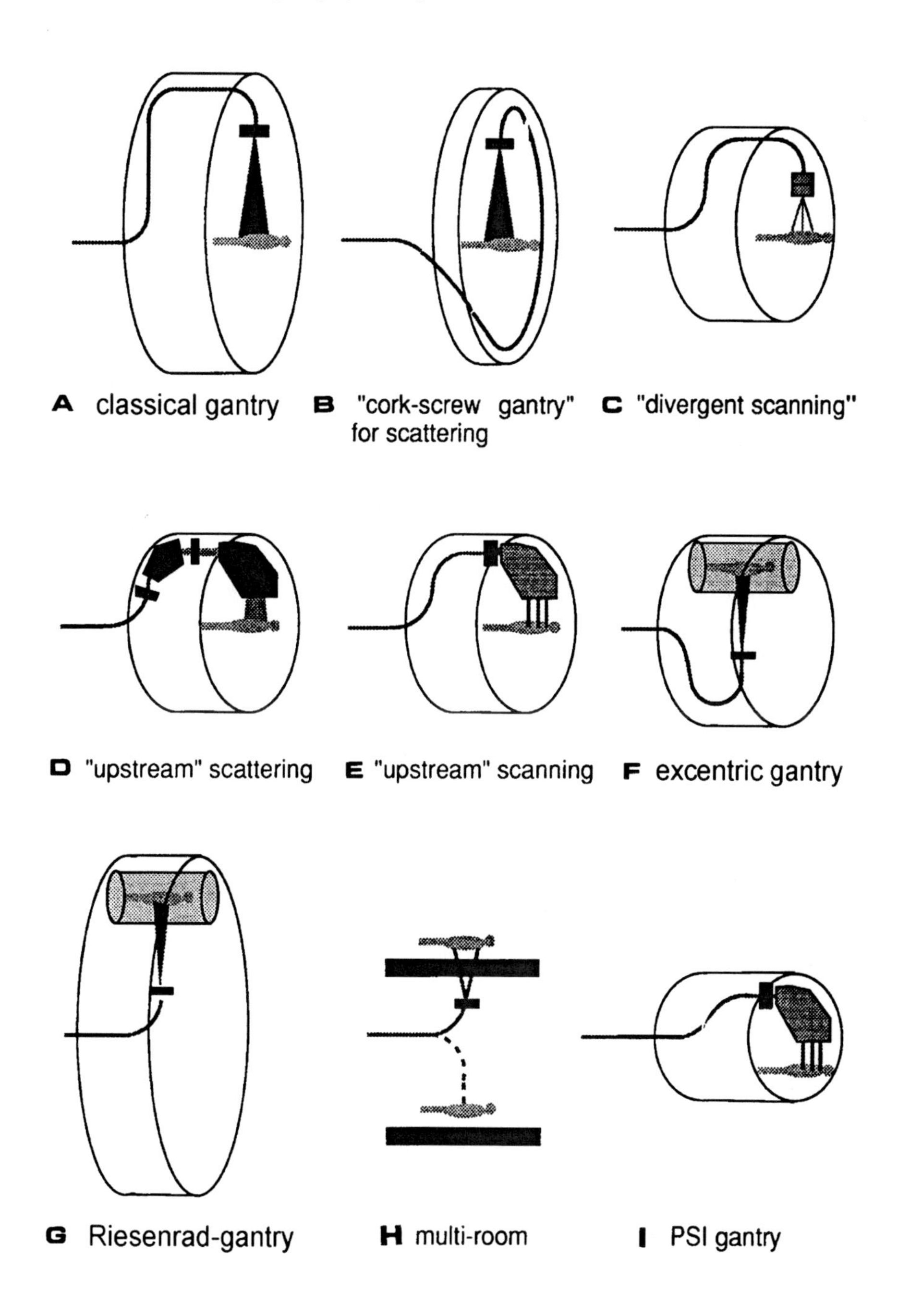

Figure 5.16. Classification of the different types of gantries proposed for particle therapy, according to [45].

a straightforward blow-up of a proton design. Recent proposals for carbon gantries at GSI use upstream scanning to reduce the gantry radius but even with this design the weight of the carbon gantry would be about 600 tons i.e. six times larger than that of a proton gantry.

Because of these difficulties the idea of a gantry is frequently given up for future heavy ion therapy units in favour of a fixed-angle beam delivery of 45° or 90°. The possibility of superconducting gantries has been discussed. At Berkeley's Bevalac, patients were also successfully treated in a seated position but this premises vertical CT scanners in order to diagnose the tumor under the same conditions, as it will be treated later on. It follows from the above that up to now there is no generally accepted solution for a gantry.

4.8 Accelerators

In the past, the therapy with heavy charged particles has mainly been promoted by physicists working at accelerators. Therefore, the question of the best accelerator for therapy was frequently in the centre of interest when it came to new facilities. However for optimum patient treatment, the discussion should start at the rear end i.e. with the patient, the number and indications to be treated and the kind of beam application. After these parameters have been decided the requirements for the necessary accelerator can be specified and given to the accelerator physicists and engineers.

For proton therapy, there seems to be no preference for cyclotrons or synchrotrons. Both types of accelerators have their benefits and shortcomings. Cyclotrons produce a very stable beam intensity that is suitable for beam scanning but the energy variation has to be performed with absorber systems. Cyclotrons are smaller but heavier than synchrotrons. Synchrotrons are generally more flexible and energy variation from pulse to pulse is easy. At present, both, cyclotrons and synchrotrons are used for proton therapy with good results.

For heavier ions like carbon, a synchrotron appears to be preferable since a cyclotron would become very heavy and probably more expensive than a synchrotron. In addition, synchrotrons are able to change the energy from pulse to pulse. Although a synchrotron is larger, the empty space inside the magnet ring can be used for power supplies. In addition, the size of the accelerator becomes relative when compared to the space needed for the therapy rooms. The use of superconducting accelerators has also been discussed, especially in the Eulima project [46] but was given up because of the extended repair times of several days in case of an accelerator failure. One must keep in mind that patients have to be treated for 20 to 30 fractions, one fraction a day. The closing of an accelerator for many days would be intolerable. Although the discussion on the best accelerators is still going on accelerator technology is presently at such a high level that any kind of accelerator will do.

4.9 Quality Assurance

A very important but frequently neglected problem of particle therapy is quality assurance. For particle therapy, quality assurance becomes the more important the more the precision and the efficiency of the beam are optimized. This was the case in the transition from conventional to particle therapy, which has a better dose profile. It is especially relevant for the target conforming beam scanning methods producing extremely sharp dose profiles. Here, the accuracy of the beam delivery has to be guaranteed by extensive test procedures [47].

Quality assurance starts with the test of the beam quality i.e. its isotopic purity and energy, checks of the beam, spatial beam stability and finally the test of standard volumes irradiated in water phantoms. In addition, each treatment plan has to be verified in a water phantom. This is not at all trivial because the density inhomogeneities in the human body yield a different dose distribution than that of water. For control, the treatment plan has to be converted into a "water-equivalent" plan since only this water-equivalent plan can be controlled. Another difficulty arises for beam scanning there the field is constructed sequentially and it is impossible to use thimble ionisation chambers that can be moved around in the field [48].

The best quality control would be the measurement of the irradiated volume inside the patient. To some extent this is possible with the positron emission inside the patient's body being measured by means of PET [19]. Although it is not yet possible to convert these PET images into dose distributions, the images indicate the stopping of the primary beam. This is especially relevant if the beam passes or stops close to critical structures.

4.10 Therapy Facilities, History and Patient Statistics

In 1954 particle therapy started at the Lawrence Berkeley National Laboratory (LBNL) with the first proton treatment (table 5.1. Later also Helium and heavier ions were used at Berkeley. Most of the patients worldwide have been treated at the Harvard cyclotron which started 1961 and operates down to the present day. There was another proton treatment facility in Uppsala, Sweden, from 1957 until the accelerator was closed. It was then restarted in 1989. In France, two centers - one at Nice and one at Orsay - have been operating since 1991. In Russia, too, a large number of patients have been treated with proton therapy. Loma Linda was the first centre that was not based on an old physics machine but set up as a medically dedicated facility. With currently more than 1,000 patients per year it outnumbers any other facility.

In addition to the treatment of deep-seated tumors with 200 MeV protons there is also a very efficient program for the treatment of eye tumors. Here, 70 MeV is enough energy to reach the sufficient range of 2 cm. The most active centre for eye treatment is Optis at PSI, Villigen, with altogether more than

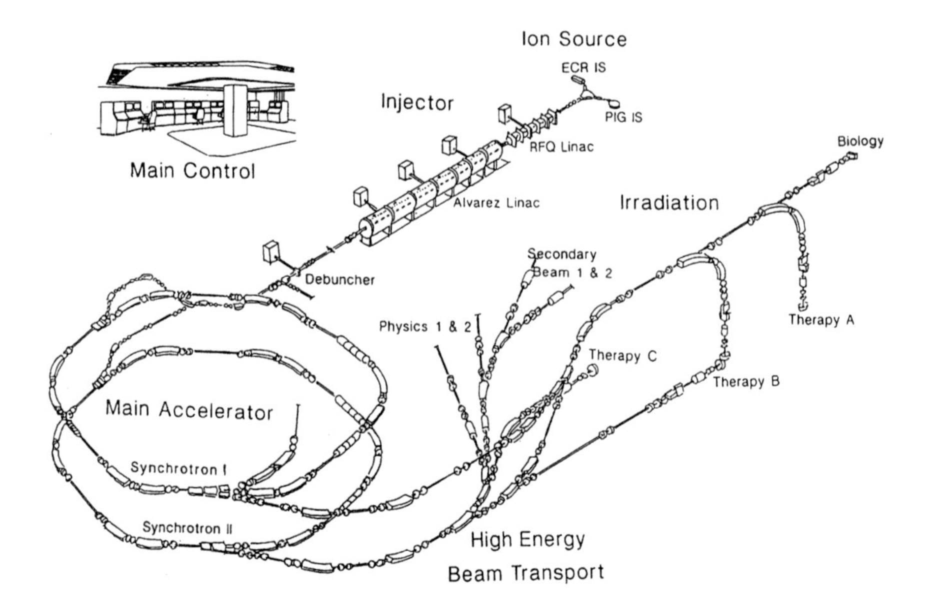

Figure 5.17. General layout of the HIMAC facility for heavy-ion treatment at Chiba, Japan, according to [49]

3,000 patients up to now. Also Clatterbridge, Nice and recently Berlin have set up very successful treatment facilities.

The development for heavy ions is much slower than for protons. On the one hand, this is because the required accelerators are more expensive to build and to run and on the other hand, because the RBE problem had to be explored in its clinical aspects first.

Due to their extreme biological efficiency Berkeley started with Argon ions but the very good tumor control was unfortunately linked to intolerable side effects. Consequently, Berkeley switched to lighter ions and treated approximately 400 patients with Neon. For radiobiological reasons, the heavy-ion therapy at NIRS, Chiba, began treatment with carbon ions right from the start, although it would have been possible to accelerate heavier ions up to Argon at the HIMAC. The therapy unit at Chiba is the first heavy-ion accelerator that is dedicated solely to therapy. It is the most complete system possible and consists of two large synchrotrons that are able to accelerate all kinds of ions from He to Ar.

Up to now, more than 700 patients have been treated at the HIMAC [30]. The heavy-ion therapy unit at GSI, Darmstadt, is a small experimental set-up where the beam is available for patient treatment only three months a year. The main purpose of the facility is to demonstrate the clinical feasibility and reliability of

Table 5.1. World wide charged particle patient totals (adapted from Sisterson [50], January 2002.)

Who	Where	What	Date first RX	Date last RX	Recent Patient Total	Date of Total
Berkeley 184	CA. USA	p	1954	1957	30	
Berkeley	CA. USA	He	1957	1992	2054	June-91
Uppsala	Sweden	p	1957	1976	73	
Harvard	MA. USA	p	1961		9067	Jan-02
Dubna	Russia	p	1967	1974	84	
Moscow	Russia	p	1969		3445	Oct-01
St. Petersburg	Russia	p	1975		1029	Jun-98
Berkeley	CA. USA	ion	1975	1992	433	June-91
Chiba	Japan	p	1979		133	Apr-00
PMRC, Tsukuba	Japan	p	1983	2000	700	Jul-00
PSI (72 MeV)	Switzerland	p	1984		3429	Dec-01
Dubna	Russia	p	1987		88	May-01
Uppsala	Sweden	p	1989		236	June-00
Clatterbridge	England	p	1989		1102	Dec-00
Loma Linda	CA. USA	p	1990		6672	Dec-01
Louvain-la-Neuve	Belgium	p	1991	1993	21	
Nice	France	p	1991		1590	June-00
Orsay	France	p	1991		1894	Jan-01
IThemba LABS	S.Africa	p	1993		408	Nov-01
MPRI	IN. USA	p	1993		34	Dec-99
UCSF - CNL	CA. USA	p	1994		284	June-00
HIMAC, Chiba	Japan	ion	1994		917	June-01
TRIUMF	Canada	p	1995		57	June-00
PSI (200 MeV)	Switzerland	p	1996		99	Dec-01
GSI Darmstadt	Germany	ion	1997		106	Jan-02
HMI, Berlin	Germany	p	1998		236	Dec-01
NCC, Kashiwa	Japan	p	1998		75	May-01
Harimac, Hyogo	Japan	p, ion	2001		30	Nov-01
PMRC, Tsukuba	Japan	p	2001		19	Jan-02
NPTC, MGH	MA. USA	p	2001		2	Jan-02
			3510	ions		
			30837	protons		
		Total	34347	all particles		

newly developed techniques such as the intensity-controlled raster scanning, the biology-based treatment planning, the on-line PET and the control and safety systems. The intention is to transfer these techniques to a clinical therapy unit as soon as possible.

4.11 Future Plans

Table 5.2. New facilities [50].

INSTITUTION	PLACE	TYPE	First RX?
INFN-LNS, Catania	Italy	p	2002
Wakasa Bay	Japan		2002
Bratislava	Slovakia	p, ion	2003
IMP, Lanzhou	PR China	C-Ar ion	2003
Shizuoka Cancer Center	Japan	p	2003
Rinecker, Munich	Germany	p	2003
PSI	Switzerland	p	2004
IThemba LABS, Somerset West	South Africa	p	2006
CGMH, Northern Taiwan	Taiwan	p	2001?
Erlangen	Germany	p	2002?
CNAO, Milan & Pavia	Italy	p, ion	2004?
M.D. Anderson Cancer Center	TX. USA	p	2004?
Heidelberg	Germany	p, ion	2005?
AUSTRON	Austria	p, ion	?
Beijing	China	p	?
Central Italy	Italy	p	?
Clatterbridge	England	p	?
TOP project ISS Rome	Italy	p	?
3 projects in Moscow	Russia	p	?
Krakow	Poland	p	?
Proton Development N.A. Inc.	IL. USA	p	?

Another proton facility will be starting at Harvard University in the near future. Most of the new therapy units for heavy charged particles are being set up in Japan. Apart from Chiba, there are 4 new facilities being planned or under construction in Tsukuba, Kashiva and Wakasa. Hyogo started treatment only recently.

In Europe, there are several strong initiatives for heavy-ion therapy, that have been funded - at least partially - from their governments, for instance in Italy (TERA), Austria (Austron) and in Etoile, France. Elsewhere new facilities will also be constructed in the near future: In Lanzhou, China, a heavy ion facility including a therapy unit is under construction. In Germany, many plans for particle therapy are virulent as for instance in Munich, Erlangen and Regensburg but only the project at Heidelberg has been given high priority from the German Scientific Council. Because particle therapy has produced very good clinical results, many of the projects of table 5.2 have a good chance to be realized.

5. Problems of Radioprotection in Space

The problems of radiation protection in space are different from those in conventional radiation protection because the radiation in space mainly consists of heavy charged particles from protons to iron ions. These particles are delivered with energies far above the nuclear reaction threshold. Therefore, not only the primary particles but also cascades of secondaries contribute to the exposure of a space craft. There are two sensitive targets that are affected by the impact of radiation - man and computer. In the following, we will focus on the consequences for man and will not elaborate on such topics as for example semiconductor response which are beyond the capacity of the author and the scope of this contribution. Comprehensive reports on those problems of space radiation can be found for instance in [51, 52]

5.1 The Radiation Field

Depending on the distance from earth the composition of the radiation field varies because particles are trapped in the geomagnetic field. At low earth orbit (LEO), trapped protons and electrons from the radiation belt predominate (fig. 5.18).

Apart from the cosmic galactic rays there are also the protons of solar eruption events, the solar flares. These solar particle events are statistically distributed and cannot be predicted to occur or not to occur during a specific space mission but in the long run, a stable period of eleven years for maximum solar activity has been observed. In solar particle events, a dose of a fraction of one Gray up to the Gray region can be delivered in a short time of a few hours or even less. Therefore, solar particle events could be life threatening in extreme cases but viewed on a long-term basis, they are so rare that they contribute only with a small fraction to the general radiation burden.

The main contribution of dose during a space mission outside the magnetic shielding of the earth originates from the galactic cosmic rays (GCR) (fig. 5.19). GCR are heavy charged particles from the most frequent protons up to iron ions. Ions heavier than iron are several orders of magnitude less frequent since they originate from super nova explosions and cannot be synthesized by exo-thermic nuclear fusion reactions like iron and the elements being lighter than iron.

The galactic cosmic radiation has an energy spectrum with a broad maximum at a few hundred MeV per nucleon and a steep and continuous decay to-wards higher energies (fig. 5.19). This high-energy tail has been measured up to the TeV region and is a big problem for an effective shielding. Because the

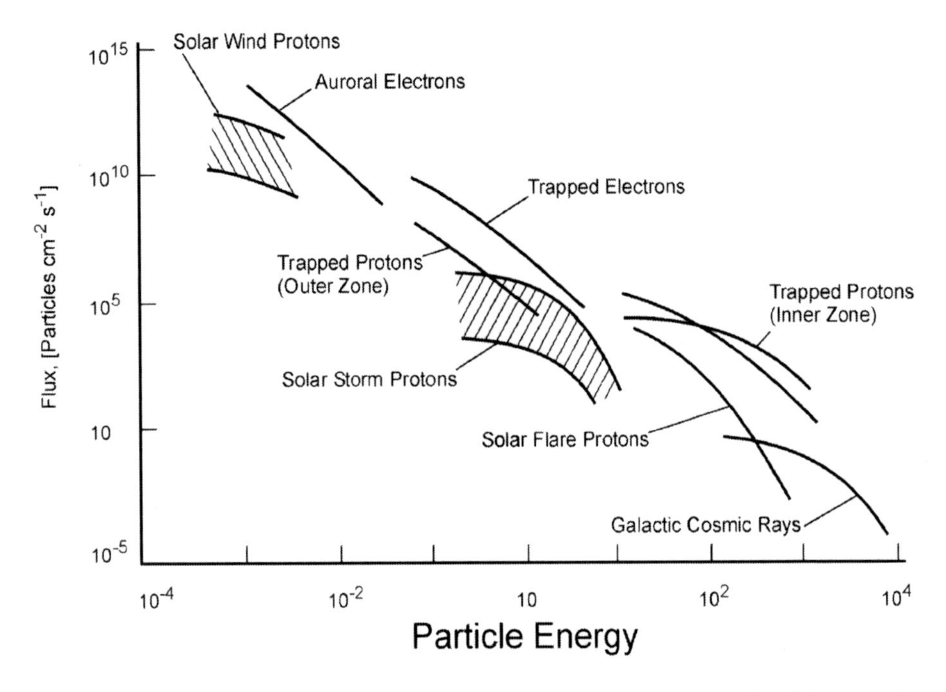

Figure 5.18. Energy spectrum of the galactic cosmic radiation as function of particle energy in MeV (redrawn from J.W. Wilson [53]).]

cross sections for nuclear fragmentation are still significant for the ions of these extremely high energies, any shielding layer introduced to stop the low-energy particles also produces showers of nuclear fragments from the high-energy particle impact.

Transport calculations for the GCR spectrum in various shielding materials like aluminium showed that after a small benefit in the first thin shielding layers, thicker absorbers do not produce a net reduction of the biological effect that correlates reasonably with the increasing mass of the shielding material. The decrease in the amount of low-energy particles is almost compensated by the increase in nuclear fragmentation. For details of this very complex but important ion transport avalanches we ask the reader to refer to the comprehensive article by Wilson *et al.* [53].

The fluence distribution of protons ranges over three orders of magnitude over Fe-ions but the energy deposition i.e. the dose of a single particle depends on the square of the atomic number. Therefore, the difference between protons and iron in their frequency contribution is nearly compensated by the dose. Taking the change in the relative biological efficiency into account the fraction of iron particles becomes as important as that of protons.

For low earth orbit (LEO), the total dose per day is about 1mSv behind a shielding of $1 g/cm^2$ of Aluminium. This is the average dose per year on earth.

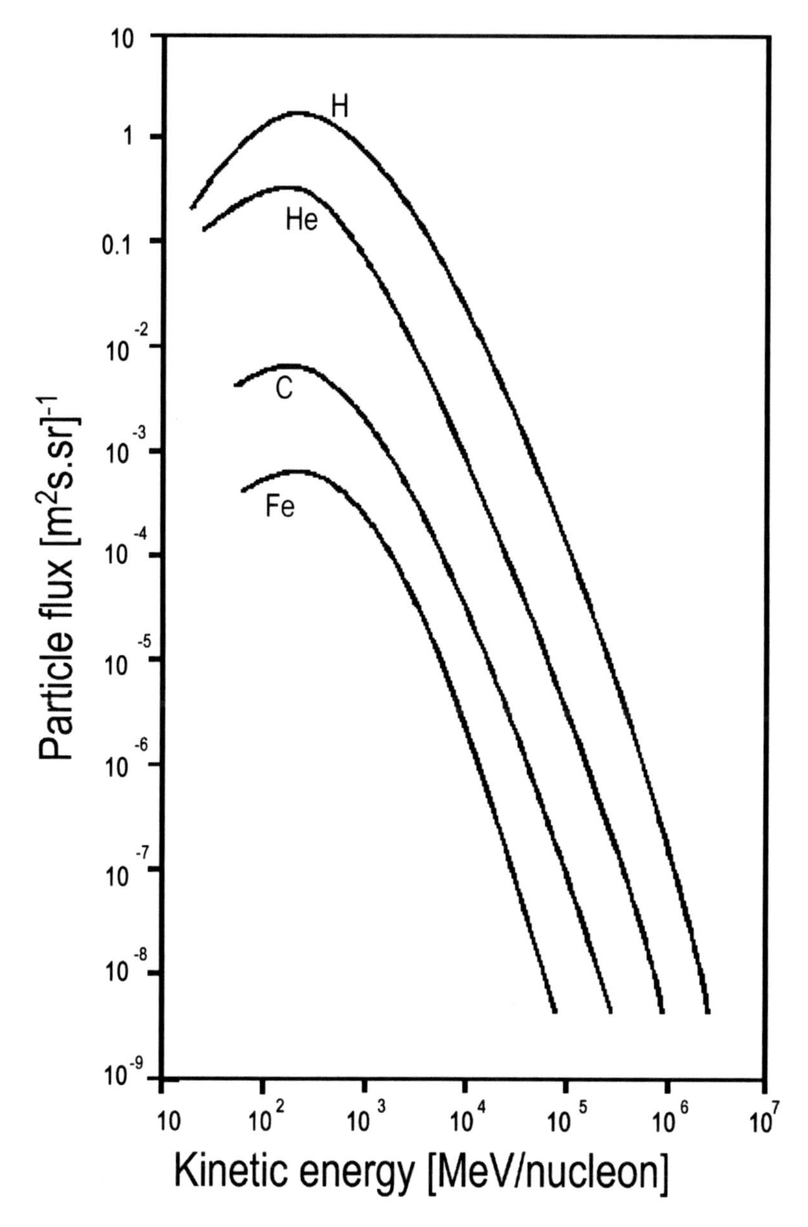

Figure 5.19. Energy spectrum of the galactic cosmic radiation (GCR) (adapted from Wilson [53])

Spaceflights are on average 300 times more exposure-intensive than our daily life but the actual value of a space mission very much depends on the altitude of the flight and on the inclination of the route. Because at the magnetic poles, shielding is drastically reduced and an orbit over the earth poles results in a greater exposure than an equatorial flight.

Outside the geomagnetic shielding the particle composition of the GCR becomes more relevant and the estimation of the biological effect is determined rather by the hit probability of the critical target inside the cells, i.e. the cell nuclei, than by a dose averaged over a long time. For example, a three years' mission to Mars was calculated to produce 400 proton and 40 He-ion traversals through the nucleus of each cell of the human body. This number decreases for light ions to a hit probability of 3% for iron ions. According to this statistic, every cell nucleus will be hit by a proton once every three days and by a He ion once every thirty days. A human body consists of about 10^{14} cells. Concerning the high hit frequency and the large number of cells at risk it is very evident that a non-protracted exposure would be lethal while the distribution over a period of three years allows repair with a good chance to survive. However, the long-term consequences like cancer induction or genetic mutation determine the risk. It is known that a macroscopic tumor originates from one transformed cell. Concerning the high number of cells at risk it is obvious that most transformed cells are eliminated by the body's imuno-system but if one mutated cell survives cancer will develop. Consequently, cancer is determined more by the biological processing than by mere induction statistics of DNA damage (fig. 5.20).

5.2 Genetic Effects and Cancer Induction

It is very difficult to make reliable and substantial statements concerning the biological consequences of human exposure to GCR. First the physical exposure cannot be predicted with sufficient accuracy because the physical i.e. the absorbed dose depends on the accidental exposure and its composition which cannot be predicted precisely. Even a retrospective analysis often lacks the necessary accuracy because it would request personal dosimeters to be carried by the astronauts that could give an analysis in atomic number and energy of all particles and the dose of other types of radiation.

On the biological side, the genetic mutation and cancer induction by radiation are very rare events in the range of 0,05 per Gray per person or less. This means for example that from a population of 10,000 people, irradiated with one Gray - which is just a third of the lethal dose - fortunately only a few hundred will develop cancer. This radiation-induced cancer has a latency of many years and becomes evident only within a long period of about 20 years. During this period many more cancer patients are diagnosed, who develop the disease due to other reasons than radiation. This was exactly the situation of the victims of the atomic bombing of Nagasaki and Hiroshima where only after some years

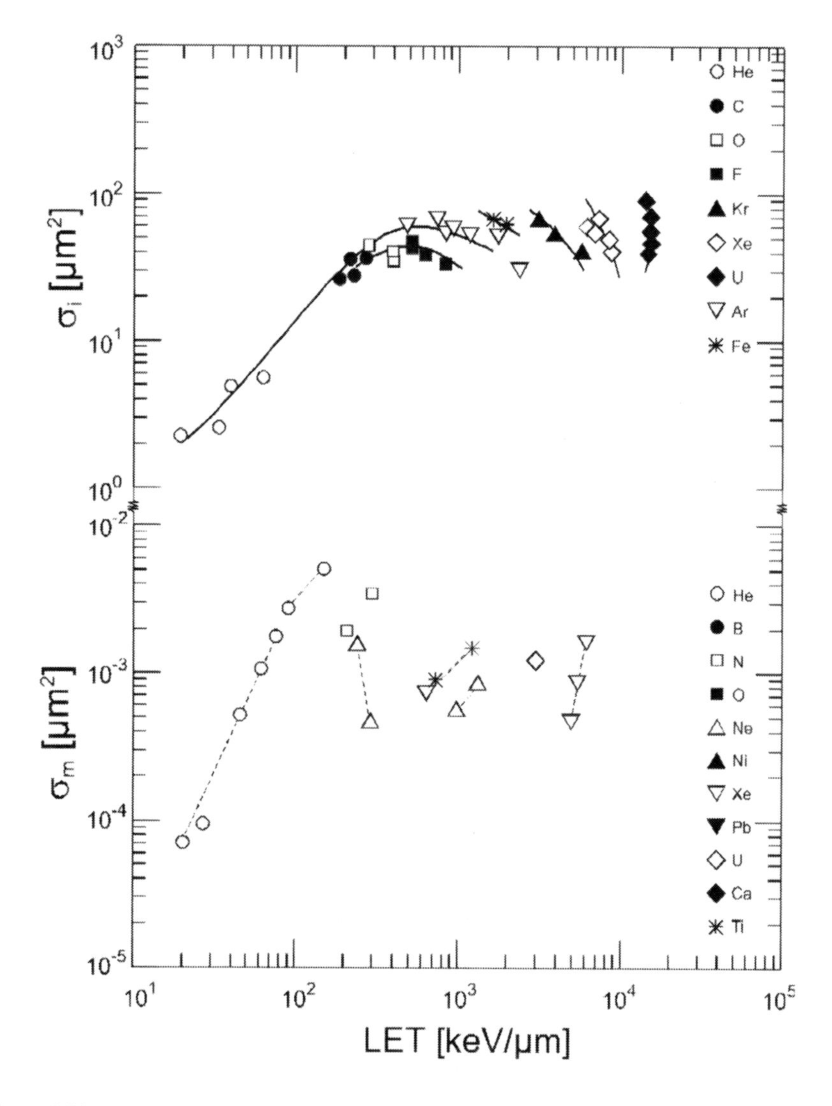

Figure 5.20. (top): Inactivation cross section of V79 Chinese hamster cells for various particles from He to U as a function of the linear energy transfer. (bottom): Mutation cross section of the HGPRT gen in Chinese hamster cells as a function of the same parameters as above, according to [54].

a measurable increase in the cancer incidence was found for leukemia but not for other cancer types because of the low incidence rate.

For ethical reasons, experiments inducing cancer in humans are not possible and even animal experiments are restricted and difficult to perform because of the large number of animals needed. Such experiments are mostly confined to sparsely ionising radiation. The data for Hiroshima and Nagasaki also refer to sparsely ionising radiation because the neutron component in both bombs contributed a few percent only.

Animal experiments with particle radiation are scarce and a straight-forward extrapolation from sparsely ionising radiation to particle exposure is yet impossible because we know that the biological response to densely ionising radiation is different from the X-ray response. An extensive synopsis of this problem was compiled by LBNL [1] and by the Nat Res. Council [52].

With regard to the modelling of inactivation as shown in 3.2 one might consider a similar theoretical procedure for the calculation of the genetic or cancer risk. Starting from X-ray dose effect curves and the radial dose distribution in the track, the expected effect for particles should be determined. Such calculations were published [53] and represent the first reliable theoretical approach. However, the uncertainty of these calculations is large because of small probabilities and the complex biological mechanisms necessary to yield stable genetic mutations. In comparison to inactivation the cross section for mutations are three to four orders of magnitude smaller. For inactivation one could attribute the higher RBE values to the greater local ionisation density that impairs a successful repair of high fidelity (fig 5.20).

Stable genetic mutations are the result of a more complex biological process, that is, of a repair that was successful in terms of survival but not in terms of the fidelity of the genetic code. This means that all essential genes are still present but at least one of them in a mutated but still meaningful way. Therefore, an increase in local ionisation density causes a local increase in DNA damage, but these clustered DNA lesions at higher energy depositions are not necessarily correlated with a higher frequency for stable mutations.

The most complete set of mutation data on cellular level and for heavy particles concern the mutations of the HPRT gene. Without going into details on the relevance and limitations of these cellular data in comparison to the risk of a human being, the cross section given in fig. 5.20 clearly shows that, first, the absolute mutation cross section are 3-4 orders of magnitude smaller than the inactivation values. Second, the mutation cross sections for heavier ions are smaller than for light ions indicating - in a still very crude way - that the mutation process is strongly determined also by other factors and not only by the primary ionisation density.

Similar results are reported for the induction of chromosome aberrations. Older measurements postulated a much lower incidence of chromosomal aberrations even for lighter ions. More recent experiments have shown that the expression time of heavy-ion-induced aberrations can be drastically delayed yielding low cross sections when analysed too early [55]. When time-integrated data are collected over longer intervals the number of chromosome aberrations as indicator of genetic effects has greater RBE values, indicating that heavy ions are more efficient in the initial stage of producing genetic damage. However in the course of time most of these aberrations lead to cell death and are therefore eliminated from the cell population. Mainly, aberrations from the translocation type i.e. aberrations where two DNA breaks are repaired with the wrong ends have a great chance to survive durably. However, only a fraction of this aberration type might have biological consequences.

In addition to these genetic changes that become visible in the very next cell division after the exposure, a long-term genetic instability has been observed after particle exposure i.e. chromosome aberrations that become visible only after about 50 cell divisions but not earlier [56].

From these considerations, the complexity of the risk estimation for GCR becomes evident but because the general risk is small, upper limits have been fixed as a worst case scenario, as for instance the figure of an incidence rate of 0.05 per Gray for leukemia as given above.

6. Conclusions

The radiobiology of highly charged ions differs from the conventional radiobiology with photons because of the great local ionisation density that is produced along a particle track. From the primary data of energy loss and electron emission, the radial dose distribution inside a track can be calculated. In the center of the track, doses up to the Mega-Gray region can be found that decay with square of the radial distance from the center up to a maximum value given by the range of the most energetic electrons. Because of this very inhomogeneous dose distribution clustered DNA damage becomes more frequent that cannot be repaired by the cell. This yields a greater relative biological efficiency of particles compared to that of sparsely ionising radiaton. For cell inactivation the particle efficiency can be calculated by folding the X-ray dose effect curve with the radial dose distribution of the particles. This method is used to calculate the killing efficiency in ion-beam treatment planning. Other parameters for treatment planning in ion-beam therapy are the pure physical data like beam lateral and longitudinal scattering and beam fragmentation.

Dose distributions for cancer treatment can be achieved from these data, which are superior to any conventional therapy. In order to transform the planning into a therapy different beam application methods have been developed that allow - in its final and most sophisticated form - to reach a perfect congruence between planned and treated volume.

Using the small amount of radioactive positron emitters that are produced inside the patient's body, the beam delivery can be monitored from outside using positron emission tomography.

Because of the superiority of these treatment methods extremely good results in tumor therapy have been achieved. Up to now, approximately 30,000 patients have been treated, mainly with protons and, a great number of projects for dedicated particle therapy centers are under way all over the world.

In space exploration, a major problem are the cosmic galactic rays that consist of highly charged ions from protons up to iron. Again, these particles have a greater biological efficiency than X-rays to induce genetic mutations and cancer. However the genetic processes are not only determined by the primary ionisation density but also by the biological response to the primary injury i.e. by the biological repair systems. Up to now, it is not possible to calculate the radiation risk in space with the desired accuracy. Because the energy spectrum of the GCR stretches up to very high values and because of the fragmentation process the application of absorber material is not only expensive but also - to a large extent - ineffective. Therefore, more accurate measurements and modelling is necessary to determine the radiation risk in space.

In general, particle radiobiology has reached a very microscopic understanding of the interaction of highly charged ions with living material. This knowledge can be used for a successful tumor therapy but for genetic changes the understanding is not sufficient to deliver exact data.

References

[1] LBNL Report, *Modelling Human Risk: Cell and Molecular Biology in Context*, LBNL Report 40278, (1997)

[2] G. Kraft, *RBE and its Interpretation*, Strahlenther. Onkol. **175**, Suppl 2, 44-47 (1999)

[3] G. Kraft and M. Krämer, *Linear Energy Transfer and Track Structure*, Advances in Radiat. Biology , **17**, 1-52 (1993)

[4] C.O. Reinhold, D.R. Schultz, U. Bechthold, G. Kraft, S. Hagmann, and H. Schmidt-Böcking, *Ternary ridge of ejected electrons from fast ion-atom collisions*, Physical Review A **58**(3), 2611-14 (1998)

[5] M. Krämer and G. Kraft, *Calculations of Heavy Ion Track Structure*, Radiat. Environm. Biophys. **33**, 91-109 (1994)

[6] G. Kraft, M. Scholz, and U. Bechthold , *Tumor Therapy and Track Structure*, Radiat.Environm.Biophys. **38**, 229-237 (1999)

[7] W. Bragg and R. Kleemann , *On the α-Particles of Radium and their Loss of Range in Passing Through Various Atoms and Molecules*, Phil. Mag. **10**, 318-340 (1905)

[8] R.R. Wilson, *Radiological Use of Fast Protons*, Radiology **47**, 487-491 (1946)

[9] U. Weber, *Volumenkonforme Bestrahlung mit Kohlenstoffionen*, PhD-Thesis, Universität Gh Kassel, (1996)

[10] H. Bethe, *Zur Theorie des Durchgangs schneller Korpuskularstrahlung durch Materie*, Ann. Phys. (Leipzig) **5**, 325-400 (1930)

[11] F. Bloch, *Bremsvermögen von Atomen mit mehreren Elektronen*, Z. Phys. **81**, 363-376 (1933)

[12] F. Bloch, *Zur Bremsung rasch bewegter Teilchen beim Durchgang durch Materie*, Ann. Phys. (Leipzig) **5**, 285-321 (1933)

[13] H.W. Barkas, *Nuclear Research Emulsions, Vol. I*, Academic Press New York and London, (1963)

[14] G. Molière, *Theorie der Streuung schneller geladener Teilchen II, Mehrfach- und Vielfachstreuung*, Z. Naturforschung **3a**, 78-97 (1948)

[15] B. Gottschalk, A.M. Koehler, R.J. Schneider, J.M. Sisterson, and M.S. Wagner, *Multiple Coulomb Scattering of 160 MeV Protons*, Nucl. Instr. and Meth. **B74**, 467-490 (1992)

[16] J. Hüfner, *Heavy Fragments Produced in Proton-Nucleus and Nucleus-Nucleus Collisions at Relativistic Energies*, Phys. Reports **125**, 129-185 (1985)

[17] E.M. Friedländer and H.H. Heckmann, Relativistic Heavy-Ion Collisions Experiment, in: *Treatise on Heavy-Ion Science, Vol. 4*, ed. D.A. Bromly, 304-365 (1985)

[18] W. Enghardt, W.D. Fromm, H. Geissel, H. Keller, G. Kraft, A. Magel, P. Manfraß, G. Münzenberg, F. Nickel, J. Pawelke, D. Schardt, C. Scheidenberger, and M. Sobiella, *Positron Emission Tomography for Dose Localization and Beam Monitoring in Light Ion Tumor Therapy*, Proc. Int. Conf. on Biological Applications of Relativistic Nuclei, Clermont-Ferrand, France, 30 Oct. (1992)

[19] W. Enghardt, J. Debus, T. Haberer, B.G. Hasch, R. Hinz, O. Jäkel, M. Krämer, K. Lauckner, and J. Pawelke, *The Application of PET to Quality Assurance of Heavy-Ion Tumor Therapy*, Strahlenther. Onkol. **175**, Suppl. II, 33-36 (1999)

[20] A. Wambersie, The Future of High-Let Radiation in Cancer Therapy, in: *Eulima Workshop on the Potential Value of Light Ion Beam Therapy*, ed. P. Chauvel and A. Wambersie, Publication No. EUR 12165, Commission of the European Community, Brussels, (1989)

[21] G. Kraft, Radiobiology of Heavy Charged Particles, in: *Advances in Hadrontherapy*, eds. U. Amaldi, B. Larsson, and Y. Lemoigne; Excerpta Medica, Int. Congr. Series 1144, Elsevier, 385-404 (1997)

[22] G. Kraft, *Radiobiological effects of very heavy ions: inactivation, induction of chromosome aberrations and strand breaks*, Nucl. Sci. Appl. **3**, 1-28 (1987)

[23] G. Kraft, W. Kraft-Weyrather, G. Taucher-Scholz, and M. Scholz, What Kind of Radiobiology should be done at a Hadron Therapy Center, *Advances in Hadrontherapy*, Proceedings of the International Week on Hadrontherapy, European Scientific Institute, Archamps, France, 20-24 November 1995 and of the Second International Symposium on Hadrontherapy, PSI and CERN, Switzerland, 9-13 September (1996)

[24] W.K. Weyrather, S. Ritter, M. Scholz, and G. Kraft, *RBE for Carbon Track Segment Irradiation in Cell Lines of Differing Repair Capacity*, Int. J. Radiat. Biol **11**, 1357-1364 (1999)

[25] W.T. Chu, B.A. Ludewigt, and T.R. Renner, *Instrumentation for Treatment of Cancer Using Proton and Light-Ion Beams*, Review of Scientific Instruments, Accelerator & Fusion Research Division, LBL-33403 UC-406 Preprint (1993)

[26] O. Jäkel and M. Krämer, *Treatment planning for heavy ion irradiation*, Physica Medica Vol XIV/1; 53-1998, (1998)

[27] E.A. Blakely, C.A. Tobias, F.Q.H. Ngo, and S.B. Curtis, Physical and Cellular Radiobiological Properties of Heavy Ions in Relation to Cancer Therapy Application, in: *Biological and Medical Research with Accelerated Heavy Ions at the Bevalac*, eds.: M.D. Pirncello, C.A. Tobias, LBL 11220, 73-88 (1980)

[28] T. Kanai, Y. Furusawa, K. Fukutsu, H. Itsukaichi, K. Eguchi-Kasai, and H. Ohara, *Irradiation of Mixed Beam and Design of Spread-out Bragg Peak for Heavy-Ion Radiotherapy*, Radiat. Res. **147**(1), 78-85 (1997)

[29] T. Kanai, *et al. Biophysical Characteristics of HIMAC Clinical Irradiation System for Heavy-Ion Radiation Therapy*, Int. J. Radiat. Oncol. Biol. Phys. **44**(1), 2001-210, (1999)

[30] H. Tsujii, *Preliminary Results of Phase I/II Carbon Ion Therapy at NIRS*, Int. Part. Therapy Meeting and XXIV. PTCOG Meeting, April 24-26, Detroit Michigan (1996)

[31] M. Scholz, A.M. Kellerer, W. Kraft-Weyrather, and G. Kraft, *Computation of Cell Survival in Heavy-Ion Beams for Therapy. The Model and its Approximation*, Radiat.Environ.Biophys. **36**, 59-66 (1997)

[32] M. Scholz and G. Kraft, *Calculation of Heavy Ion Inactivation Probabilities Based on Track Structure, X-ray sensitivity and target size*, Radiat. Prot. Dosimetry, **52**, Nos 1-4, 29-33 (1994)

[33] J. Debus, T. Haberer, D. Schulz-Ertner, O. Jäkel, F. Wenz, W. Enghardt, W. Schlegel, G. Kraft, and M. Wannenmacher, *Fractionated Carbon Ion Irradiation of Skull Base Tumors at GSI. First Clinical Results and Future Perspectives*, Strahlenther. Onkol. **176**(5), 211-216 (2000)

[34] H. Blattmann, G. Munkel, E. Pedroni, T. Böhringer, A. Coray, S. Lin, A. Lomax, and B. Kaser-Hotz Conformal proton radiotherapy with a dynamic application technique at PSI, Progress in: *Radio-Oncology V*, ed. H.D. Kogelnik, Monduzzi Editore, Bologna, 347-352 (1995)

[35] G. Kraft, W. Becher, K. Blasche, D. Böhne, B. Fischer, G. Gademann, H. Geissel, Th. Haberer, J. Klabunde, W. K.-Weyrather, B. Langenbeck, G. Münzenberg, S. Ritter, W. Rösch, D. Schardt, H. Stelzer, and Th. Schwab, *The Heavy Ion Project at GSI*, Nucl. Tracks Radiat. Meas. **19**, 911-914, (Int. J. Radiat. Appl. Instrum. Part D) (1991)

[36] Th. Haberer, W. Becher, D. Schardt, and G. Kraft, *Magnetic scanning system for heavy ion therapy*, Nucl. Instr. and Meth. in Phys. Res.A **330**, 296-305 (1993)

[37] H. Brand, H.G. Essel, H. Herdel, J. Hoffmann, N. Kurz, W. Ott, and M. Richter, *Therapy: Slow Control System, Data Analysis and on-line Monitoring*, GSI Report, 146-189 (1998)

[38] L.J. Verhey, Comparison of achievable Dose Distributions Using Gamma Knife, Protons and IMRT, in: it Intensity-Modulated Radiation Therapy, ed. E.S. Sternik, Advanced Med. Pub., 127-142 (1997)

[39] W. Schlegel, O. Pastyr, T. Bortfeld, G. Gademann, M. Menke, and W. Maier-Borst, *Stereotactically Guided Fractionated Radiotherapy: Technical Aspects*, Radiother. Oncol. **29**, 197-204 (1993)

[40] K.M. Langen and T.L. Jones, *Organ Motion and its Management*, Int.J.Radiation Oncology Biol. Phys. **50**, 265-278 (2001)

[41] J.R. Alonso, Synchrotrons: the American Experience, in: *Hadrontherapy in Oncology*, eds.: U. Amaldi, B. Larsson, Excerpta Medica Intercongress Series 1077, Elsevier, 266-281 (1994)

[42] T. Renner, M. Nymann, and R.P. Singh, Control Systems for Ion Beam Radiotherapy Facilities, in: *Ion Beams in Tumor Therapy*, ed. U. Linz, Chapman & Hall, 156-265 (1995)

[43] V. Wieszczycka, *Preliminary Project of a Polish Dedicated Proton Therapy Facility*, Ph.D Thesis, Warsaw, 194-198 (1999)

[44] G. Kraft, *Tumor Therapy with Heavy Charged Ions*, Progr. Part. Nucl. Phys. **45**, 473-544 (2000)

[45] E. Pedroni, *Beam Delivery in Hadrontherapy in Oncology*, eds. U. Amaldi, B. Larsson, Excerpta Medica, 434-452 (1994)

[46] Eulima: *Cancer Treatment with Light Ions in Europe*, Final Report, Geneva (1991)

[47] P. Heeg, G.H. Hartmann, O. Jäkel, C. Karger, and G. Kraft, *Quality Assurance at the Heavy-Ion Therapy Facility at GSI*, Strahlenther. Onkol. **175**, 36-39, (1998)

[48] C. Brusasco, *A Detector System for the Verification of Three-Dimensional Dose Distributions in the Tumor Therapy with Heavy Ions*, PhD Thesis Kassel (1999)

[49] U. Amaldi, *The National Centre for Oncological Hadrontherapy at Mirasole*, INFN, Frascati (1997)

[50] Sisterson J.: Particles, A Newsletter for those Interested in Proton Light Ion and Heavy Charged Particle Radiotherapy, 28/2001, 13-14 (2001)

[51] Majima H.J. and Fujitaka K. (eds), *Exploring Future Research Strategies in Space Radiation Science*, Proceedings of the 2nd International Space Workshop Feb. 16-18, 2000, Chiba NIRS (2000)

[52] National Research Council: *Radiation Hazard to Crews of Interplanetary Missions - Biological Issues and Research Strategies*, National Academy Press, Washington (1996)

[53] J.W. Wilson, *et al.*, *Transport Methods and Interactions for Space Radiations*, NASA Publication 1257 (1991)

[54] J. Kiefer, *Heavy ion effects on cells: chromosomal aberrations, mutations and neoplastic transformation*, Radiat. Environm. Biophys. **31**, 279-288 (1992)

[55] S. Ritter, E. Nasonova, E. Gudowska-Nowak, M. Scholz, and G. Kraft, *High-LET-induced chromosome aberrations in V79 cells analysed in first and second post-irradiation metaphases*, Int. J. Radiat. Biol. **76**, 149-161 (2000)

[56] M.A. Kadhim, D.A. Macdonald, D.T. Goodhead, S.A. Lorimore, S.A. Marsden, and E.G. Wright, *Transmission of chromosomal instability after plutonium alpha-particle irradiation*, Nature **355**, 738-740 (1992)

III

FUNDAMENTAL INTERACTIONS

Chapter 6

PHOTOIONIZATION OF ATOMIC IONS

N. J. Wilson and K. L. Bell
School of Mathematics and Physics,
The Queen's University of Belfast,
Belfast BT7 1NN, UK.
kl.bell@qub.ac.uk

Abstract This chapter presents the theory of the R-matrix method as applied to photoionization of atomic ions. Two examples of the application of the theory are given and the results discussed in some detail. The discussion include comparison with experiment, resonances, the effect of including damping and recombination processes.

1. The Theory of Photoionization

The interaction of atoms with electromagnetic radiation has been of interest to physicists since Hertz's discovery in 1885 that the passage of a spark was facilitated by ultraviolet light falling on a metallic electrode. This phenomenon became known as the photoelectric effect and was further investigated experimentally by Hallwachs, Stoletov, Lenard and others. Experimental activity revealed that

- For any given atom, there is a minimum frequency, below which no emission of electrons can take place.
- Electrons can emerge with a range of velocities. The maximum kinetic energy is directly proportional to the frequency of the radiation but is independent of its intensity.
- The number of electrons emitted per unit time is proportional to the intensity of the incident radiation.
- Electron emission occurs immediately; there is no detectable time delay.

Since Newton's time, light had been considered to consist of particles, known as corpuscles. At the turn of the twentieth century, the new quantum physics was

F.J. Currell (ed.), The Physics of Multiply and Highly Charged Ions, Vol. 1, 199-229.

developing and, in 1905, Einstein extended Planck's idea of the quantization of black-body radiation. He postulated that electromagnetic fields could be quantized and that light consisted of corpuscles called photons, each travelling at the velocity of light c and having associated with it a quantum of energy of magnitude

$$E = h\nu = \frac{hc}{\lambda} \tag{6.1}$$

in accordance with the findings of de Broglie, where h is Planck's constant and ν and λ are the frequency and wavelength of the photon, respectively.

1.1 Theory of Interaction of Light with Atoms

For ease of explanation, it is convenient to consider the interaction of light with atoms. The theory, however, is also applicable to the interaction of electromagnetic radiation with positive ions.

It is well known that atoms can only exist in a number of discrete states, each having associated with it a fixed energy level. When a photon comes into contact with an atom, the whole quantum of the photon's energy can be absorbed by the atom at one time. The atom thus has more energy and as a consequence can be transferred from the ground state, which is the state of lowest energy, to a state of higher energy, known as an excited state. The absorption lines corresponding to transitions from a given initial state form a spectral series, in which the wavelength difference between successive members decreases regularly toward shorter wavelengths. The series converges on a series limit, at which theoretically an infinite number of lines lie. Beyond this limit, there exists a region of continuous absorption, called the continuum. In this region, the photon uses the entire energy to eject an electron from the atom. This process can be represented by

$$A + h\nu \longrightarrow A^{+} + e^{-} \tag{6.2}$$

and is called photoionization, where A is an atom with $(N+1)$ electrons and A^{+} is a positive ion, which is formed after one of the atom's electrons has been removed. Because of the interaction of the ejected electron with other electrons, a minimum energy is required for the electron to escape from the atom. This minimum energy is known as the ionization energy E_{I}. Any excess energy can be removed as kinetic energy. To conserve momentum, almost all the kinetic energy must be taken by the electron, since the ion is always much heavier than the electron. Thus, the maximum kinetic energy of the photoelectron is given by

$$\frac{1}{2}mv_{\mathrm{max}}^{2} = h\nu - E_{\mathrm{I}} \tag{6.3}$$

where v_{max} is its maximum velocity and m its mass.

In the visible and near-ultraviolet regions, the absorption spectrum of the atom is that of the electron which is furthest from the nucleus. This electron is known as the valence electron. It is, however, also possible to excite electrons from inner shells. It is intuitive that the photon energies required for inner-shell excitation are greater than those required for excitation of valence electrons, since these inner-shell electrons are closer to the nucleus and so have a greater attractive force. Absorption lines representing these transitions are found in the vacuum ultraviolet region of the spectrum, and converge to series limits corresponding to excited ionic states. Part or all of the inner-electron series may overlap the normal photoionization continuum, in which case two atomic states with the same energy may exist and thus two mechanisms for photoionization exist.

Firstly, the incident photon may result in the direct photoionization of the atom, in which case a valence electron may be excited with sufficient kinetic energy to remove it from the atom, as represented by equation (6.2). Secondly, it is possible that an inner electron may be excited. In this case the atom then exists in a discrete superexcited state A^*. If there is no interaction between the electron configurations involved in the discrete states and the continuous spectrum, the discrete absorption lines observed do not exhibit any special properties.

However, if both discrete and continuum states at the same energy have the same angular momentum, the eigenfunctions and the properties of the two are mixed. The unquantized continuum of states is obtained from the solution of Schrödinger's equation for positive energy values. The interaction between the continuum and the discrete states leads to the property that the discrete states are unstable. Physically this means that one of the electrons moves in an orbit which extends to infinity. The superexcited state can therefore spontaneously ionize. This property is known as autoionization. This process can be better represented as

$$A + h\nu \longrightarrow A^* \longrightarrow A^+ + e^- \tag{6.4}$$

where A^* represents the atom in a superexcited state.

The inner-shell vacancy in the superexcited state may be filled by an electron from an outer shell. This results in the emission of a photon, representing the excess energy. It is also possible that there may be a series of superexcited states, in which the outer electron in each member of the series has the same orbital angular momentum number, but has a different principal quantum number, becoming closer towards a limit. Such a series is termed a Rydberg series and is often observed in photoionization processes, manifesting itself as resonant features in a photoionization cross section. Successive terms in the Rydberg series can be related in a regular fashion. This will be explored in more depth in Section 6.1.3.4.

1.2 The Wave Equation

To calculate atomic data, it is necessary to have a knowledge of the wave functions Ψ_n $(n = 1, 2, \ldots)$ of the relevant atomic states. These functions should be single-valued, finite and continuous, and should vanish at infinity. If relativistic effects are neglected, then these functions are solutions of the time-dependent Schrödinger equation

$$H\Psi_n(\mathbf{r}, t) = i\hbar \frac{\partial \Psi_n}{\partial t}(\mathbf{r}, t) \tag{6.5}$$

where H is the quantum mechanical Hamiltonian, which will be defined in Section 6.1.3.

It is usual to consider stationary states of atoms, and the solutions of the time-dependent Schrödinger equation can be written as

$$\Psi_n(\mathbf{r}, t) = \Psi_n(\mathbf{r}) e^{-iE_n t} \tag{6.6}$$

where E_n is a constant, so that

$$|\Psi_n(\mathbf{r}, t)|^2 = |\Psi_n(\mathbf{r})|^2 . \tag{6.7}$$

Substitution of (6.6) into the Schrödinger equation (6.5) yields the equation

$$H\Psi_n = E_n \Psi_n \tag{6.8}$$

which is referred to as the stationary-state or time-independent form of the Schrödinger equation. Equation (6.8) is an eigenvalue equation; the wave-functions Ψ_n are the eigenfunctions of the Hamiltonian operator H, while the constants E_n are given by the eigenvalues associated with the wavefunction Ψ_n and represent an upper bound to the energy of the system for that particular state.

1.3 The Hamiltonian Operator

The Hamiltonian operator H represents the total energy of the system under consideration. The kinetic energy of the electrons and the potential energy, due to both the Coulomb repulsion between each pair of electrons and the Coulomb interaction between the nucleus and the orbiting electrons, are included. An algebraic expression for the Hamiltonian operator in atomic units is

$$H = \sum_i \left(-\frac{1}{2}\nabla_i^2 - \frac{Z}{r_i} \right) + \sum_{i<j} \frac{1}{r_{ij}} \tag{6.9}$$

in which the summation is over the total number of electrons. In equation (6.9), the nucleus is taken as the origin of coordinates, Z is the nuclear charge and the interelectronic distance r_{ij} is given by

$$r_{ij} = |\mathbf{r}_i - \mathbf{r}_j| . \tag{6.10}$$

The substitution of the expression for the Hamiltonian operator given in equation (6.9) into the stationary-state form of Schrödinger's equation (6.8) presents a number of complications. In particular, the presence of the term involving r_{ij} prevents a solution by the separation-of-variables technique. Exact solutions are thus only possible for one-electron or hydrogenic systems. However, approximate solutions can be obtained by simplification of either the Hamiltonian operator or the form of the wavefunctions. To systemize these approximations, perturbation theory or the variational principle have been used.

1.3.1 Relativistic Effects. On occasion, the wavefunctions and results obtained by using the Hamiltonian operator (6.9) may not compare favourably with experiment or do not display particular properties inherent to the problem being solved. One possible explanation may be that the discussion so far has neglected relativistic effects. The non-relativistic Hamiltonian given by equation (6.9) can be augmented by terms representing various effects, leading to the calculation of different wavefunctions. The Breit-Pauli Hamiltonian H_{BP} includes one-body correction terms and can be derived from the Dirac equation. It can be written

$$H_{\mathrm{BP}} = H_{\mathrm{NR}} + H_{\mathrm{REL}} \tag{6.11}$$

where H_{NR} is the non-relativistic Hamiltonian given by equation (6.9) and H_{REL} is the sum of the one-body correction terms and is given by

$$H_{\mathrm{REL}} = H_{\mathrm{mass}} + H_{\mathrm{D}_1} + H_{\mathrm{SO}} \tag{6.12}$$

where H_{mass} is the mass-correction term given by

$$H_{\mathrm{mass}} = -\frac{1}{8}\alpha^2 \sum_i \nabla_i{}^4 \tag{6.13}$$

H_{D_1} is the one-body Darwin term

$$H_{\mathrm{D}_1} = -\frac{1}{8}\alpha^2 Z \sum_i \nabla_i{}^2 \left(\frac{1}{r_i}\right) \tag{6.14}$$

and H_{SO} is the spin-orbit term

$$H_{\mathrm{SO}} = \frac{1}{2}\alpha^2 Z \sum_i \frac{\mathbf{l}_i.\mathbf{s}_i}{r_i^3} \tag{6.15}$$

representing the spin-orbit interaction of the ith electron in the field of the nucleus.

It is also possible to define a number of two-body terms, which can be classified as fine-structure terms and non-fine-structure terms. One of the two-body fine-structure terms represents the spin-other-orbit interaction and is defined by

$$H_{\mathrm{SOO}} = -\alpha^2 \sum_{i<j} \left\{ \left(\frac{\mathbf{r}_{ij}}{r_{ij}^3} \times \mathbf{p}_i \right) \cdot \mathbf{s}_j + \left(\frac{\mathbf{r}_{ij}}{r_{ij}^3} \times \mathbf{p}_j \right) \cdot \mathbf{s}_i \right\} \tag{6.16}$$

whilst the spin-spin interaction is given by

$$H_{\mathrm{SS}} = \alpha^2 \sum_{i<j} \left\{ \frac{\mathbf{s}_i \cdot \mathbf{s}_j}{r_{ij}^3} - 3 \frac{(\mathbf{s}_i \cdot \mathbf{r}_{ij})(\mathbf{s}_j \cdot \mathbf{r}_{ij})}{r_{ij}^5} \right\}' . \tag{6.17}$$

It is also possible to define the mutual spin-orbit interaction, which comes from a reduction of the many-body generalization of the Dirac equation and is given by

$$H_{\mathrm{MSO}} = -\frac{1}{2}\alpha^2 \sum_{i<j} \left\{ \left(\frac{\mathbf{r}_{ij}}{r_{ij}^3} \times \mathbf{p}_i \right) \cdot \mathbf{s}_i + \left(\frac{\mathbf{r}_{ij}}{r_{ij}^3} \times \mathbf{p}_j \right) \cdot \mathbf{s}_j \right\} . \tag{6.18}$$

The two-body non-fine-structure terms can be classified into terms representing the spin-spin contact interaction and the orbit-orbit interaction, and the two-body Darwin term. For definitions of these terms, the reader is referred to Bethe and Salpeter [1].

In equations (6.13) – (6.18), α is the dimensionless fine-structure constant, with a value of approximately $\frac{1}{137}$ and $\mathbf{l}_i$ and $\mathbf{s}_i$ are the orbital angular momentum and spin of the electron, respectively. When the wavefunction has been obtained, it will become possible to calculate a variety of observables, including photoionization cross sections, which shall be considered in the remainder of this chapter.

1.3.2 Photoionization Cross Sections. As described earlier, it is clear that photoionization processes can take place at a range of energies greater than or equal to the ionization energy of the atom. Furthermore, for any range of photon energies, a number of transitions from a fixed lower state to other states of higher energy, including continuum states, are possible.

The photoionization cross section σ_ν associated with a particular frequency ν is defined per unit frequency interval as

$$\sigma_\nu = \frac{\pi e^2}{mc^2} \frac{df}{d\nu} \tag{6.19}$$

where f is the oscillator strength. It is possible to define two forms of the photoionization cross section. The length form of the photoionization cross section σ_ν^{l} is given by

$$\sigma_\nu^{\mathrm{l}} = \frac{4\pi^2 \alpha a_0^2}{3} \frac{\omega}{g_i} \sum_{i,j} \left| \left\langle \psi_j \middle| \sum_n \mathbf{r}_n \middle| \psi_i \right\rangle \right|^2 \tag{6.20}$$

whilst the velocity form σ_ν^{V} is given by

$$\sigma_\nu^{\mathrm{V}} = \frac{4\pi^2\alpha a_0^2}{3}\frac{1}{\omega g_i}\sum_{i,j}\left|\left\langle \psi_j \left| \sum_n \nabla_n \right| \psi_i \right\rangle\right|^2 \tag{6.21}$$

where α is the fine structure constant, a_0 is the Bohr radius of the hydrogen atom and ω is the incident photon energy.

The magnitude of the photoionization cross section at any given frequency is related to the probability of photoionization occurring at that particular energy.

1.3.3 Selection Rules. There are restrictions on the use of the dipole approximation in calculations. These are known as the selection rules and are dependent on the type of coupling employed.

(i) LS coupling

- The dipole operator is independent of spin. Clearly for a non-zero oscillator strength, it is a requirement that ψ_i and ψ_j have the same total spin.
- The operator is a vector. As a consequence, it is required that the orbital angular momenta L_i and L_j differ by no more than 1. The result for $L_i = L_j = 0$ is zero.
- Since **r** and **p** are operators of odd parity, ψ_i and ψ_j must be of opposite parity.

(ii) Intermediate (or LSJ) coupling

- J_i and J_j must differ by no more than 1 subject to the exception that $J_i = J_j = 0$ is not an allowed transition.
- ψ_i and ψ_j must be of opposite parity.

1.3.4 Autoionizing Resonances. In general a plot of the photoionization cross section will reveal it to be a smooth function of energy. However, the excitation of inner-shell electrons and the interaction between the discrete states and the continuum as discussed in Section 6.1.1 may result in the cross section varying rapidly within a small change in the incident photon energy. Such a feature is termed a resonance and in terms of inner-shell photoexcitation represents the existence of an atom in a superexcited state. A general expression for the cross section in the vicinity of a resonance has been shown by Fano [2] and Fano and Cooper [3] to be

$$\sigma = \sigma_a \frac{(q+\varepsilon)^2}{1+\varepsilon^2} + \sigma_b \tag{6.22}$$

where q is the line profile index or shape parameter, σ_a and σ_b are constants to be determined and ε is defined by the equation

$$\varepsilon = \frac{E - E_r}{\frac{\Gamma}{2}} \tag{6.23}$$

in which E is the photon energy, E_r the photon energy at the resonance position and Γ the width of the resonance at half-height. The constants σ_a and σ_b represent respectively those parts of the continuum which do and do not interact with the discrete autoionizing state.

This work was extended by Gersbacher and Broad [4] to express the cross section near a resonance as

$$\sigma = a_0 + a_1\varepsilon + a_2\frac{(q+\varepsilon)^2}{1+\varepsilon^2} \tag{6.24}$$

which allows for a background cross section to be linear as compared with the constant background given by equation (6.22). The lifetime of the resonance is given by

$$\tau = \frac{\hbar}{\Gamma} \tag{6.25}$$

and so narrower resonances have longer lifetimes.

These autoionizing resonances are a special case of Feshbach resonances, which are typically very narrow and are characterized by a peak followed by a trough at a cross section lower than that of the background.

It may be the case that the photoionization spectrum exhibits a series of Feshbach resonances. These resonances represent states, in which an inner-shell electron has been excited to an orbital with the same orbital angular momentum l but different principal quantum numbers n. Such a series is called a Rydberg series. Successive members of the series are observed in the photoabsorption spectrum at decreasing frequency intervals, converging to the energy of the target state to which the autoionizing states ionize.

It is often possible to use eigenvector components to identify the first member of the series. Subsequent members can be identified by using a quantum defect analysis using the equation

$$E_{nl} = -\frac{1}{2}\frac{z^2}{(n-\nu_l)^2} \tag{6.26}$$

where $z = Z - N$ and is the residual target charge, E_{nl} is the energy of the resonance with respect to the ionization limit and ν_l is the quantum defect, which varies slightly along the series but for identification purposes can be assumed to be constant.

2. R-matrix Theory

In Section 6.1.3.2, expressions for the photoionization cross section were introduced. Whilst cross sections could be obtained by use of pseudo-orbitals to model the continuum, a better method is to use a method which treats the continuum and discrete states in two different ways. One such model is the R-matrix method, which produces cross sections of a high degree of accuracy.

The R-matrix method provides an excellent basis for describing many types of atomic process. The theory was first introduced in 1947 by Wigner and Eisenbud [5] and was first adapted for atomic structure problems by Burke and Seaton [6] and by Burke *et al.* [7]. A useful summary of the non-relativistic theory is provided by Burke and Robb [8].

Together with the mathematical development of the theory, computer programs have been developed. The first publication was by Berrington *et al.* [9] in 1974. A modified package was published by Berrington *et al.* [10] in 1978.

The theory was extended to calculation of the Breit-Pauli Hamiltonian by Scott and Burke [11] and Breit-Pauli R-matrix programs were described by Scott and Taylor [12]. More recently, the R-matrix programs were reviewed by Berrington *et al.* [13], who described the Belfast atomic R-matrix package (RMATRX1) for the internal region problem. The preferred approach in the external region, though, is to use the description offered by the earlier paper of Berrington *et al.* [14], which separates STG4 (the program in this region) into a suite of modules.

2.1 Notation

As described in Section 6.1, the photoionization process is described by the equation

$$A + h\nu \longrightarrow A^{+} + e^{-}. \tag{6.27}$$

For the purpose of the R-matrix method the final state A^{+} is referred to as the target and has N electrons. The initial state A is thus a bound state of the $(N+1)$-electron system. The method is independent of the mechanism by which the target state forms.

In photoionization processes, the initial and target states are clearly distinct atomic systems. The only difference between the two systems is the status of the $(N+1)$th electron, which is bound in the initial state and free in the target. A quantity r_a is chosen to be the R-matrix radius. The value should be chosen so that a sphere of radius r_a just encloses the atomic charge distribution for the target state. It is then straightforward to define two distinct regions.

Taking the nucleus as the centre of coordinates, the internal region is defined as that region in which $r \leqslant r_a$. In this region the $(N \ + \ 1)$th electron is bound. This means that it is necessary to take account of electron exchange and correlation between the free electron and the target. This can be achieved by adopting a configuration interaction approach. The external region is obviously the region where $r > r_a$. The $(N+1)$th electron is free and so electron exchange can be neglected in the external region.

Clearly, obtaining a good description of the target is essential to solving the problems in both the internal and external regions. In LS coupling, the target is described using a number of $LS\pi$ target states, where L is the total orbital angular momentum, S is the total spin angular momentum, and π is the parity. In LSJ coupling, the target is described using $LS\pi$ target state energies split into fine structure levels with symmetry $J\pi$, where J is the total angular momentum of the system.

2.1.1 Channels. A state of the $(N + 1)$-electron system can be considered by allowing a continuum electron with orbital angular momentum l_i to approach the target and couple to it. This coupling may happen vectorially in several different ways to yield the same total L but must always obey the rules of Russell-Saunders coupling. The different possibilities are called the channels. If E is the total energy and ε_i the energy of the target state coupled to the ith channel, then the channel energy of the continuum electron, k_i^2, is given by

$$k_i^2 = 2(E - \varepsilon_i). \tag{6.28}$$

The different possible values of k_i^2 leads to three possible descriptions for any channel. If $k_i^2 > 0$, the channel is termed open; if $k_i^2 < 0$, the channel is considered to be closed; if $k_i^2 = 0$, the continuum electron possesses an energy which is the threshold energy.

2.2 Description of target

It has already been stated that an adequate description of the target is essential to determining solutions in the internal and external regions. The representation of the target using $LS\pi$ target state wavefunctions is achieved by solving the time-independent Schrödinger equation given by

$$H^N \Phi_i = \varepsilon_i \Phi_i \tag{6.29}$$

where H^N is the non-relativistic Hamiltonian given by equation (6.9) and ε_i is the energy of the corresponding target state. The $\{\Phi_i\}$ are the eigenfunctions and the $\{\varepsilon_i\}$ are the eigenvalues of equation (6.29). Since H^N is the non-relativistic Hamiltonian, the Φ_i are antisymmetric $LS\pi$ states and are also eigenfunctions of the operators $\mathbf{L^2}$ and $\mathbf{S^2}$. It is usual to use the method of configuration interaction for describing the target in an R-matrix calculation.

Each $LS\pi$ target state can be written in terms of basis configurations ϕ_k using a configuration-interaction expansion such as

$$\Phi_i(x_1, \ldots, x_N) = \sum_{k=1}^{M} a_{ik}\phi_k(x_1, \ldots, x_N) \tag{6.30}$$

where $x_i = (\mathbf{r}_i, \sigma_i)$ refers to both the space ($\mathbf{r}_i = r_i\hat{\mathbf{r}}_i$) and spin ($\sigma_i$) coordinates of the ith electron and $\{a_{ik}\}$ are the configuration mixing coefficients, which are unique for each target state. The basis configurations are constructed from a bound orbital basis consisting of a set of real and possibly pseudo orbitals, whilst the expansion coefficients $\{a_{ik}\}$ are determined by diagonalizing the N-electron Hamiltonian matrix. This results in expressions for the target-state functions and upper bounds to their energies.

The orbital set is chosen so that it is sufficient for the representation of both the target and the $(N+1)$-electron system. The radial functions are determined self-consistently. It is common to use the CIV3 code [15] to determine the required radial functions before they are input into the R-matrix codes. The radial functions also determine the R-matrix radius. The value of r_a is chosen as the critical value at which the magnitude of the radial function becomes less than a pre-specified number δ,

$$|P_{nl}| \leqslant \delta \tag{6.31}$$

and is usually chosen as the value of r at which the orbitals have decreased to about 0.1% of their maximum value.

2.3 The R-matrix Basis

It is necessary to define a suitable zero-order basis for expanding the $(N+1)$-electron wavefunction. This basis is constructed from the real, pseudo and continuum orbitals. The continuum orbitals are included to represent the motion of the free electron. For an electron with orbital angular momentum l_i, the set of continuum orbitals $\nu_{ij}(r)$ are the eigensolutions of the equation

$$\left(\frac{d^2}{dr^2} - \frac{l_i(l_i+1)}{r^2} + V(r) + k_{ij}^2\right)\nu_{ij}(r) = \sum_n \lambda_{ijn}P_{nl_i}(r) \tag{6.32}$$

subject to the R-matrix boundary conditions

$$\nu_{ij}(0) = 0 \tag{6.33}$$

and

$$\left(\frac{r_a}{\nu_{ij}(r_a)}\right)\left(\frac{d\nu_{ij}}{dr}\right)_{r=r_a} = b \tag{6.34}$$

and the orthonormality condition

$$\int_{o}^{r_a} \nu_{ij}\nu_{ik}dr = \delta_{jk} \tag{6.35}$$

for all j,k. In equation (6.34), b is an arbitrary constant known as the logarithmic derivative and is normally taken to be zero since, if enough terms are retained in the R-matrix basis, the result is independent of b. The eigenvalues of the equation, k_{ij}^2, are the channel energies introduced in equation (6.28). The λ_{ijn} are Lagrange undetermined multipliers that ensure the continuum orbitals are orthogonal to the bound orbitals of the same angular symmetry,

$$\int_{0}^{r_a} \nu_{ij}(r)P_{nl_i}(r)dr = 0 \tag{6.36}$$

but not to all bound orbitals, whilst $V(r)$ in equation (6.32) is a zero-order potential, chosen to represent the static potential of the target.

2.4 The Internal Region

The key to finding information about the photoionization process is finding a wavefunction Ψ which is an eigensolution of the stationary-state or time-independent Schrödinger equation

$$H^{N+1}\Psi = E\Psi \tag{6.37}$$

subject to appropriate boundary conditions where H^{N+1} is the $(N+1)$-electron Hamiltonian described by equation (6.9) in which N has been replaced by $(N+1)$ and where E is the total energy of the system. Equation (6.37) needs to be solved in both the internal and external regions. As previously described in Section 6.2.1, different physical problems exist in the two regions.

In the internal region, strong interaction between the bound states and the continuum may require a summation to approximate the continuum using pseudo states, which satisfy the equations introduced to describe the target states but are constructed from a combination of real and pseudo orbitals. A configuration interaction expansion can be used to represent the total wavefunction. That is,

$$\Psi = \sum_k A_{Ek}\psi_k \tag{6.38}$$

where A_{Ek} are energy-dependent coefficients and ψ_k are states which form a basis for the total wavefunction in the internal region $(r < r_a)$, are energy-independent and are given by the expansion

$$\begin{aligned}\psi_k(x_1,\ldots,x_{N+1}) &= \mathcal{A}\sum_{i,j} c_{ijk}\overline{\Phi_i}(x_1,\ldots,x_N;\hat{\mathbf{r}}_{N+1}\sigma_{N+1})\frac{1}{r_{N+1}}\nu_{ij}(r_{N+1}) \\ &+ \sum_j d_{jk}\chi_j(x_1,\ldots,x_{N+1}) \end{aligned} \tag{6.39}$$

where $\overline{\Phi_i}$ are the channel functions obtained by coupling the target states Φ_i with the angular and spin functions of the continuum electron to form states of the same total angular momentum and parity. $\mathcal{A}$ is the antisymmetrization operator, which ensures the Pauli exclusion principle is satisfied when electron exchange between the target electrons and the free electron is taken into account, whilst the χ_j represent the quadratically integrable (L^2) functions (or $(N+1)$-electron configurations) which vanish at the surface of the internal region, are formed from the bound orbitals and are included to ensure completeness of the total wavefunction. The ν_{ij} are the continuum orbitals corresponding to the appropriate angular momentum obtained from equation (6.32) and are the only terms in equation (6.39) that are non-zero on the surface of the internal region. The coefficients c_{ijk} and d_{jk} are determined by diagonalizing H^{N+1} in the finite space $r \leqslant r_a$ by writing

$$\left(\psi_k \left| H^{N+1} \right| \psi_{k'}\right) = E_k^{N+1} \delta_{kk'} \tag{6.40}$$

where the round brackets indicate that the radial integrals are limited to the region $r \leqslant r_a$.

If the total wavefunction Ψ_E at any energy E can be expanded in terms of the basis ψ_k as

$$\Psi_E = \sum_k A_{Ek} \psi_k \tag{6.41}$$

then the coefficients A_{Ek} can be determined by considering the relation

$$\left(\psi_k \left| H^{N+1} \right| \Psi_E\right) - \left(\Psi_E \left| H^{N+1} \right| \psi_k\right) = \left(E - E_k^{N+1}\right) \left(\psi_k \mid \Psi_E\right) \tag{6.42}$$

where it is assumed that the complete basis used in equation (6.41) diagonalizes the Hamiltonian H^{N+1}. Since only the kinetic energy operator contributes to the left-hand side of equation (6.42) it simplifies to

$$-\frac{1}{2}(N+1)\left\{ \left(\psi_k \left| \nabla_{N+1}^2 \right| \Psi_E\right) - \left(\Psi_E \left| \nabla_{N+1}^2 \right| \psi_k\right) \right\} = \left(E - E_k^{N+1}\right) \left(\psi_k \mid \Psi_E\right) \tag{6.43}$$

A further simplification can be made, since the only non-zero contribution can take place when the operator ∇_{N+1}^2 acts on the continuum orbitals. Writing

$$w_{ik} = \sum_j c_{ijk} \nu_j \tag{6.44}$$

and using equation (6.41) yields

$$-\frac{1}{2} \sum_{ijk'} A_{Ek'} \left\{ \left(\Phi_i w_{ik} \left| \nabla_{N+1}^2 \right| \Phi_j w_{jk'}\right) - \left(\Phi_j w_{jk'} \left| \nabla_{N+1}^2 \right| \Phi_i w_{ik}\right) \right\} = \left(E - E_k^{N+1}\right) \left(\psi_k \mid \Psi_E\right). \tag{6.45}$$

Defining the reduced radial wavefunction of the continuum electron in channel i with energy E by

$$y_i = \sum_k A_{Ek} w_{ik} \tag{6.46}$$

and recalling that the channel functions Ψ_i are orthonormal the equation

$$-\frac{1}{2}\sum_i \left\{ \left(w_{ik} \left| \frac{d^2}{dr^2} \right| y_i \right) - \left(y_i \left| \frac{d^2}{dr^2} \right| w_{ik} \right) \right\} = \left(E - E_k^{N+1} \right) A_{Ek} \tag{6.47}$$

follows. Applying Green's theorem and using the R-matrix boundary conditions (6.33) and (6.34) at $r = r_a$ satisfied by the w_{ik} gives

$$\sum_i w_{ik}(r_a) \left(r_a \frac{dy_i}{dr} - b y_i \right)_{r=r_a} = 2r_a \left(E_k^{N+1} - E \right) A_{Ek} \tag{6.48}$$

which can be rewritten in terms of the R-matrix which has elements defined by

$$R_{ij} = \frac{1}{2r_a} \sum_k \frac{w_{ik}(r_a) w_{jk}(r_a)}{E_k^{N+1} - E} \tag{6.49}$$

in the form

$$y_i(r_a) = \sum_j R_{ij} \left(r_a \frac{dy_j}{dr} - b y_j \right)_{r=r_a} . \tag{6.50}$$

The two unknowns in equations (6.49) and (6.50), the surface amplitudes w_{ik} and the R-matrix poles, E_k^{N+1}, are easily obtained from the eigenvalues and eigenvectors of the Hamiltonian matrix. So equations (6.50) and (6.49) are the basic equations from which the wavefunctions for the internal region can be obtained. The infinite sum in equation (6.50) is not a practical course of action and so the expansion is truncated to a finite number of terms using the Buttle correction.

2.4.1 The Buttle Correction. The contribution from high-lying levels is given by the expression derived by Buttle [16]

$$R_{ii}^c = \left[\frac{r_a}{\nu_i} \frac{d\nu_i}{dr} - b \right]_{r=r_a}^{-1} - \frac{1}{r_a} \sum_j \frac{[\nu_{ij}(r_a)]^2}{k_{ij}^2 - k_i^2} \tag{6.51}$$

which is diagonal in the channel quantum numbers. The first term in equation (6.51) is obtained by solving equation (6.32) at the energy in question, whilst in the second term k_i^2 is twice the difference between the total energy and the channel energy and thus the summation subtracts those low-lying levels obtained by diagonalizing H^{N+1} and which are already included in equation (6.49).

2.5 The External Region

It is now necessary to find a solution for the external region, similar to that obtained in the internal region. The total wavefunction at energy E can be expressed as

$$\Psi_E = \mathcal{A}\sum_i \Phi_i y_i + \sum_j d_j \phi_j \tag{6.52}$$

and using Kohn's variational principle [17] and taking arbitrary variations of the functions y_i and coefficients d_j, the close coupling equations

$$\langle \Phi_i \left| H^{N+1} - E \right| \Phi_E \rangle = 0 \text{ for } i = 1, \ldots, n \tag{6.53}$$

and

$$\langle \phi_j \left| H^{N+1} - E \right| \Phi_E \rangle = 0 \text{ for } j = 1, \ldots, m \tag{6.54}$$

follow where it is assumed that n channel functions have been retained in (6.53) and m $(N+1)$-electron bound functions in (6.54). These equations reduce to n coupled integro-differential equations for the functions y_i, which can be solved in the external region $(r > r_a)$. The choice of the R-matrix radius ensures that exchange effects between the free electron and the target vanish and so the close-coupling equations reduce to coupled differential equations of the form

$$\left(\frac{d^2}{dr^2} - \frac{l_i(l_i+1)}{r^2} + \frac{2Z}{r} + k_i^2 \right) y_i(r) = 2\sum_{j=1}^{n} V_{ij}(r) y_j(r) \tag{6.55}$$

for $i = 1, \ldots, n \quad (r \geq r_a)$. The notation l_i is used to represent the channel angular momentum; $V_{ij}(r)$ is the potential matrix given by

$$V_{ij}(r) = \left\langle \overline{\Phi}_i \left| \sum_{k=1}^{N} \frac{1}{r_{k,N+1}} \right| \overline{\Phi}_j \right\rangle; \tag{6.56}$$

Z is the nuclear charge and k_i^2 are the channel energies. The long-range potential coefficients a_{ij}^{λ} are defined by the equation

$$a_{ij}^{\lambda} = \left\langle \overline{\Phi}_i \left| \sum_{k=1}^{N} r_k^{\lambda} P_{\lambda}(\cos\theta_{k,N+1}) \right| \overline{\Phi}_j \right\rangle. \tag{6.57}$$

Due to the orthonormality of the channel functions,

$$a_{ij}^{0} = N\delta_{ij} \tag{6.58}$$

which, combined with the following expansion

$$\sum_{k=1}^{N} \frac{1}{r_{k,N+1}} = \sum_{\lambda=0}^{\infty} \frac{1}{r_{N+1}^{\lambda+1}} \sum_{k=1}^{N} r_k^{\lambda} P_{\lambda}(\cos\theta_{k,N+1}) \tag{6.59}$$

for $r_{N+1} > r_k$ and $k = 1, \ldots, N$, reduces the differential equations of (6.55) to

$$\left(\frac{d^2}{dr^2} - \frac{l_i(l_i+1)}{r^2} + \frac{2z}{r} + k_i^2\right) y_i(r) = 2\sum_{\lambda=1}^{\lambda_{max}} \sum_{j=1}^{n} \frac{a_{ij}^{\lambda}}{r^{\lambda+1}} y_j(r) \tag{6.60}$$

for $i = 1, \ldots, n; r > r_a$where $z = Z - N$ is the residual target charge. It is important that solutions obtained in the external region match those already obtained in the internal region, so appropriate boundary conditions need to be set up.

2.6 Boundary Conditions

The status of the $(N+1)$th electron is crucial in choosing appropriate boundary conditions. If it is free, $k_i^2 > 0$ and so the associated channel is open. If, however, the $(N+1)$th electron is bound, then $k_i^2 < 0$ and so the channels are closed.

2.6.1 Open Channels. If there are n channels and n_a of these channels are open then as $r \to \infty$

$$y_{ij} \sim k_i^{-\frac{1}{2}} (\sin\theta_i \delta_{ij} + \cos\theta_i K_{ij}) \tag{6.61}$$

$$y_{ij} \sim O(r^{-2}) \tag{6.62}$$

for $i = 1, \ldots, n_a; j = 1, \ldots, n_a$ and $i = n_a + 1, \ldots, n; j = 1, \ldots, n_a$ respectively. Equation (6.61) defines the elements of the reactance matrix K_{ij} and the other quantities are given by

$$\theta_i = k_i r - \frac{1}{2} l_i \pi - \eta_i \ln 2k_i r + \sigma_{l_i} \tag{6.63}$$

$$\eta_i = -\frac{Z-N}{k_i} \tag{6.64}$$

$$\sigma_{l_i} = \arg\Gamma(l_i + 1 + i\eta_i) \tag{6.65}$$

To relate the $n \times n$ dimensional R-matrix to the $n_a \times n_a$ dimensional K-matrix defined by equation (6.61) it is possible to introduce $n+n_a$ linearly independent solutions ν_{ij} of equation (6.60) which satisfy the following boundary conditions as $n \to \infty$

$$\nu_{ij} \sim k_i^{-\frac{1}{2}} \sin\theta_i \delta_{ij} + O(r^{-1}) \tag{6.66}$$

$$\nu_{ij} \sim k_i^{-\frac{1}{2}} \cos\theta_i \delta_{ij-n_a} + O(r^{-1}) \tag{6.67}$$

$$\nu_{ij} \sim \exp(-\phi_i)\delta_{ij-n_a} + O(r^{-1}\exp(-\phi_{j-n_a})) \tag{6.68}$$

for $j = 1, \ldots, n_a$, $j = n_a+1, \ldots, 2n_a$ and $j = 2n_a+1, \ldots, n+n_a$ respectively and where

$$\phi_i = |k_i|r - \frac{Z-N}{|k_i|}\ln(2|k_i|r). \tag{6.69}$$

Hence the continuum orbitals in the external region can be obtained and the reduced radial wavefunctions can be written

$$y_{ij}(r) = \sum_{l=1}^{n+n_a} \nu_{il}(r)x_{lj} \tag{6.70}$$

for $i = 1, \ldots, n$, $j = 1, \ldots, n_a$ where x_{lj} satisfy the n_a equations

$$x_{lj} = \delta_{lj} \tag{6.71}$$

for $l = 1, \ldots, n_a$, together with the n equations given by

$$\sum_{l=1}^{n+n_a}\left\{\nu_{il}(r_a) - \sum_{m=1}^{n} R_{im}\left(r_a\frac{d\nu_{ml}}{dr} - b\nu_{ml}\right)_{r=r_a}\right\}x_{lj} = 0 \tag{6.72}$$

for $i = 1, \ldots, n$. Equations (6.71) and (6.72) must be solved for each $j = 1, \ldots, n_a$. The K-matrix can thus be obtained from

$$K_{ij} = x_{i+n_a j} \tag{6.73}$$

which completes the solution for the case of open channels.

2.6.2 Closed Channels. If all the channels are closed, the wavefunction Φ_E corresponds to a bound state of the electron-plus-atom and the problem reduces to finding a discrete eigenvalue and the corresponding eigenvector of H^{N+1}. It is possible to define n linearly independent solutions of equation (6.60), which satisfy the boundary conditions

$$\nu_{ij} \sim \exp(-|k_i|r)\delta_{ij} \tag{6.74}$$

as $r \to \infty$ for $i, j = 1, \ldots, n$. The required solution can be expanded in terms of

$$y_{ij} = \sum_{j=1}^{n} \nu_{ij}x_j \tag{6.75}$$

for $i = 1, \ldots, n$. The coefficients x_j can be determined by substituting equation (6.75) into equation (6.50) which leads to the n homogeneous equations given by

$$\sum_{j=1}^{n}\left\{\nu_{ij}(r_a) - \sum_{k=1}^{n} R_{ik}\left(r_a\frac{d\nu_{kj}}{dr} - b\nu_{kj}\right)_{r=r_a}\right\}x_j = 0 \tag{6.76}$$

for $i = 1, \ldots, n$ which can be expressed in the more convenient form

$$\sum_{j=1}^{n} B_{ij} x_j = 0 \tag{6.77}$$

for $i = 1, \ldots, n$. The only non-trivial solutions occur at the negative-energy eigenvalues corresponding to the final bound state. The condition for a solution is

$$\det \mathbf{B} = 0. \tag{6.78}$$

A solution can be found by setting $x_1 = 1$ and solving the equations

$$\sum_{j=2}^{n} B_{ij} x_j = B_{i1} \tag{6.79}$$

for $i = 2, \ldots, n$, and then looking for the zeros of

$$f(E) = \sum_{j=1}^{n} B_{1j} x_j \tag{6.80}$$

as a function of energy, which are determined using Newton's iteration method.

2.7 Photoionization Cross Sections

Photoionization cross sections are a direct measure of the probability that light of various energies will ionize a particular system. They involve the sum of the absorption cross-sections that represent transitions from one fixed lower state to all other states of higher energy. The information obtained from the solutions in the internal and external regions is required to calculate the photoionization cross section.

The differential cross section for the photoionization of an $(N+1)$-electron atom or ion, with the ejection of an electron in the $\mathbf{k}$-direction, is given by

$$\frac{d\sigma_L}{d\mathbf{k}} = 8\pi^2 \alpha a_o^2 \omega \left| \left\langle \Psi_f^-(\mathbf{k}) \left| \sum_{j=1}^{N+1} z_j \right| \Psi_i \right\rangle \right|^2 \tag{6.81}$$

in the dipole-length approximation and by

$$\frac{d\sigma_V}{d\mathbf{k}} = \frac{8\pi^2 \alpha a_o^2}{\omega} \left| \left\langle \Psi_f^-(\mathbf{k}) \left| \sum_{j=1}^{N+1} \frac{\partial}{\partial z_j} \right| \Psi_i \right\rangle \right|^2 \tag{6.82}$$

in the dipole-velocity approximation. In these equations, ω is the incident photon energy in atomic units, α is the fine-structure constant, a_0 is the Bohr radius of the hydrogen atom and $\Psi_f^-(\mathbf{k})$ and Ψ_i are the final and initial continuum wavefunctions. The boundary condition satisfied by $\Psi_f^-(\mathbf{k})$ corresponds to a plane wave in direction $\mathbf{k}$ incident on the final state of the ion, with ingoing waves in all open channels. In LS coupling, $\Psi_f^-(\mathbf{k})$ can be expanded as

$$\Psi_f^-(\mathbf{k}) = \sum_{l_f, m_{l_f}, L} i^{l_f} \exp(-i\sigma_{l_f}) Y_{l_f}^{m_{l_f}\star}(\mathbf{k}) C(L_f l_f L; M_{L_f} m_{l_f} M_L) \times C(S_f \tfrac{1}{2} S; M_{S_f} \tfrac{1}{2} M_S) \Psi_f^- \quad (6.83)$$

where Ψ_f^- is an eigenstate of total angular momentum L and total spin angular momentum S, the $C(j_1 j_2 j_{12}; m_1 m_2 m_{12})$ are Clebsch-Gordan coefficients, $L_f S_f M_{L_f} M_{S_f}$ are the quantum numbers describing the final state of the ion, $Y_{l_f}^{m_{l_f}}$ is the spherical harmonic of Rose [18] and σ_{l_f} is defined by equation (6.65).

It is possible to expand Ψ_i in equations (6.81) and (6.82) and Ψ_f^- in equation (6.83) in the form given by (6.39) and a solution can be obtained using the method outlined in the preceding sections. It is convenient to define the reduced matrix elements by

$$\langle \Psi_f^- || M_1 || \Psi_i \rangle = \frac{\sqrt{2L+1}}{C(l_i l L; M_{L_i} \mu M_L)} \langle \Psi_f^- | M_1^\mu | \Psi_i \rangle \quad (6.84)$$

where the operator M_1^μ reduces to the dipole length operator in equation (6.81) or the dipole velocity operator in (6.82) and where it is assumed that the only significant contribution to the reduced matrix element comes from the internal region. If the reduced matrix $\mathbf{M}$ is defined by

$$M_{kk'} = \langle \Phi_k || M_1 || \Phi_{k'} \rangle \quad (6.85)$$

where Φ_k comes from a rewriting of equation (6.39) in the form

$$\Psi_k = \sum_{k'} \Phi_{k'} V_{kk'} \quad (6.86)$$

where the Φ_k denote collectively the basis functions $\mathcal{A}\overline{\Phi}_i \frac{\nu_{ij}}{r_{N+1}}$ and χ_j, and the $V_{kk'}$ denote collectively the coefficients c_{ijk} and d_{jk} then the $\mathbf{V}$ matrices are given by the eigenvalue equation

$$\sum_{kk''} (\Phi_k | H | \Phi_{k'}) V_{k'k'''} = E_{k''} \delta_{k''k'''} \quad (6.87)$$

An expression for the cross section is given by

$$\sigma = \frac{2\pi^2 \alpha a_0^2 C}{3 r_a^2 (2L_i + 1)} \sum_{l_f L} \mathbf{y}_f^{-T} \mathbf{R}_f^{-1} \mathbf{w}_f \mathbf{G}_f \mathbf{V}_f^T \mathbf{M} \mathbf{V}_i \mathbf{G}_i \mathbf{w}_i^T \mathbf{R}_i^{-1} \mathbf{y}_i \quad (6.88)$$

where $C = \omega$ in the length approximation and $C = \frac{1}{\omega}$ in the velocity approximation.

The $\mathbf{w}$ matrices are defined by equation (6.44) and the diagonal matrices $\mathbf{G}_i$ and $\mathbf{G}_f$ are defined by

$$G_{ki} = \frac{1}{E_k - E_i} \tag{6.89}$$

and

$$G_{kf} = \frac{1}{E_k - E_i - \omega} \tag{6.90}$$

respectively. The $\mathbf{R}$ matrices are defined by equation (6.49) and the $\mathbf{y}$ matrices satisfy equations (6.70) and (6.75).

2.8 Relativistic Effects

The R-matrix theory outlined in the preceding sections of this chapter has related to the solution of Schrödinger's equation with a non-relativistic Hamiltonian. This theory has been extended by Scott and Burke [11] to include relativistic effects in calculations. This was achieved by using the Breit-Pauli Hamiltonian incorporating the one-body mass correction, Darwin and spin-orbit terms. The relativistic Hamiltonian has already been described in Section 1.3.1. The one-body mass correction and Darwin terms are non-fine-structure interactions and as a consequence commute with $\mathbf{L}^2$, $\mathbf{S}^2$, $\mathbf{L}_z$ and $\mathbf{S}_z$. However, the spin-orbit term is a fine-structure interaction and so commutes with $\mathbf{J}^2$ and $\mathbf{J}_z$. It is therefore necessary to use a representation which is diagonal in $\mathbf{J}^2$ and $\mathbf{J}_z$.

The solution of the bound-state problem in the internal region requires the evaluation of the matrix elements of the Breit-Pauli Hamiltonian. In the external region, the relativistic effects manifest themselves only through the change of representation and through their effect on the channel energies.

3. Applications

3.1 Photoabsorption of K II

One application of the foregoing theory is the photoexcitation of a 3s electron in singly-ionized potassium, an argon-like ion.

The autoionizing series $3s^23p^6\ ^1S_0 - 3s3p^6np\ ^1P^o_1$ was observed for neutral argon by Madden and Codling [19]. The autoionizing state $3s3p^64p\ ^1P^o$ in singly ionized potassium was identified by Aizawa *et al.* [20] using the experimental technique of ejected electron spectroscopy. An absolute photoionization cross section of K II was measured by Peart and Lyon [21] using the experimental technique of merged ion and synchotron electron beams. Theoretically, Nasreen *et al.* [22] have performed calculations at the relativistic random-phase approximation level for K II.

More recently, high resolution inner shell photoabsorption spectra have been produced for K II by van Kampen *et al.* [23] using the dual laser plasma technique. In addition, calculations associated with $3s \rightarrow np$ excitations have been performed by the same authors using the random phase approximation with exchange for calculation of photoionization amplitudes, the Dyson equation method for energy shifts and corrections to wavefunctions of ground and discrete excited states and many-body perturbation theory for corrections to the photoamplitudes due to double-electron processes.

Lagutin *et al.* [24] have also calculated absolute photoionization cross sections for K II. Their calculations used the Pauli-Fock atomic orbitals within the configuration interaction technique.

Here we illustrate the application of the *R*-matrix method to obtain accurate absolute photoabsorption cross sections for K II.

3.1.1 Details of the Calculation. The *R*-matrix method (Berrington *et al.* [10],[13],[14], Burke and Robb [8], Seaton [25]) is used to obtain accurate absolute photoabsorption cross sections and to enable a study of resonances.

Our calculations neglect relativistic effects and are done in *LS* coupling. The initial and final states are treated consistently in that the initial state of K II is represented as a bound state of the electron-plus-target system. The twelve lowest target states of K III are included in the wavefunction expansion.

The target states are represented by configuration interaction wavefunctions based on the orbital set: 1s, 2s, 2p, 3s, 3p, 3d, 4s, 4p, $\overline{4d}$. The 1s, 2s, 2p, 3s and 3p orbitals are taken to be the Hartree-Fock orbitals of the ground state of the target as determined by Clementi and Roetti [26]. The 3d orbital is obtained by using the CIV3 code [15] to optimize on the energy of the $3s^2 3p^4 (^3P) 3d\ ^2F$ state of the target. The 4s and 4p orbitals are obtained by optimizing on the energies of the $3s^2 3p^4 (^1S) 4s\ ^2S$ and $3s^2 3p^4 (^3P) 4p\ ^2D^o$ states respectively. The 3d orbital obtained, being incapable of representing the 3d orbital in all the various states considered, is corrected by employing a $\overline{4d}$ pseudo-orbital derived by optimising on the energy of the $3s^2 3p^4 (^3P) 3d\ ^2D$ state using two configurations $(3s^2 3p^4 3d + 3s^2 3p^4 \overline{4d})$. Although this state was eventually not included as a target state, the optimal 3d orbital for this state is radially the most different from that of the other 3d states. The optimization of $\overline{4d}$ in this way thus gives the greatest flexibility in the radial representation of the outer orbital for all the states labelled "... 3d".

Using this set of orbitals, the configuration interaction wavefunctions for the target states are generated by considering the 1s, 2s and 2p orbitals to form a closed core and then considering one-electron replacement on the basis distributions $3s^2 3p^5$, $3s 3p^6$, $3s^2 3p^4 3d$, $3s^2 3p^4 4s$, $3s^2 3p^4 4p$, $3s 3p^5 3d$ and $3s 3p^5 4s$.

In addition, the distributions $3s^23p^23d^3$ and $3s^23p^23d^2\overline{4d}$ are included. A total of 632 configurations are employed in representing the target states.

The target state energies found for K III are compared with the available experimental data, compiled by Corliss and Sugar [27]. Such a comparison is useful in assessing the accuracy of the target-state wavefunctions. It is only possible to compare data for the first three target states as the energy level values provided by Corliss and Sugar for higher states are unreliable. The agreement is very satisfactory and we therefore feel that the target state representations are sufficiently sophisticated for the subsequent photoionization calculations.

Using these target state data, the *R*-matrix method is employed for K II with 20 continuum orbitals for each continuum electron orbital angular momentum $0 \leqslant l \leqslant 5$. The *R*-matrix radius is taken to be 11.8 atomic units. To maintain the balance between the initial ground state and the target-plus-electron system, the $(N+1)$-electron correlation configurations are constructed by the addition of one electron from the orbital set to the target-state distributions. A total of 195 and 437 configurations are obtained for the ^{1}S and ^{1}P^o total system symmetries, respectively. The photoionization cross sections are calculated in both the length and velocity formulations. The length formulation normally gives the most accurate values within the framework of the procedures adopted, since the length formulation is less sensitive to electron correlation in the core.

To enable a complete resolution of the resonances for each ion, a small energy mesh for the free-electron is taken, namely 7.0750×10^{-3} eV. It is customary, in employing the *R*-matrix method, to adjust the target state energies to their experimental values prior to diagonalization of the $(N+1)$-electron Hamiltonian. This ensures the correct positioning of resonances in the photoabsorption spectrum. In our calculation of the photoionization cross section of K II, we are restricted to adjustment of energies of the first three K III target states to their experimental values.

3.1.2 Results and Discussion. Figure 6.1 shows a plot of the total photoionization cross section together with the experimental data of van Kampen *et al.* [23]. The cross section is observed to exhibit many of the features of the experimental cross section. Two methods have been used to identify the resonances in the photoabsorption spectrum. First, the *R*-matrix eigenvectors were used to identify the $3s3p^64p\ ^1P^o$ and $3s^2(3p^4\ ^1D)3d(^2D)4p\ ^1P^o$ resonances. Second, a quantum defect analysis was utilized to identify the remaining resonances of the $3s3p^6np\ ^1P^o$series. The quantum defect for the series was determined from the $3s3p^64p\ ^1P^o$ resonance to be 1.41482 and from the positions of the other resonances the value of n was calculated. The value of n in each case was found to be close to an integer value, indicating that the relative positions of the resonances in the Rydberg series were quite accurate.

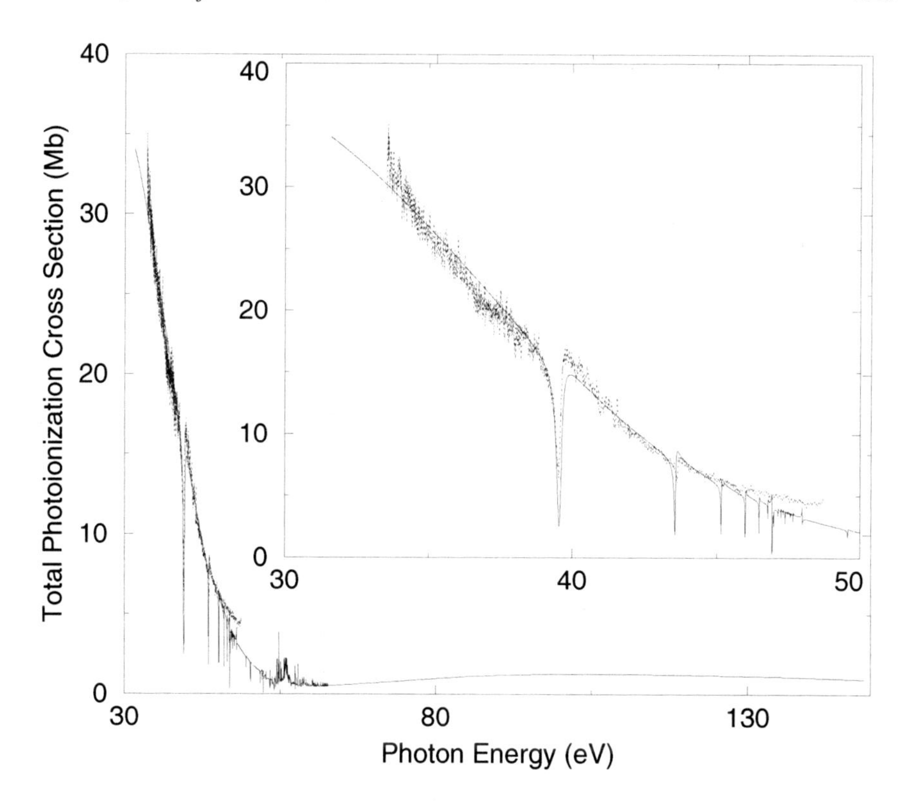

Figure 6.1. The total photoionization cross section for K II calculated using the R-matrix approach (solid line) compared with the experimental data of van Kampen *et al.* (dashed line).

The positions of the resonances have been compared with previously published experimental and theoretical positions (Aizawa *et al.* [20], van Kampen *et al.* [23]). The positions of the $3s3p^6 4p\ ^1P^o$ and $3s3p^6 5p\ ^1P^o$ resonances are in excellent accord with experiment. The present calculation replicates experiment more closely than the calculation of Lagutin *et al.* [24], particularly at lower photon energies and around the 4p resonance.

Our calculation does not indicate the presence of any structure in the photoabsorption spectrum around 35.5 eV (Peart and Lyon [21]) and so we support the interpretation of Nasreen *et al.* [22] that any possible structure is simply an artefact of this particular experiment. The presence of a $3s^2 3p^4 3d4p\ ^1P^o$ resonance at a photon energy of 46.89 eV, above the $3s3p^6 4p\ ^1P^o$ resonance at 39.56 eV is not unexpected since two-electron processes are significant in the photoionization continuum of argon.

Table 6.1. Parameters for the 3s → np transitions in K II compared with those of van Kampen *et al.* [23].

Transition	van Kampen *et al.* Theory		van Kampen *et al.* Experiment		This Calculation	
	q	Γ(eV)	q	Γ(eV)	q	Γ(eV)
3s → 4p	0.37	0.23	0.4±0.1	0.19+0.02/−0.08	0.143	0.202
3s → 5p	0.53	0.056	0.5±0.2	0.05±0.03	0.357	0.055

The $3s3p^6 4p\ ^1P^o$ and $3s3p^6 5p\ ^1P^o$ resonances were fitted using the formula

$$Q(\varepsilon) = a_0 + a_1\varepsilon + a_2\frac{(q+\varepsilon)^2}{1+\varepsilon^2} \tag{6.91}$$

with

$$\varepsilon = \frac{2}{\Gamma}(E - E_r) \tag{6.92}$$

where Q is the photoionization cross section measured in megabarns, Γ is the width of the resonance, q is the Fano shape parameter, E_r is the position of the resonance measured in Rydbergs and a_0, a_1 and a_2 are constants to be determined. We note that equation (6.91) allows for a background cross section to be non-constant.

The shape parameters q and the widths of both the $3s3p^6 4p\ ^1P^o$ and $3s3p^6 5p\ ^1P^o$ resonances are compared with those obtained by van Kampen *et al.* [23] in Table 6.1. Our calculations of the widths of the two resonances are in good agreement with the literature values. However there is poor accord between the calculated and experimental value of q. This may be due to the higher resolution used in the present calculation, permitting a more accurate definition of the resonance.

The length and velocity forms of the total photoionization cross section are compared in Figure 6.2. The agreement between the two forms of the cross section is very satisfactory. We are therefore confident in the accuracy of our results.

3.2 Photoabsorption of Fe XVI and Fe XVII Recombination Rate Coefficients

This second example illustrates the use of the R-matrix method for a more highly ionized, heavier species. For such an element it is important that relativistic effects are included, and these effects have been discussed from the theoretical viewpoint in Sections 6.1.3.1 and 6.2.8.

Line emission from Fe L-shell ions has been extensively detected in the soft X-ray spectra of coronal plasmas such as the Sun (Phillips *et al.* [28]), and from accretion-powered sources including active galactic nuclei, cataclysmic

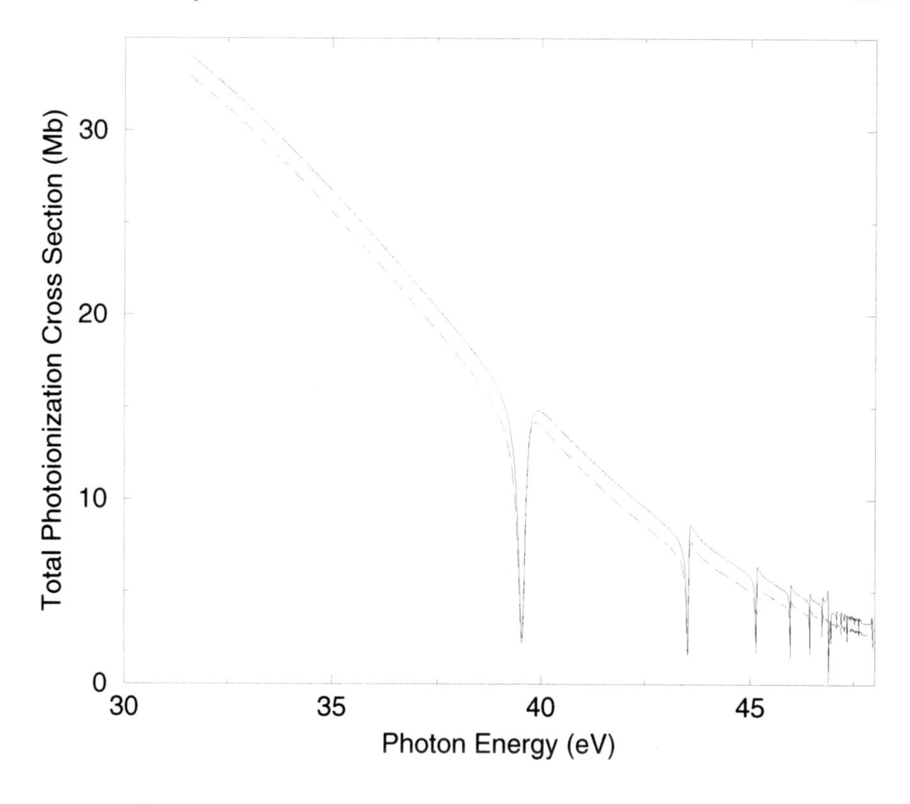

Figure 6.2. A comparison of the length (solid curve) and velocity (dashed curve) forms of the total photoionization cross section for K II.

variables and X-ray binaries (Liedahl *et al.* [29]). To date, the diagnostic applications of Fe L-shell spectra have been primarily restricted to the solar corona (see, for example, Keenan [30] and references therein), due to the availability of high spectral resolution observations. By contrast, current data for cosmic X-ray sources (provided for example by the BBXRT and ASCA satellites; Yaqoob *et al.* [31]; Audley *et al.* [32]), are at best of moderate spectral resolution, thereby limiting the analysis and interpretation (Liedahl *et al.* [29]). However high resolution spectra for such objects will soon become available from the Chandra and XMM satellites (Weisskopf [33]; Brinkman [34]), which will also provide observations for the coronal regions of late-type stars (Haisch and Schmitt [35]).

The modelling of Fe L-shell emission from all of the above astronomical objects requires reliable atomic data, with photoionization and recombination being particularly important for the accretion-powered X-ray sources (Savin *et al.* [36]).

3.2.1 Details of the Calculation. We use the relativistic codes based on the theory of Scott and Burke [11], and therefore include in the Hamiltonian the one-body mass correction, Darwin and spin-orbit terms, as previously discussed in Section 6.1.3.1. We initally represent the residual ion by the following 21 $LS\pi$ target states: $2s^22p^6\ ^1S$; $2s^22p^53s\ ^1P^o, ^3P^o$; $2s^22p^53p\ ^1S, ^3S, ^1P, ^3P, ^1D, ^3D$; $2s^22p^53d\ ^1P^o, ^3P^o, ^1D^o, ^3D^o, ^1F^o, ^3F^o$; $2s2p^63s\ ^1S, ^3S$; $2s2p^63p\ ^1P^o, ^3P^o$; $2s2p^63d\ ^1D, ^3D$. Each of these target states is represented using a configuration-interaction type wavefunction, as shown by equation 6.30, with the following set of one-electron orbitals $\{1s,2s,2p,3s,3p,3d,\overline{4p},\overline{4d}\}$ whose radial parts are given by linear combinations of Slater-type orbitals

$$P_{nl}(r) = \sum_{j=1}^{k} c_{jnl} \left[\frac{(2\zeta_{jnl})^{2I_{jnl}+1}}{(2I_{jnl})!} \right]^{\frac{1}{2}} r^{I_{jnl}} e^{-\zeta_{jnl} r}. \qquad (6.93)$$

The 1s, 2s and 2p orbitals are those calculated for the Hartree-Fock ground state of Fe XVII by Clementi and Roetti [26]. The remaining orbitals are generated using the CIV3 code [15] in the following manner. The 3s and 3p orbitals are obtained by treating the exponents, ζ_{jnl}, as variational parameters in the minimization of the energies of the $2s^22p^53s\ ^3P^o$ and $2s^22p^53p\ ^3S$ states, respectively. The 3d orbital is obtained by optimizing on the $2s^22p^53d\ ^3P^o$ state but both exponents and coefficients are allowed to vary. Further valence shell correlation is introduced by using pseudo orbitals: a $\overline{4p}$ optimized on $2s^22p^53p\ ^3P$ using two configurations $[2s^22p^53p + 2s^22p^5\overline{4p}]\ ^3P$ and a $\overline{4d}$ optimized on $2s^22p^53d\ ^3D^o$ using two configurations $[2s^22p^53d + 2s^22p^5\overline{4d}]\ ^3D^o$. Using this set of orbitals, the configuration-interaction wavefunctions for the target states are generated by allowing two electron replacement in the $2s^22p^6$ basis set with any other two electrons from the orbital set now available to us. However, the resulting 825 configurations are too many for available computing resources. Unimportant configurations are therefore eliminated using the following procedure. Examining the eigenvectors generated by the CIV3 computer code in determining the target state energies in LS-coupling, only configurations which have an expansion coefficient, a_{ik} in equation 6.30, greater than 0.0075 for at least one of the target state wavefunctions are included. This procedure reduces the number of N-electron configurations to 171. This choice of cut-off for a_{ik} was made, noting that the wavefunctions must contain high accuracy. The recoupling method described by Scott and Burke [11] is then employed to give the fine-structure energies and expansion coefficients. Comparison with experimental values compiled by NIST (obtained from the NIST atomic spectroscopic database at `http://physics.nist.gov`) for the energies of the fine-structure levels shows excellent accord and this suggests that the target state wavefunctions are adequately represented by the reduced configuration set. Using basis states of the form given by equation 6.39, the present calculation ensures that the Fe XVI and Fe XVII systems are balanced, by using $(N+1)$-electron configurations that are obtained by adding one electron from the orbital set to the 171 configurations used in the target state representation. Effectively, Fe XVI is being treated as a bound state of the (Fe XVII+e$^-$) system, while the motion and energy of the free electron is represented by the set of continuum orbitals. Twenty-five such orbitals are used for each value of $l \leqslant 6$ and an R-matrix radius of 4.6 atomic units is employed.

Selection rules, discussed briefly in Section 6.1.3.3, indicate that only two transitions need be considered in the present calculation : $\frac{1}{2}^{\mathrm{e}} \rightarrow \frac{1}{2}^{\mathrm{o}}$ and $\frac{1}{2}^{\mathrm{e}} \rightarrow \frac{3}{2}^{\mathrm{o}}$. Hamiltonian matrices are calculated for the three symmetries involved in these transitions where all the $LS\pi$ symmetries which contribute to them are included. An indication of the accuracy of the approach thus far used is given in the calculation of the $2\mathrm{s}^2 2\mathrm{p}^6 3\mathrm{s}\ {}^2\mathrm{S}_{1/2}$ Fe XVI bound state wavefunction which provides a value for the ionization energy. A value of 18.06 atomic units is obtained which is in excellent agreement with the value of 17.98 atomic units tabulated by NIST.

We note that for certain elements an effect referred to as radiation damping may affect photoionization cross sections and recombination rates. Pradhan and Zhang [37] have performed investigations into the effect of the inclusion of radiation damping and found that the effect on larger resonances was minimal and while weak resonances may be damped out, the omission of damping is unlikely to cause serious inaccuracies in the photoionization cross sections. Radiation damping of autoionization resonances has thus been omitted in the present investigation.

3.2.2 Results and Discussion. Total photoionization cross sections are calculated in the $18 \rightarrow 65$ atomic unit photon energy range using a step size of 5×10^{-3} atomic units. This small step size ensures that the resonance structure is well resolved. The results of the length formulation calculation are displayed in Figure 6.3, since they are expected to be more accurate than the velocity formulation values within the framework of the procedures adopted.

We note the sharp jump in the cross section at photon energies of 44.77 atomic units and 49.82 atomic units, which correspond to the 2p and 2s thresholds respectively, and the extensive resonance structure included due to the high resolution. A comparison with the Opacity Project LS-coupled calculation (Cunto *et al.* [38]) is made. Agreement with the Opacity Project data is highly satisfactory although it is noted that 2p and 2s photoionization was not included in that calculation and thus comparison above the 2p threshold is not meaningful. In addition, the Opacity Project results have been shifted by 1.5 atomic units in order to match the position of the 3s threshold with that of the present work.

To investigate the effect of including relativity in the present case, the LS-coupled cross sections have been calculated for the present scenario. The results

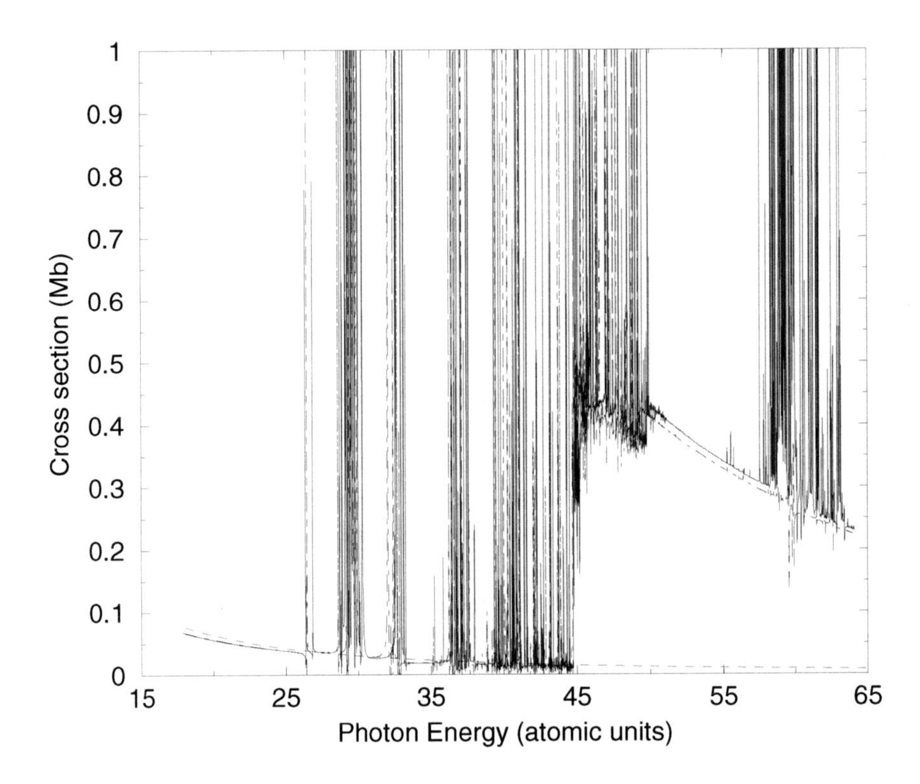

Figure 6.3. Total photoionization cross section for the photoionization of the Fe XVI ground state (solid curve) compared with the present calculation performed in LS-coupling (dot-dashed curve) and with the Opacity Project calculation (dashed curve).

are also provided in Figure 6.3. In the region of 3s photoionization, very little difference is observed between the two calculations, which is consistent with the observed purity of the target ground state in both coupling schemes. Significant mixing between the $2s^22p^53s\ ^1P_1^o$ and $^3P_1^o$ target states produces a slight increase in the background cross section shortly following the 2p ionization threshold. (The question of whether or not relativistic effects increase or decrease the background cross section is dependent solely on the magnitude of the partial cross sections involved.) The mixing in the remaining target states, however, has little effect on the magnitude of the total cross section, due to the relatively small magnitude of the partial cross sections for these target states.

3.2.3 Recombination Rates for Fe XVII. The recombination rate coefficient, α_{ji}, for electrons combining with Fe XVII in state i to form Fe XVI in state j is related to the partial photoionization cross section σ_{ij} by the Milne relation (Milne [39])

$$\alpha_{ji}(T) = \frac{\omega_j}{\omega_i}\sqrt{\frac{2}{\pi}}\frac{1}{c^2(mkT)^{\frac{3}{2}}}\int_{I_{ji}}^{\infty}(h\nu)^2\sigma_{ij}(h\nu)\exp\left[-\frac{h\nu - I_{ji}}{kT}\right]dh\nu \tag{6.94}$$

where I_{ji} is the ionization potential from the jth state of Fe XVI to the ith state of Fe XVII; T is the temperature of the free electrons, and m is the electron mass; ω_i and ω_j are the statistical weights of the respective ionic states.

The set of partial photoionization cross sections obtained from the calculation discussed above are utilized to obtain the recombination rates for the process of recombination to each of the Fe XVII target states forming the ground state of Fe XVI. The results for the five lowest target states, which contribute significantly to the total photoionization cross section, are given in Table 6.2 for a variety of temperatures but it is noted that the $2s2p^63s\ ^3S_1$ and $2s2p^63s\ ^1S_0$ states also make a significant contribution to the total photoionization cross section. We also note that due to the presence of numerous autoionization resonances the recombination rates calculated in this procedure also include low-energy dielectronic recombination in addition to radiative recombination. A general pattern is seen, namely that as the temperature increases, the recombination rate from the excited states decreases, whereas for the Fe XVII ground state, a minimum occurs at a temperature of around 5×10^5K.

Table 6.2. Recombination rate coefficients for the recombination of states of Fe XVII with an electron to form Fe XVI in its ground state.

Temperature	Recombination rates (10^{-12} cm^{3}s^{-1}) from ...				
(K)	$2p^6\ ^1S_0$	$2p^53s\ ^3P^o_2$	$2p^53s\ ^1P^o_1$	$2p^53s\ ^3P^o_0$	$2p^53s\ ^3P^o_1$
5×10^4	1.02	4.15	4.35	3.76	3.61
1×10^5	0.72	2.87	3.03	2.70	2.67
5×10^5	0.43	1.17	1.25	1.18	1.22
1×10^6	1.76	0.79	0.85	0.82	0.84
5×10^6	3.35	0.25	0.26	0.25	0.27
1×10^7	1.81	0.11	0.12	0.11	0.12

References

[1] H.A. Bethe and E.E. Salpeter, *Quantum mechanics of one- and two-electron atoms*, Springer-Verlag, Berlin (1957)

[2] U. Fano, Phys. Rev. **124** 1866 (1961)

[3] U. Fano and J.W. Cooper, Phys. Rev. A **137** 1364 (1965)

[4] R. Gersbacher and J.T. Broad, J. Phys. B: At. Mol. Opt. Phys. **23** 365 (1990)

[5] Wigner E P and Eisenbud L, Phys. Rev. **72** 29 (1947)

[6] Burke P G and Seaton M J, *Methods Comput. Phys.* **10** 1 (1971)

[7] Burke P G, Hibbert A and Robb W D J. Phys. B: At. Mol. Phys. **4** 153 (1971)

[8] Burke P G and Robb W D, *Adv. At. Mol. Phys.* **11** 143 (1975)

[9] Berrington K A, Burke P G, Chang J J, Chivers A T, Robb W D and Taylor K T, Comput. Phys. Comm. **8** 149 (1974)

[10] Berrington K A, Burke P G, Le Dourneuf M, Robb W D, Taylor K T and Vo Ky Lan, Comput. Phys. Comm. **14** 367 (1978)

[11] Scott N S and Burke P G, J. Phys. B: At. Mol. Phys. **13** 4299 (1980)

[12] Scott N S and Taylor K T, Comput. Phys. Comm. **25** 347 (1982)

[13] Berrington K A, Eissner W B and Norrington P H, Comput. Phys. Comm. **92** 290 (1995)

[14] Berrington K A, Burke P G, Butler K, Seaton M J, Storey P J, Taylor K T and Yu Yan, J. Phys. B: At. Mol. Opt. Phys. **20** 6379 (1987)

[15] Hibbert A, Comput. Phys. Comm. **9** 141 (1975)

[16] Buttle P J A, Phys. Rev. **160** 719 (1967)

[17] Kohn W, Phys. Rev. **74** 1763 (1948)

[18] Rose M E, *Elementary Theory of Angular Momentum*, Wiley, New York (1957)

[19] Madden R P and Codling K, Phys. Rev. Lett. **10** 516 (1963)

[20] Aizawa H, Wakiya K, Suzuki H, Koike F and Sasaki F, J. Phys. B: At. Mol. Phys. **18** 289 (1985)

[21] Peart B and Lyon I C, J. Phys. B: At. Mol. Phys. **20** L673 (1987)

[22] Nasreen G, Deshmukh P C and Manson S T, J. Phys. B: At. Mol. Opt. Phys. **21** L281 (1988)

[23] van Kampen P, O'Sullivan G, Ivanov V K, Ipatov A N, Costello J T and Kennedy E T, Phys. Rev. Lett. **78** 3082 (1997)

[24] Lagutin B M, Demekhin Ph V, Petrov I D, Sukhorukov V L, Lauer S, Liebel H, Vollweiler F, Schmoranzer H, Wilhelmi O, Mentzel G and Schartner K-H, J. Phys. B: At. Mol. Opt. Phys. **32** 1795 (1999)

[25] Seaton M J, J. Phys. B: At. Mol. Phys. **20** 6363 (1987)

[26] Clementi E and Roetti C, At. Data Nucl. Data Tables **14** 177 (1974)

[27] Corliss C and Sugar J, *J. Phys. Chem. Ref. Data* **8** 1109 (1979)

[28] Phillips K J H, Leibacher J W, Wolfson C J, Parkinson J H, Fawcett B C, Kent B J, Mason H E, Acton L W, Culhane J L and Gabriel A H, *Astrophys. J.* **256** 774 (1982)

[29] Liedahl D A, Osterheld A L and Goldstein W H, *Astrophys. J.* **438** L115 (1995)

[30] Keenan F P, *Space Sci. Rev.* **75** 537 (1996)

[31] Yaqoob T, Serlemitsos P J, Ptak A, Mushotzky R, Kunieda H and Terashima Y, *Astrophys. J.* **455** 508 (1995)

[32] Audley M D, Kelley R L, Boldt E A, Jahoda K M, Marshall F E, Petre R, Serlemitsos P J, Smale A P, Swank J H and Weaver K A, *Astrophys. J.* **457** 397 (1996)

[33] Weisskopf M C, *Space Sci. Rev.* **47** 47 (1988)

[34] Brinkman A C, in *Proc. SPIE, EUV, X-Ray and Gamma-Ray Instrumentation for Astronomy and Astrophysics*, ed. C J Hailey and O H W Siegmund **1159** 495 (1989)

[35] Haisch B and Schmitt J H M M, *Proc. Astron. Soc. Pacific* **108** 113 (1996)

[36] Savin D W, Bartsch T, Chen M H, Kahn S M, Liedahl D A, Linkemann J, Muller A, Schippers S, Schmitt M, Schwalm D and Wolf A, *Astrophys. J.* **489** L115 (1997)

[37] Pradhan A K and Zhang H L, *J. Phys. B: At. Mol. Opt. Phys.* **30** L571 (1997)

[38] Cunto W, Mendoza C, Ochsenbein F and Zeippen C J, *Astrophys. J.* **275** L5 (1993)

[39] Milne E A, *Phil. Mag.* **47** 209 (1924)

Chapter 7

RECOMBINATION OF COOLED HIGHLY CHARGED IONS WITH LOW-ENERGY ELECTRONS

E. Lindroth and R. Schuch
Department of Atomic Physics, Fysikum, Stockholm University, S-106 91 Stockholm, Sweden
lindroth@physto.se, schuch@physto.se

Abstract The basic processes of recombination between free electrons and atomic ions are reviewed. We concentrate particularly here on spectroscopic studies of few-electron atomic systems by dielectronic recombination at the electron cooler of the CRYRING heavy-ion storage ring facility. Recent measurements of dielectronic recombination resonances with Li-like, Na-like, and Cu-like ions are shown, where a resolution in the order of 10^{-3} eV was obtained just above the first ionization threshold. Theories for recombination are discussed. From the spectra of the dielectronic resonances very accurate values for energy splittings and resonance strengths are derived. These allow crucial tests of relativistic, correlation, and quantum electrodynamical effects in these systems.

Keywords: Electron ion impact, Electron ion recombination and electron attachment, Relativistic and quantum electrodynamic effects in atoms and molecules, Electron correlation calculations for atoms and ions: excited states, Cooler storage rings

1. Introduction

It was in the middle of the 80's when the experimental techniques for studying electron-ion collision processes started a rapid development which then occurred during the past 15 years. Before that it was barely possible to observe the most fundamental electron-ion reactions in the laboratory. Experiments with conventional, merged beams or crossed beams of electrons and ions, by so called 'single passage' arrangements were done in different laboratories. Then, in the end of the 80's the advent of heavy-ion cooler storage rings[1, 2] and electron beam ion traps(EBIT)[3] caused revolutionary developments. These devices allow investigations of reactions between electrons and ions in almost

F.J. Currell (ed.), The Physics of Multiply and Highly Charged Ions, Vol. 1, 231-268.

any charge state with high resolution, signal-to-background ratio, and luminosity. Primarily radiative recombination, dielectronic recombination, laser induced recombination, and dissociative recombination [1, 4] were studied. Some work on electron impact excitation and ionization has been done as well.

The most fundamental process in the interaction of free electrons with ions is radiative recombination (RR). The photon is directly emitted with the capture of an electron into the quantum state n of the ion $A^{(q-1)}$ (process 7.1).

$$A^{q+} + e^- \longrightarrow A^{(q-1)+}(n) + h\nu \tag{7.1}$$

$$+ \; e^- \longrightarrow A^{(q-1)+}(n, n') \longrightarrow A^{(q-1)+}(n'') + h\nu \,, (q < Z) \tag{7.2}$$

$$+ \; 2e^- \longrightarrow A^{(q-1)+}(n) + e^- \,. \tag{7.3}$$

Where Z is the nuclear charge and q is the charge state.

In process 7.2, called dielectronic recombination (DR), a free electron is captured simultaneously with the excitation of a bound electron in the projectile. Due to energy conservation, the binding energy plus the kinetic energy of the captured electron must equal the excitation energy of the bound electron. The resulting doubly excited state (n, n') will have a very large probability to autoionize and loose the electron again, but it may emit a photon and end up in a singly excited state. This last step completes dielectronic recombination.

For completeness, we mention three-body recombination (TBR, process 7.3), where one electron recombines in the vicinity of another electron by transferring energy and momentum to it[5]. TBR can be viewed as the time reverse of electron-impact ionization; RR and the first step of DR is the time reverse of photoionization and of the Auger effect, respectively. Due to its dependence on the electron temperature and density, TBR can play an important role for the recombination rate in cold, dense electron plasmas. Replacing the atomic ion A^{q+} by a molecular ion AB^+ one can have additionally dissociative recombination. In this process a low energy electron collides with a stable molecular ion and causes its dissociation by neutralizing it.

The interest in these electron impact phenomena with bare or few electron ions and molecular ions is due to several practical applications and to very fundamental questions. Electron-ion recombination processes appear as important phenomena in astrophysical plasmas[6], the chemistry of interstellar clouds[7], modelling of supernova reminiscence, and earth atmosphere, and in fusion plasma[8]. Recombination can cause significant energy losses from plasma as being a source for radiation. This radiation has proven on the other hand to be a valuable tool for plasma diagnostics. In fact, dielectronic recombination was postulated to explain discrepancies in the ionization balance of the solar corona[9, 10]. Much of the energy transport and reactions in these plasma and media occur as electrons collide with atomic and molecular ions. In such collisions, the ions can be excited or further ionized, or the electrons can recombine leading to emission of photons, excitation of ions or dissociation of molecules. Electron-ion recombination occurring in fusion or astrophysical plasma are needed for their spectroscopic investigations.

On the fundamental level, studies of electron-ion impact phenomena can be very useful for developing our understanding of the structure and decay modes in many electron systems. In recent years recombination studies obtained a decisive role also in fundamental atomic spectroscopy [11–17]. This is due to the new possibilities given by coolers in storage rings and by electron-beam ion traps which allow measurements of electron-impact ionization and recombination with unprecedented resolution and luminosity. They can serve e.g. as testing grounds for highly accurate calculations of energy levels in few electron ions. As we will see in this article the present day most accurate determination of energy splittings in few-electron ions are done by electron-impact experiments. These allow the tests of the treatment of relativistic, quantum electrodynamical (QED), and correlation effects in many-body atomic systems.

2. Experimental Methods

The reasons that EBITs and cooler-storage ring facilities offer unique properties for the study of electron-ion reactions are manifold: In these devices, ions can be accumulated and confined under excellent vacuum for bombardment with electrons. The devices for production of ions in EBIT and for cooling ions with electrons in a storage ring provide intense, high-quality electron beams. This results in a high luminosity with low background and excellent resolution. With an EBIT, in particular, photon spectra from the electron ion reactions can be observed in high resolution. It has additionally the advantage of allowing high electron impact energies. Storage rings, in contrary, have the strength in low impact energies. The disadvantage of not being able to perform photon spectroscopy of recombination in storage rings at present, is partly compensated by an excellent energy resolution and by the selectivity in ion charge states as well as the easy detection of recombined ions. Otherwise, EBITs have, in comparison, compact designs with low investment and operation costs and in many aspects of electron-ion collisions they are complementary to storage rings.

In storage rings, ions are kept rotating with high velocity (up to $\sim 50\%$ of the speed of light) within a 50 to 100 meter circumference vacuum tube guided by magnetic fields. Storage times from seconds to several days (depending on the electronic structure of the ions and the vacuum in the ring) have been observed for high circulating ion currents ranging from μA to mA. In these machines one has thus the possibility to work with a well known number of ions (typically 10^3 to 10^{10}) at well defined speed and charge state. With the capability of storing ions one can also cool them i.e. make the beam monoenergetic and reducing its angular divergence and geometrical size. This gives access to enormous improvements in spectroscopy of transitions and resolution in reaction channels. Stored particles can be cooled, in general, by stochastic or resistive cooling, collisional cooling (buffer gas cooling, electron cooling, sympathetic cooling), laser cooling, and synchrotron radiation cooling. With highly-charged heavy ions only electron cooling is practical, and it is a very powerful method for cooling. The electron cooler is also the device used for studies of electron-ion recombination. The high velocity has the advantage of a kinematic expansion of the energy scale in the electron-ion center-of-mass system. Thus leading to a high energy resolution. The disadvantage is a limitation to quite low collision energies. A recent development of using expanded electron beams has additionally reduced their velocity spread. The ability to store ions for seconds or more can in some cases be exploited to reduce molecular vibrational states, to eliminate atomic metastable states, or to measure their lifetimes.

Heavy ion cooler storage rings at present operating are; ASTRID in Aarhus [18], CRYRING in Stockholm [19], ESR at GSI Darmstadt [20, 21], and TSR in Heidelberg[22]. For storing heavy ions, the vacuum has to be in the low 10^{-11} mbar pressure region in order to obtain an adequate beam lifetime and low background counts. For example at CRYRING, of the Manne Siegbahn Laboratory, Stockholm, ions from a plasmatron ion source or for high charge states (q) from an electron-beam ion source (EBIS) are injected into the storage ring. Starting at 5 - 50 keV x q, they can either be injected directly into the ring or after preacceleration to 300 keV/amu in a radio-frequency quadrupole (RFQ). After storing the ions, they are further accelerated to $96(q/A)^2 MeV/amu$ maximum energy (A is the atomic or molecular weight number). The motion of ions in the ring is governed by magnetic fields which provide the confining forces on the ions. Usually, a periodic structure of magnetic elements: dipole magnets provide the centripetal force, and higher order magnetic multi-pole fields the focussing force for radial confinement[23]. Between the magnets there are drift lengths containing the beam injection elements, the device for acceleration, the electron cooler, and targets, as well as detector installations for experiments. This structure, which repeats itself periodically, is called superperiod. For example CRYRING has a (bend, focus, defocus, focus, bend, drift) - superperiod, with six of such superperiods, a sixfold symmetry. After injection, the ions fill the whole aperture of the ring. Acceleration decreases the beam emittance somewhat, but it is the possibility of cooling the stored ions which makes storage rings to powerful tools providing intense beams of low emittance and high energy definition.

A schematic view of the electron cooler installed at CRYRING[24] is shown in Fig. 7.1. The electron beam is made by thermal emission of the electrons from a cathode at $T_{\mathrm{K}} = 1000^o K$ which corresponds to an electron energy spread of $kT_{\mathrm{K}} \approx 0.1\ eV$ (k is here the Boltzmann constant) . Acceleration to the energy $E_{\mathrm{e}} = (m_{\mathrm{e}}/2)v_{\mathrm{e}}^2$ reduces the longitudinal temperature by a large

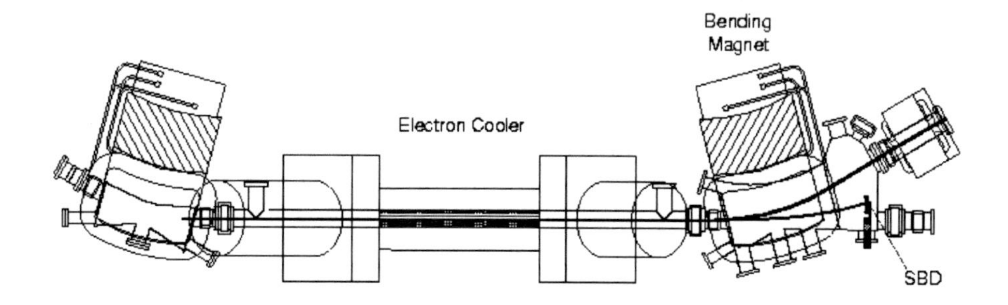

Figure 7.1. The experimental setup showing schematically the section of CRYRING with the electron cooler and actuators for placing detectors into the ultra-high vacuum system. The recombined atoms or ions are separated from the stored ion beam in the ring by the first bending magnet after the electron cooler and detected with surface-barrier detectors (SBD), channel plates, gas counters, or scintillators.

factor $T_{e\|} = \frac{kT_K^2}{4E_e}$. The effective longitudinal temperature is, due to space charge relaxation between electrons in the beam, the so-called longitudinal-longitudinal relaxation[24, 25], increased to $T_{e\|} = 1.6e^2 n_e^{1/3}/k(4\pi\epsilon_0)$. The electron gun in the present version of the CRYRING cooler uses a two-stage acceleration: After the first anode, the electrons go through a 25 cm long metal tube before they are accelerated to the final energy. Inside this tube, the electron density is around $10^9\ cm^{-3}$. Relaxation should then give $T_{e\|} = 2 \cdot 10^{-4}$ eV$/k$. At the second acceleration stage, from e.g. 500 eV to final energy, the longitudinal energy spread becomes reduced by a factor equal to the ratio between the two energies to $T_{e\|} \leq 0.1$ meV$/k$. A further relaxation takes place after the second acceleration, increasing $T_{e\|}$ by an almost negligible amount. Almost all electron coolers have typical values of $T_{e\|} \sim 0.1 - 1$ meV$/k$ for the longitudinal temperature. The transverse electron temperature would be $T_{e\perp} \sim 0.1$ eV$/k$. Recently, a big step towards lower transverse energy spread of the electron beam was achieved[24]. In the interaction region a magnetic field with a strength B in the order of $0.02 - 0.05$ T prevents the electrons from diverging. By immersing the electron gun in a strong magnetic field, of up to factor 100 higher value, the electron beam can be expanded adiabatically by the same factor in the guiding magnetic field. The transverse temperature decreases by the same factor. Almost all electron coolers have expanded electron beams nowadays. At CRYRING the expansion factor is up to 100 (which can result in a decrease of $T_{e\perp}$ to 1 meV/k), TSR has expansion of up to 30, ASTRID has expansion of up to 20.

The electron velocity distribution $f(\vec{v}_e)$ is described by a flattened Maxwellian distribution:

$$f(\vec{v}_e) = Cexp\left(-\frac{m_e}{2}\frac{v_{e\perp}^2}{kT_{e\perp}} - \frac{m_e}{2}\frac{(v_{e\parallel} - v_{rel})^2}{kT_{e\parallel}}\right), \tag{7.4}$$

where $C^{-1} = (\frac{2\pi}{m_e})^{\frac{3}{2}} kT_\perp (kT_\parallel)^{\frac{1}{2}}$ is a normalization constant and v_{rel} is the average longitudinal center-of-mass velocity (see below).

The ion beam is then completely immersed in the constant density electron beam. Typical electron densities are on the order of $n_e \sim 10^7\ cm^{-3}$. Cooling occurs by collisions with the low-temperature electrons as the ions pass through the cooler $\sim$ a million times per second as they circulate in the ring. Thus, at thermal equilibrium, the ion-beam energy spread will be reduced from MeV to a few eV. At CRYRING electron cooling has been applied both to atomic and molecular ions at energies between 290 keV/u (the injection energy) and 24 MeV/u. The relative momentum spread of the ion beam after cooling is usually $5 - 10 \cdot 10^{-5}$ and occasionally somewhat smaller, dependent on the ion density, charge, and mass.

The reduction of the momentum spread during cooling is seen in the longitudinal Schottky noise spectrum[23]. The area of the Schottky peak remains nearly constant, whereas its width decreases with increasing signal height, demonstrating directly the concentration of the particles in momentum space by cooling. For electron-ion recombination experiments at storage rings, the longitudinal Schottky noise spectrum is essential for the calibration of the energy scale. The difficulty in the calibration lies in the determination of the actual electron-ion relative energy in the collision. With the peak in the Schottky spectrum one has the mean ion circulation frequency in the ring. If one knows the circumference, the ions take in the ring, also their velocity is known. Under the condition of electron cooling the ions and electrons are forced to have the same velocity and thus the electron velocity is also known.

In the recombination experiments the recombined atoms or ions are separated from the stored ion beam in the ring by the first bending magnet after the electron cooler and detected (see Fig. 7.1). This is done with surface-barrier detectors (SBD), channel plates, gas counters, or scintillators. Fig. 7.1 indicates also a manipulator used to insert the detectors in the ring vacuum. The advantage of SBD is that they have detection efficiency unity and reasonable good energy resolution at the high energy of the ions. Channel plates, gas counters, and scintillators have a detection efficiency dependent on the ion charge and energy and do not have a good energy resolution; however, they do have the advantage of high irradiation thresholds. Position sensitive channel plates are useful to monitor the spatial distribution of detected atoms. In that way, valuable information on beam cooling can be acquired and used in optimizing the alignment of electron and ion beams [26].

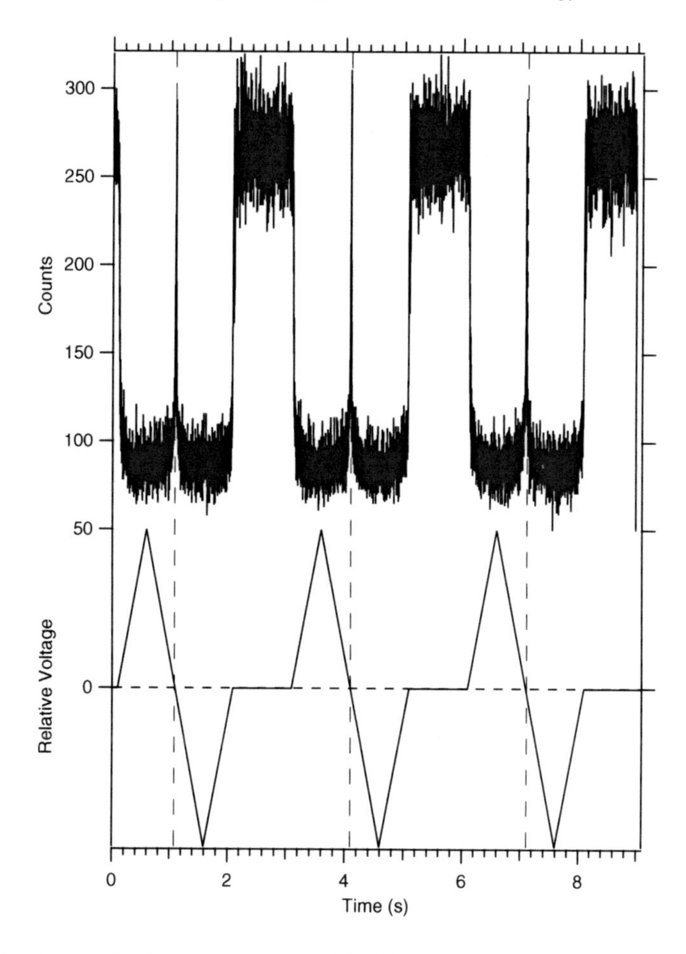

Figure 7.2. Recombination rate measured with 8 Mev/amu D^+ (upper part) when the cathode voltage U_{cath} of the electron gun is scanned from its value at cooling to higher and lower values as function of time (lower part). Relative voltage = 0 corresponds to the value of U_{cath} at cooling.

At cooling condition the mean ion and electron velocities are equal, which defines the electron cooling energy as $E_{cool} = (m_e/m_i)E_i$. Here E_i is the energy of the stored ions, whereas m_e and m_i are the electron and ion mass, respectively. After cooling the ion beam, the electron energy can be changed by a certain amount ΔE which results in a center-of-mass energy: $E_{\rm cm} \approx \frac{\Delta E^2}{4E_{\rm e}}$. An example for the recombination rate measured with 8 Mev/amu D^+ when ΔE, i.e. the cathode voltage (U_{cath}) is scanned from its value at cooling to higher and lower values as function of time is shown in Fig. 7.2. It is obvious that the recombination rate maximizes very sharply when electron and ion velocity matches. At the crossing points of U_{cath} with the cooling values

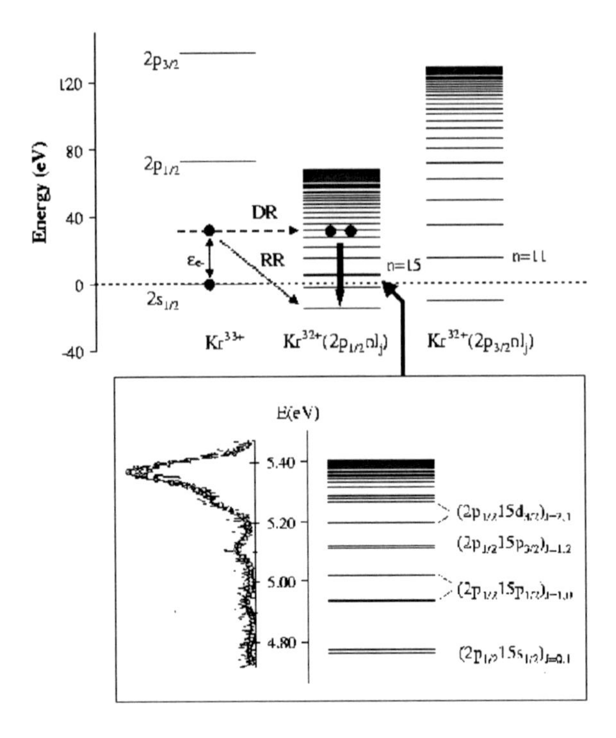

Figure 7.3. Top: The upper part shows schematically the RR and DR processes of a free electron and a Li-like Kr^{33+} ion. In both cases the final state is the electron bound to the ion and an emitted photon. The energy levels of Li-like Kr^{33+} (initial ion) and Be-like Kr^{32+} (the recombined ion) are shown on the same energy scale. The doubly excited states $Kr^{32+}(2p_{1/2}n\ell_j)$, with $n \geq 15$, as well as $Kr^{32+}(2p_{3/2}n\ell_j)$, with $n \geq 11$, are situated above the ground state of Kr^{33+}. The states in these manifolds have been accurately calculated while energies of other states are only indicated with Rydberg's formulae. This gives rise to a slightly thicker energy-level line and a small deviation of the $1/n^2$ distribution of the energy levels. Bottom:In the lower panel the doubly excited states $Kr^{32+}(2p_{1/2}15\ell_j)$ states are shown in more detail together with the experimental spectrum.

(relative voltage = 0) the count rate of D^+ peaks at the same value reached at cooling. From this sensitivity of the recombination rate on the relative energy one can also understand that it can be exploited effectively for optimizing the alignment of electron and ion beams.

For obtaining such spectra as shown in Fig. 7.2, each detected recombination event triggers the recording of U_{cath}, the time t related to the scanning trigger, the particle detector pulse height, stored ion number, and so on in event mode. From U_{cath} and the Schottky frequency one can obtain the parameters E_e and E_i, respectively. When derived from such values E_cm has, however, not a high accuracy. Different ways to get E_{cm} with high accuracy, which were developed in experiments at CRYRING, are described in Sec. 7.4 below.

3. Theory

In this section we will discuss how the cross sections for the two recombination processes, RR and DR, depend on the collision energy and the properties of the ion. We will also outline how recombination can be treated theoretically. In Sec.7.5 below examples of calculated spectra are shown and compared to experimental results.

The two processes are schematically depicted in Fig. 7.3 for the case of recombination of Li-like Kr^{33+} where the active bound electron is in $2s$ (see also Sec. 7.5.2). Fig. 7.3 shows the energy levels of Li-like Kr^{33+} (initial ion) and Be-like Kr^{32+} (recombined ion) on the same energy scale. When the electron collides with Kr^{33+} a certain amount of kinetic energy, ε_{e-}, is brought into the system. If the energy of the electron-ion system now matches a state in Kr^{32+} it can transform into this state. The new state is a very short lived, autoionizing state, and in most cases it will decay back the same way it was formed. If a photon is emitted, however, the system may end up below the ionization limit of Kr^{32+} (thick arrow) and then recombination is completed. This is the DR process. In the RR process the systems emits a photon and proceeds directly to a bound state of Kr^{32+}.

It is worth noting that the RR process as well as the DR process starts with a system of a free electron and an ion and ends with a system where the electron is bound to the ion and a photon has been emitted. The only difference between the processes is the intermediate step present in the DR process. From first principles it can then be argued that RR and DR should be treated as one single process and that the division into two processes is artificial. Eventually it will indeed be necessary to treat them as one process, as we will discuss below, but in fact no system has yet been found where it makes a visible difference with today's experimental resolution. In most practical cases the division into two processes is on the contrary natural. The reason is that RR and DR often contribute for rather different center of mass energies and that they prefer different final states. We will now discuss these differences.

3.1 Radiative Recombination

RR is a non-resonant process, possible also for bare ions, which scales as the inverse of the relative energy between the ions and the electrons and as the inverse of the n-quantum number of the state to which the electron recombine. This was shown by Kramers [27], who treated recombination of bare ions with a semi-classical approach already in the 20's. Bethe and Salpeter derived the radiative recombination cross section under basically the same assumptions using quantum mechanics and they obtained the same relation as Kramers, which for recombination into state n is:

$$\sigma_n^K(\varepsilon_r) = 2.11 \cdot 10^{-22}[cm^2]\frac{E_{1s}^2}{n(E_{1s}+n^2\varepsilon_r)\varepsilon_r}, \tag{7.5}$$

where E_{1s} stands for the ground-state ionization energy of the hydrogen-like ion. A fully quantum mechanical derivation of the radiative recombination cross section with non-relativistic wave functions was done by Stobbe [28]. In all calculations the emission of the photon was described within the dipole approximation, which should be reasonable for the cases considered here, as long as $h\nu \ll 2Z \cdot Ry/\alpha$ (α the fine structure constant), or differently expressed; as long as the wavelength of the photon is much larger than the size of the ion. The expressions given by Stobbe for the radiative recombination cross sections can lead to rather tedious numerical calculations if a sum over many n up to high quantum states needs to be performed. One can then use the so-called Gaunt-factor which connects Kramers results with Stobbe's quantum mechanically correct expression $\sigma_{rr}^{nl}(E_r)$: $g_{nl}(E_r) = \frac{\sigma_{rr}^{nl}(E_r)}{\sigma_{rr}^{K}(E_r)}$. For most cases the l averaged approximative Gaunt-factor $g_n = \frac{1}{n^2}\sum_l (2l+1)g_{nl} = 1 + 0.1728n^{-2/3}(x-1)(x+1)^{-2/3} - 0.0496n^{-4/3}(x+1)^{-4/3}(x^2+4x/3+1)+...$, where $x = E_r/E_{nl}$, from Ref.[29], is a useful approximation (deviation from Stobbe $< 1\%$).

An example for observation of RR is demonstrated in Fig. 7.4, where the recombination rate coefficient for D^+ is shown. It was measured with three different transversal temperatures of the electron beam (100 meV, 10 meV, and 1 meV). Some of this data (1 meV) is derived from the raw spectra shown in Fig. 7.2. The RR cross section (Eq. 7.5 with Gaunt-factor as just discussed) is folded with the electron velocity distribution (Eq. 7.4) as will be discussed below, Eq. 7.14 . Both theory and experiment are in absolute scale. They agree well except for the 1 meV temperature at very low energy ($<$ 1 meV) where the experimental rate coefficient is here enhanced by nearly a factor 2. This enhancement at low relative energy is observed for every ion, although its size varies to some extent [30–32]. Its origin is not clear at present.

As seen in Eq. 7.5 and the data of Fig. 7.2 and Fig. 7.4, the RR cross section peaks at zero relative energy and decays smoothly when the energy is increased. The first observation is not surprising; the probability for recombination should approach unity when the electron and ion have no relative velocity. The second observation tells us that the ground state of the recombined ion is the dominating final state. RR occurs for bare and non-bare ions. The situation is very different for the DR process.

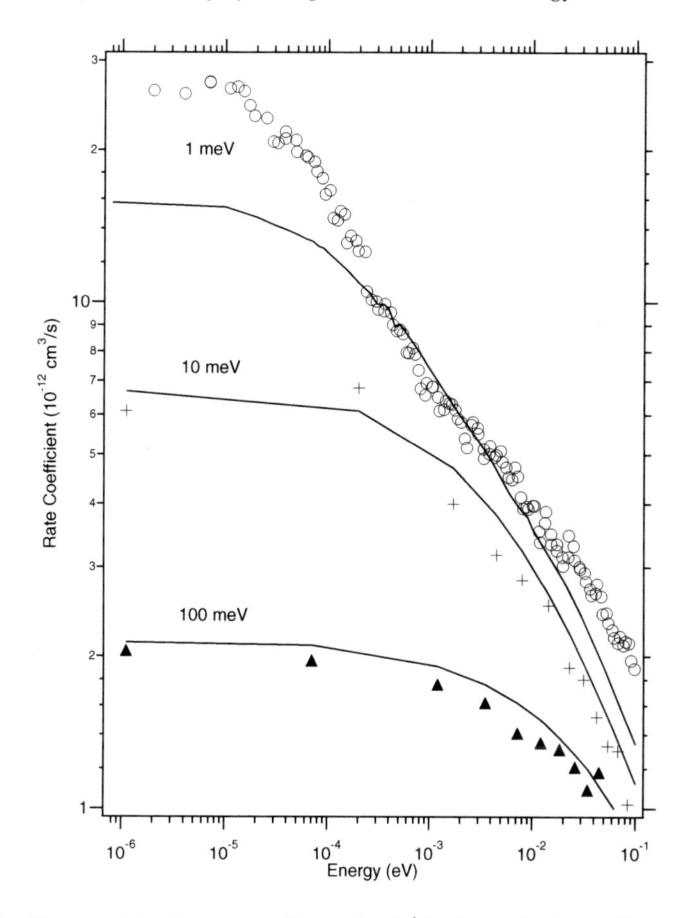

Figure 7.4. The recombination rate coefficient for D^+ is shown for three different transversal temperatures of the electron beam (100 meV, 10 meV, and 1 meV). The lines represent the RR cross section folded with the electron velocity distribution (see text).

3.2 Dielectronic Recombination

DR is a resonant process. When the energy of the free electron is tuned through a doubly excited state this state will be formed with large probability. Although the lifetime of this state, governed by Auger decay, is short it prolongs the interaction time between the ion and the electrons drastically and increases the probability for photon emission leading to stabilization. A doubly excited state has generally much larger overlap with some singly excited bound states then with the ground state and recombination into excited states will dominate.

If dielectronic recombination is regarded as a two-step process its probability will be the product of the probability of the two steps and the integrated cross section, or the strength, S, of the process can be written as;

$$S = \int \sigma(\varepsilon_r) d\varepsilon_r = \frac{\hbar^3 \pi^2}{2m_e \left(E_d - E_{ion}\right)} \frac{g_d}{g_i} \frac{A^a_{i\to d} \sum_s A^{rad}_{d\to s}}{A^a + \sum_s A^{rad}_{d\to s}}. \tag{7.6}$$

The strength is proportional to the capture rate $A^a_{i\to d}$ into the doubly excited state d, and to the probability of state d to decay radiatively to a non-autoionizing level. The multiplicity of the intermediate doubly excited state is given by g_d and that of the initial target state by g_i. If the ground state of the studied ion is of $S_{1/2}$-symmetry, as in all the cases discussed here, then $g_i = 2$. $A^a = \Gamma^a/\hbar$ is the total autoionization rate from the doubly excited state d, and $A^{rad}_{d\to s}$ is the radiative transition rate from level d to a level s below the ionization threshold.

The cross section itself should (we are still regarding DR as a two-step process) be in the form of a Lorentz profile

$$\sigma(\varepsilon_e) = \frac{1}{\pi} S \frac{\Gamma/2}{(E_d - E_{ion} - \varepsilon_e)^2 + \Gamma^2/4}, \tag{7.7}$$

where $E_d - E_{ion}$ is the position of the doubly excited state compared to the ground state in the initial ion, cf. Fig. 7.3 and Γ is the natural life time of the doubly excited state

The radiative transition rate is in the dipole approximation is given by

$$A^{rad}_{d\to s} = \frac{1}{g_d} \sum_{M_s, M_d} \frac{e^2}{4\pi\varepsilon_0} \frac{4}{3\hbar} \left(\frac{\omega}{c}\right)^3 \langle \Psi_s^{J_s M_s} \mid \mathbf{r} \mid \Psi_d^{J_d M_d} \rangle^2 \tag{7.8}$$

with $\omega = \left(E_d - E_s\right)/\hbar$, where E_s is the energy of a specific final state. The sum in Eq. (7.8) is over all magnetic substates of the two levels d and s. The last denominator in Eq. (7.6) equals the total transition rate, $\Gamma/\hbar$, from the doubly excited state. If this state can autoionize only to the ground state of the target ion then the autoionization rate, A^a, in the denominator equals the capture rate, $A^a_{i\to d}$, in the nominator. As can be seen in Eq. (7.6), the strength is inversely proportional to the position of the resonance relative to threshold (i.e. relative to the initial state in the target ion), $(E_d - E_{ion})$, and depends crucially on the slowest type of decay of the doubly excited state through the ratio $A^a A^{rad}/(A^a + A^{rad})$. For light to medium heavy systems the radiative rate is usually the slowest, and it is thus this rate which determines the recombination rate. Exceptions to this rule may occur when the doubly excited state is very asymmetrically excited. Lithium-like krypton, presented below, is an interesting case where the radiative and autoionization rate are of equal importance. The recombination rate is then more directly dependent on the autoionization rate.

3.3 Computational Method for Dielectronic Recombination Resonances

To describe the DR process several properties of doubly excited states have to be predicted. Such states are generally highly correlated and thus rather demanding to calculate. They are also autoionizing, i.e. they are embedded in the continuum, and this property requires special considerations. An experimental spectrum covers in addition often several hundred doubly excited states, especially near the series limit, which all have to be accounted for to reproduce the spectrum. Several computational approaches, with slightly different goals, have been developed for DR calculations. There is an obvious interest to be able to describe most ions on a wide energy range, even if the prediction of energy positions is of modest accuracy. Other methods focus on an accurate many-body description suitable for comparison with high accuracy data showing resolved resonances. In Sec. 7.5.1 below examples of calculations with different methods will be shown. Here we will only give a brief background to the many-body method we have developed in connection with the CRYRING experiments.

Many-body perturbation theory (MBPT), is a method with a high potential for accuracy, especially in its all order formulation. It is also very well suited for calculations on positive ions. With the approach, implemented for bound states by Salomonson and Öster, the non-relativistic two-electron problem [33] is solved essentially exactly in an iterative procedure. The relativistic counterpart (RMBPT) [34] starts with the Dirac equation , neglects radiative corrections but includes all relativistic effects within the so called no-virtual pair approximation (i.e. neglecting virtual electron-positron pairs). This procedure includes all effects to order α^2 Ry. For systems with more than two-electrons Salomonson and Öster have implemented the coupled-cluster single- and double excitation scheme (CCSD) [35], which is an approximation to the full many-particle problem in that it neglects true three- and four etc particle excitations. This method is well suited for lithium- or beryllium-like systems where the two $1s$-electrons provide a relatively inert core. This CCSD approach has also been used for relativistic calculations, e.g. by Lindroth and Hvarfner [36] for beryllium-like iron and molybdenum. In Ref. [36] it is also explained how the Breit interaction, accounting for magnetic interaction and the retardation of the electron-electron interaction, can be treated on equal footing with the ordinary Coulomb interaction between the electrons. This approach is now customary.

In order to be able to use many-body perturbation theory also for autoionizing states it has been combined with complex rotation both in the non-relativistic case [37, 14] as well as in the relativistic case [38, 39, 17]. Complex rotation has since a long time been used by many groups to account for the instability of autoionizing states, see e.g. Refs. [40–42]. Through the scaling of the radial variable $r \to re^{i\theta}$ the Hamiltonian is transformed or *rotated* into the complex plane. Solution of this transformed Hamiltonian gives directly the autoionization width of the doubly excited state as the imaginary part of a complex binding energy. The real part of the energy corresponds to the position of the state. RMBPT combined with complex rotation will thus provide energy positions of the resonances as well as the autoionization rate. The radiative rate can be calculated from Eq. 7.8 and finally the strength can be obtained from Eq. 7.6.

3.4 Symmetric or Asymmetric Resonances?

As mentioned above, Eq. 7.6 is an approximation in that we regard the recombination process as a two-step process. In principle we do not know if the systems passed through the doubly excited state or if it went directly from the initial state to the final state, i.e. the two pathways are indistinguishable and their probability amplitudes have to be added before we calculate the probability to proceed from the initial to the final state. The probability amplitudes will generally interfere with each other. This is in analogy with the time reverse process where direct photoionization interfere with ionization through a doubly excited state and where the signature of interference is asymmetric line profiles, so called Fano profiles [43]. In principle asymmetric resonances are expected for the recombination process as well, although there is no conclusive experimental evidence for it so far. The observation is hindered by several reasons. Significant interference requires the two pathways to be of similar importance and although it is easy to find asymmetric DR-profiles in calculations these appear often when the intermediate to final state transition, is weak, i.e. where direct radiative recombination is relatively more important. Since most doubly excited states also have stronger radiative decay channels and recombination through these occur for the same electron energy the weak asymmetric resonance will be just a small correction on top of a large symmetric resonance. Only a small fraction of all ions have yet been studied in high precision recombination experiments, eventually it will certainly be possible to perform an experiment which can demonstrate asymmetric recombination profiles.

3.5 Unified Treatment of RR and DR

The cross section for the RR process is for bare ion very well described by the classical formulae by Kramers, Eq. 7.5, modified with the Gaunt-factor as discussed above. The formulae can be used also for non-bare ions if the hydrogen-like energies assumed in Eq. 7.5 are modified. Many experiments are well described when such an approximation of the RR cross sections is added to a DR cross section obtained from Eq. 7.7, see e.g. N^{4+} below. For systems which have resonances very close to zero relative energy it is, however, logical to employ a method which treats both processes in a unified way. Interference between the processes is then automatically accounted for. In Ref. [44] it is described how a discretized continuum can be calculated together with the doubly excited states and how the cross section for recombination into a specific bound state, s, can be written as:

$$\sigma_s(\varepsilon_r) = \frac{1}{g_i}\frac{e^2}{4\pi\varepsilon_0}\frac{4\pi}{3}\frac{\omega}{c}\left(\frac{\hbar\omega}{c}\right)^2\frac{1}{2m_e\varepsilon_r}\times$$

$$\mathrm{Im}\left(\sum_n \frac{\langle\Psi_s \mid \mathbf{r}e^{i\theta} \mid \Psi_n\rangle\langle\Psi_n \mid \mathbf{r}e^{i\theta} \mid \Psi_s\rangle}{E_n - E_{ion} - \varepsilon_r}\right). \tag{7.9}$$

The sum over n runs over all states, the doubly excited states as well as those which represent the continua. In this way both recombination through resonances as well as through direct radiative recombination are included. The energies E and state vectors, Ψ, are eigenvalues and eigenvectors to the complex rotated Hamiltonian. Both the denominator and the nominator of the last term in Eq.7.9 are thus generally complex numbers.

Note that the contribution from a specific term in the sum in Eq. 7.9 may very well be in the form of a Lorentz profile. This will be the case for a doubly excited state n, with $E_n = E_d - i\Gamma/2$, when the matrix element in Eq. 7.9 has a vanishingly small imaginary part. An integration over ε_e will then reproduce Eq. 7.6 with $A^a A^{rad}/(A^a + A^{rad})$ replaced by A^{rad}. When the imaginary part of the matrix element is comparable in size to the real part the cross section profile will instead be asymmetric and this would be a sign of interference between radiative recombination and dielectronic recombination.

4. Rate coefficients as function of energy in absolute scale

4.1 Absolute rate coefficients

The recombination rate coefficient α_{exp}, derived from first principles, for a number of stored ions N^i reacting with an electron target is:

$$\alpha_{exp}(t) = \gamma^2 \frac{N_t^{corr}/\delta t}{n_e\, N^i\, (\ell/L)}, \tag{7.10}$$

where γ is the Lorentz factor and L stands for the ring circumference. The detected ion rate per time unit δt at time t of the scan ($N_t^{corr}/\delta t$) is correcting for the electron capture background and detection efficiency. The background amounts to typically a few percent at the pressure of 10^{-11} Torr. The number of circulating ions is calculated from the relation $N^i = \mathcal{I}/(e\, f_s)$. One gets the ion circulation frequency f_s from the Schottky noise detector and the current $\mathcal{I}$

of the ion beam is measured with the current transformer. In CRYRING e.g., the ions are merged with electrons over a length of $1m$. The effective length of the interaction in recombination experiments is $\ell = 0.82 \pm 0.02\,\mathrm{m}$. This is determined in the following way: The deflection of the electron beam at both ends of the beam overlap-region in the electron cooler induces a dependence of the relative energy upon the position in the field fringe regions. Considering the mapping of the magnetic field [45] in the electron cooler, we have presently arrived at the above estimate for the effective cooler length ℓ. For the ion orbit length L one can take here the nominal length (from the ring construction) of 51.6 m. The values of α_{exp} have a systematic error of typically around 10%, originating mainly from the uncertainty in the ion current.

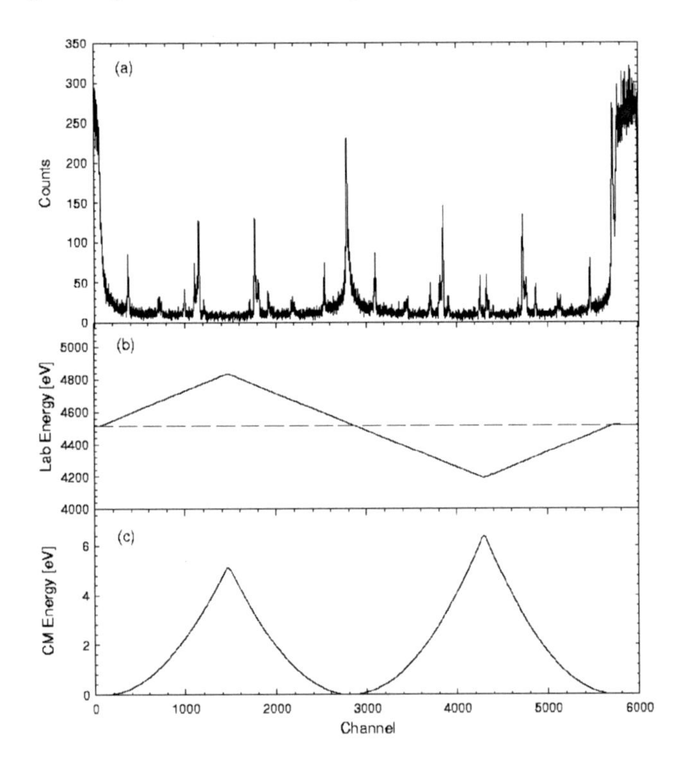

Figure 7.5. Scans of the recombination rates for the stored sodium-like Ni ions. Top: counts of recombined ions with two spectra where the electrons moved faster and two where they move slower then the ions. Note that the high peak at the center, which is due to non-resonant recombination, defines precisely the time and electron energy at which the electron velocity equals the ion velocity. middle: variation of cathode voltage, bottom: center-of-mass energy as function of scanning time.

In Fig. 7.5 the scheme of data taking is shown for the case of Ni^{17+}[46]; (middle) the detuning of the cathode voltage from transformed cooling is shown.

The lower part of this figure displays the corresponding CM energies. At the top the four spectra are displayed, two where the electrons moved faster and two where they move slower then the ions. One does not record the recombined ion spectrum as a function of cathode voltage, instead in a histogram as function of time ("time spectrum") (see also Fig. 7.2). The "time spectrum" is superior to the former for the following reasons: First, the time signal is digitally generated which is free of pickup noise and the total number of channels in a time spectrum is independent of the energy scan range. This allows one to preserve the high resolution gained from the low electron temperature. Second, the time spectra contain information revealing the variation of the ion energy during the scan of the cathode voltage, which is used in the drag force correction (see below). Finally, it is convenient to derive the rate from it as each channel has equal time length (see Eq. 7.10).

4.2 Absolute energy scale

In the first part of this section we describe how the absolute energy scale of recombination spectra is obtained by a standard calibration method (method I). Then in the second part a more accurate calibration method (method II) is described. In method II the time spectra are converted into energy spectra by converting each time channel into the corresponding cathode voltage, which is derived by averaging the recorded cathode voltages over all individual events associated with that time channel. Due to the distorted response of the cathode to its power supply, an inadequate error will be introduced when deriving these energy values simply from the function generator that drives the power supply. The energy spectra derived from the time spectrum are checked with the energy spectra obtained directly from the data by comparing the corresponding resonance peaks. Mismatches are found to be within one time channel, so that the extra error introduced by the conversion is negligible. The total energy error caused by the conversion from time channel to cathode voltage is estimated to be 0.4% within the energy range of 1 eV to 100 eV in the center of mass frame.

One still has to correct the electron kinetic energy for the space charge of the electron beam:

$$E_{\rm e} = eU_{cath} - (1-\zeta)I_c r_c m_c c^2/ev_{c0}[1+2ln(r_1/r_2)-(r/r_2)^2], \quad (7.11)$$

where r_1 and r_2 are the radii of the beam tube and of the electron beam, respectively. The classical electron radius is denoted by r_e and r is the ion beam displacement from the electron beam axis. When the two beams are coaxial then r=0. The parameter ζ represents here the contribution of trapped ions to the space charge potential. Those ions can be produced by ionization both from the electrons as well as by the ion beam. Even if those ions are not trapped (i.e. they are "cleared" by electrodes), their steady-state density can influence the absolute energy calibration. Generally the density of the trapped ions can be a function of the electron energy and time because their creation rate is electron energy dependent and their trapping potential depends on the electron density. However, in the case where the energy scan covers only a narrow range like in most of these experiment, ζ can be regarded as being constant. Another source of uncertainty for the electron-ion energy scale is the above-mentioned exact position of the ion beam relative to the electron beam axis. The condition r=0 is assumed in most experiments with the argument that the ion beam is only 1 mm in size after cooling and is aligned to the axis of the electron beam by minimizing the Schottky frequency.

The compensation parameter ζ can be determined, in principle, by inserting the cathode voltage U_{cath} and electron energy E_{e} at cooling into the above formula(Eq. 7.11). The value of E_{e} can be determined from the ion velocity at cooling and the latter can be derived from the Schottky frequency and the length of the ion orbit L. Thus, accurate determination of the ion velocity relies on the knowledge of the orbit length. In the model used here, the orbit length and trapped ion density are interconnected. If the trapped ion density is proportional to the space charge potential that confines the ions, the value of ζ should not differ very much for two measurements where the electron energies at cooling are equal. One can thus check the values of ζ and get an estimate of the value of L and its systematic error. For CRYRING it was found that L=5168.5 cm is an adequate estimate. This value is longer than the geometrical length of the ring along its axis by about 4 cm which can be within the possible variation of the orbit in the ring apertures. In experiments of laser-induced recombination [47] this value was checked by the well known $2p$ binding energy in hydrogen. It was found there to be higher then the geometrical length by the same amount. Typical values of ζ are found here to be varying from 0.7 to 0.8 for $I_e = 250$ mA to 50 mA, respectively, i. e. 20 to 30 $\%$ compensation of the space charge by trapped ions.

The very low temperature of the electron beam introduces a strong drag force on the ions at small detuning velocities. This disturbs the transformation from laboratory to CM energies. For its correction, we have calculated the change of the velocity of the ions as a function of time during the scans by numerical solution of the differential equation describing the beam acceleration due to multiple Coulomb collisions in the electron beam (for details see Ref.[48]). With the inclusion of this correction the energy scans from the laboratory system can be transformed into the center-of-mass system. The four spectra obtained from the zig-zag scans (see Fig. 7.5) should be identical in the C.M. frame. In order to check these corrections, one selects the two best pronounced peaks in the spectra in the low energy region where the drag force effect is strong. Mostly it is found that the initial big discrepancy in energy positions among the same resonance peaks from the four different parts of the zig-zag scan is removed by the drag force correction. With these corrections for the transformation an agreement of the spectra to around 10 meV at $E_{\rm cm} = 5\,{\rm eV}$ is possible. With the four spectra overlapping so well, after drag force correction, it is possible to combine them to obtain better statistics. Good alignment in energy is of course essential for adding spectra. A mismatch, much less than the obtained resolution is conditional, in order not to obscure the energy resolution and the observed features. A systematic error in the absolute energy calibration by this method is estimated to $\pm 10\,{\rm meV}$ at $E_{\rm cm} \sim 1$ eV, and somewhat more for higher energies.

A higher accuracy in the absolute energy calibration is obtained by the following calibration method (II) [49, 38, 50]: One selects calibration points in the scan (usually the maxima of prominent resonance peaks in a spectrum). The corresponding cathode voltage of the cooler needs then to be recorded. This is done by setting square voltage pulses around the peak. For each step, the recombination rate is recorded in the given time window. Then the energy shifts due to the drag force, i.e. the changing Schottky frequency is measured in the same time window. The cathode voltage where the normalized rate has a maximum can be taken as the calibration point. The cathode voltage is then set to this recorded value and the ion energy is adjusted until cooling at this new voltage is reached. Under cooling condition, the mean velocity of the electrons matches the mean velocity of the ions ($v_{\rm i}$). Since the latter can be derived from the Schottky frequency, f_s, at cooling conditions by $v_i = f_s L$, thus the velocity and the energy of the electrons can be readily deduced. It should be pointed out that the electron energy deduced in this approach is absolute, the effect of the space charge and drag force is automatically included and need not to be corrected for. The electron energies of the calibration points deduced by this method are converted to the C.M. frame (see below) and the DR spectrum is calibrated by aligning the calibration points to the obtained values. The systematic error is mainly determined by an uncertainty in the ion trajectory length around the ring. It is assumed to be in the order of centimeter. The absolute energy calibration by this method is estimated to be uncertain to 5 meV at $E_{\rm cm} \sim 1\,{\rm eV}$, and somewhat more for higher energies, at present. A better determination of L will reduce this error accordingly. We will present here, see Sec. 7.5.2 below, data with Kr^{33+}, calibrated by this method [50]. In these experiments we determined the 2s Lamb shift in Li-like Kr to an accuracy of 0.5%.

The exact relativistic expression for the mean center-of-mass energy, also called relative energy, is:

$$E_{\rm rel} = \left[\begin{array}{c} (E_{\rm e} + E_{\rm i} + m_{\rm e}c^2 + m_{\rm i}c^2)^2 - \dots \\ -(\sqrt{E_{\rm e}^2 + 2m_{\rm e}c^2E_{\rm e}} + \sqrt{E_{\rm i}^2 + 2m_{\rm i}c^2E_{\rm i}})^2 \end{array} \right]^{1/2} - \dots \\ -m_{\rm e}c^2 - m_{\rm i}c^2. \tag{7.12}$$

The energies E_e and E_i are determined as described above. The center-of-mass velocity v_{cm} agrees therefore only at high detuning energies with v_{rel}, in that case $E_{cm} = E_{rel}$.

The energy resolution for measuring resonances with an electron cooler is determined by the following parameters. Firstly, the velocity spread of the cooled ion beam is small compared to that of the electron beam. At small collision energies, the cathode temperature and the beam expansion factor set the energy-resolution ΔE_{rel}. At larger collision energies the longitudinal energy spread of the electron beam gets more important:

$$\Delta E_{rel} = kT_{e\perp} ln2 + 4\sqrt{(E_{rel} kT_{e\parallel} ln2)}. \tag{7.13}$$

Additional factors in the energy resolution are due to space-charge effects: (i) The space-charge-induced potential has a parabolic shape across the ion beam which enhances the effective longitudinal electron temperature. But one has to take into account that the cooled ion beam has a diameter ($\sim$ 1 mm) considerably smaller than that of the electron beam ($\sim$ 50 mm). (ii) A misalignment between the cooled ion beam and the electron beam will enhance the effective transverse electron temperature. Examples for the energy resolution that was obtained in some of the dielectronic recombination experiments are given in Sec. 7.5.

5. Comparison of selected results with theory

In this section we will look at the comparison between experiment and theory for a some selected ions and discuss what can be learnt from the different studies.

Before the different examples are discussed we stress that for direct comparison with the experimental rate coefficient the calculated cross section has to be folded with the electron velocity distribution, f, see Eq. 7.4, from the experiment in question:

$$\alpha(E_{rel}) = \int v_{cm} \sigma(v_{cm}) f(\vec{v}_e) d^3 v_e, \tag{7.14}$$

where α is the recombination coefficient and σ is the recombination cross section.

5.1 A test of theoretical recombination models by experimental data for Li-like N ions

The interest in Li-like ions stems from their rather simple electronic structure with two tightly bound $1s$-electrons and one more loosely bound $2s$-electron. Because of this fairly simple electronic structure the Li-like ion beams are free of any metastable ion fraction, the recombination spectra contain a limited number of resonances, and the formed doubly excited states in the corresponding Be-like ions comprise a pseudo-two-electron system with two active electrons outside a closed shell. Therefore, the calculations may be performed with pure ab initio methods. However, the electronic structure is still complicated enough to require a full many-body treatment and to match the experimental precision achieved today it is necessary to account for electron correlation to high orders as well as for relativistic and radiative effects.

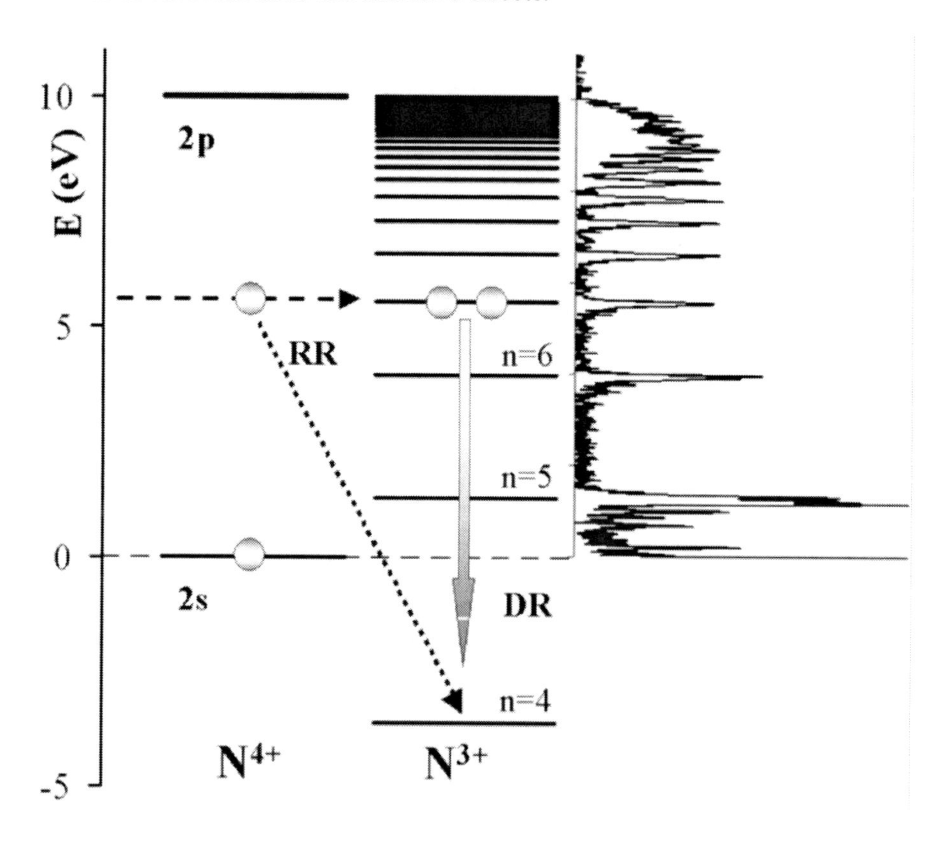

Figure 7.6. Schematic description of the RR and DR process of a free electron and a Li-like N^{4+} ion that ends in a system where the electron is bound to the ion and a photon has been emitted. The energy levels of Li-like N^{4+} (initial ion) and Be-like N^{3+} (the recombined ion) are shown on the same energy scale. The outer electron is here, for clarity, described by Rydberg's formulae. The doubly excited states $N^{3+}(2pn\ell)$, with $n \geq 5$, are situated above the ground state of N^{4+}. When the electron collides with N^{4+} a certain amount of kinetic energy, is brought into the system. If the energy of the electron-ion system now matches a state in N^{3+} it can transform into this state. On the right-hand side the experimental recombination spectrum is shown. The sharp peak at zero energy is due to RR (it is cut in its hight). At higher energies the maximum of each DR peak, coming from the higher ℓ-states in each n-manifold, clearly matches the N^{3+} energy levels. The low energy region, where $2p5\ell$ DR resonances contribute, is shown in more detail in Fig. 7.8.

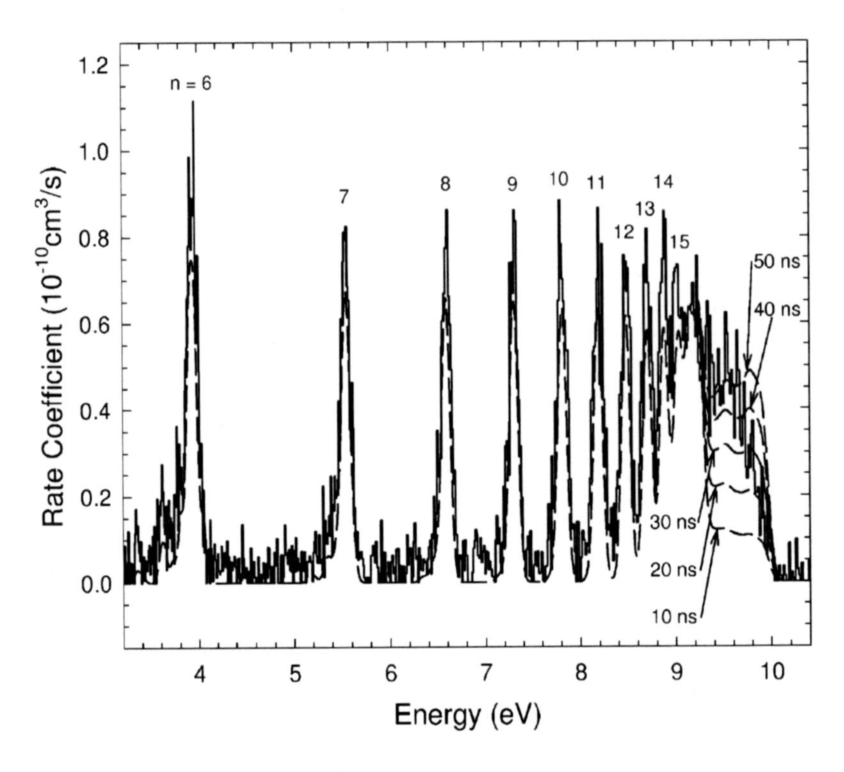

Figure 7.7. The DR spectrum of N^{4+} ions in the energy region covering the $2pn\ell$ resonances from $n = 6$ to the series limits ($n = \infty$). The dashed lines are obtained from the AS calculation using flight times of 10, 20, 30, 40, and 50 ns. The solid line is the experimental spectrum. The experimental energy scale has been adjusted slightly so that the resonance positions fit the theoretical ones near the series limits. See the text for more details.

Fig. 7.6 shows the energy levels of the $2pn\ell$ levels of N^{3+} populated by the first step of DR from the ground state of N^{4+}, $1s^2 2s$. The energy levels are calculated using the Rydberg formulae as $E_{2p} - E_{2s} - 13.6\,eV \cdot Z^2/n^2$ and compares rather well with the experimental spectrum shown on the same energy scale. Note that the radiative recombination appears in the spectrum as a sharp peak near threshold. The measurements with the N^{4+} ions were performed as described in Sec. 7.2. The rate coefficients and energy calibration were obtained as described in Sec. 7.4, using calibration method (I) see 7.4.2). An error in the energy scale of typically less than a few percent is obtained. The fact that linear corrections were sufficient to achieve a very good overlap between the four spectral parts indicates that the error in the energy scale can be substantially reduced by a linear correction. This was done by using the $(2pn\ell)$ resonances towards the series limits for energy calibration. In the DR spectrum the positions of the resonances at the series limit are accurately known from optical data[51]. The position of the series limit is given by the $2p_{1/2} - 2s_{1/2}$ and $2p_{3/2} - 2s_{1/2}$ splittings in N^{4+}, which are 9.97617 eV and 10.00824 eV, respectively. Then the spectrum was taken from the AUTOSTRUCTURE (AS)[52] calculation. Fig. 7.7 shows how the DR spectrum, from the wide data-set, compares with the theoretical AS spectrum, in the energy region where DR resonances into high Rydberg states ($n \geq 6$) contribute, after the experimental energy scale was linearly stretched. After this calibration we estimate the uncertainty of the experimental energies to be less than 1% and not more than 5 meV in the region where the $2p5\ell$ resonances occur.

Fig. 7.7 shows that the calculated rate coefficients for the $n \geq 6$ resonances appear to be about 25% lower than the experimental values. The systematic error in the experiment is, as discussed in Sec. 7.4.1, estimated to be $< 20\%$. The uncertainties in the calculated rate coefficients due to systematic errors are of the same order. Thus, the observed difference can be attributed to the total systematic errors.

Field ionization eliminates Rydberg electrons, in the motional electric field of the dipole magnet above a maximum n (n_{max}) and the ion is then not detected. In this experiment n_{max} was estimated to be 16. However, recombined ions with $n > n_{max}$ can contribute to the DR spectrum if they decay radiatively to states with $n \leq n_{max}$ before the dipole magnet. How much the states with $n > n_{max}$ contribute will thus depend on the flight time of the ions from the interaction region to the dipole magnet. The distance between the center of the interaction region and the dipole magnet is about 1.6 m and the ion velocity in the experiments was 3.9×10^7 m/s, which gives an average flight time of about 40 ns.

A "delayed cut-off" model accounting for the radiative decay of the Rydberg states was used in the AS calculations. The model is described in reference [53]. The calculations in Fig. 7.7 are made with flight times of 10, 20, 30, 40, and 50 ns. For the Rydberg resonances with $n > n_{max}$ a fairly good agreement between the experiment and the calculation using the estimated flight time of 40 ns is found. The delayed cut-off model thus seems to work quite well in this case.

For the studied system, the DR resonances closest to threshold, at low relative energies, are due to $2p5\ell$ states. Those resonances appear in the spectrum between 0.18 eV and 1.5 eV. All the three data-sets measured cover this energy region and they yield rate coefficient spectra which are in excellent agreement with each other. Fig. 7.8 shows the experimental spectrum derived from one of the narrow data-sets compared to the results from calculations using different methods. In the two uppermost panels the comparison is made with the results

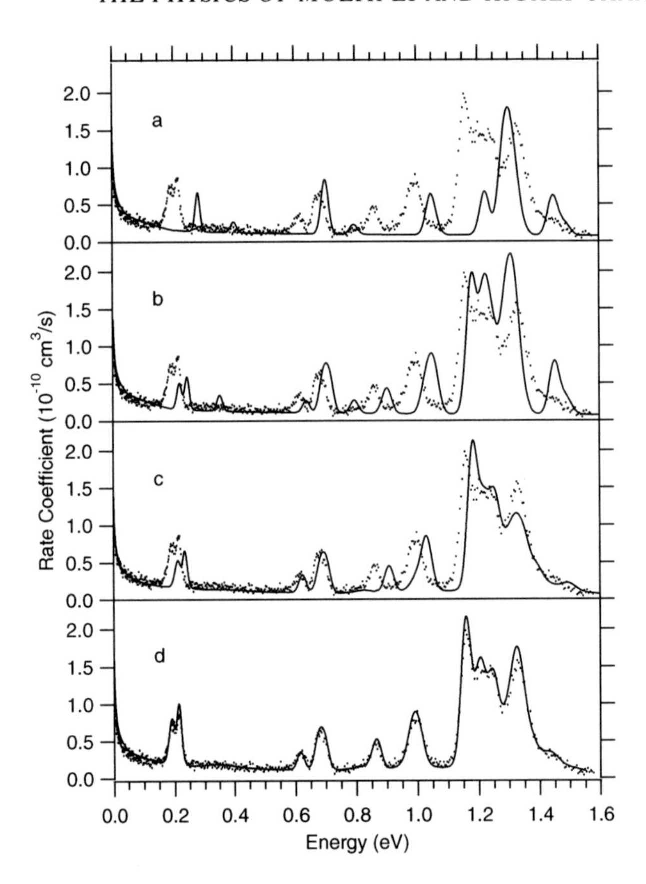

Figure 7.8. Comparison of the experimental spectrum of N^{4+} in the low-energy region, where the $2p5\ell$ DR resonances contribute, with calculations using different models and methods. The two uppermost figures show the comparison with theoretical spectra (solid line) obtained using the AUTOSTRUCTURE method, discussed in Section 7.5.1. In the upper graph the calculations were performed within the LS-coupling scheme and in the second the Breit-Pauli approximation was used. The third graph shows the comparison with an R-matrix calculation and the fourth with relativistic many-body perturbation theory combined with complex rotation.

from the two different calculations using the AUTOSTRUCTURE method, in the third panel with results from the R-matrix method, and in the lowest panel with results from a relativistic many-body calculation the complex-rotation. More details about the different methods can be found in Ref. [54].

The AUTOSTRUCTURE calculation performed within the LS-coupling scheme yields a result which is in rather poor agreement with experiment (see the uppermost panel of Fig. 7.8). In LS-coupling, transitions to the $2p5p\ ^{1,3}P$, $2p5d\ ^{1,3}D$, $2p5f\ ^{1,3}F$, and $2p5g\ ^{1,3}G$ doubly excited states are forbidden, since these states cannot be reached from the initial state conserving both parity and the orbital angular momentum L. However, the LS-forbidden transitions can give a significant contribution, which for example was shown in a previous measurement on C^{3+} [17] where they, in fact, gave rise to the strongest DR resonances. For N^{4+} it turns out that roughly one third of the DR strength for the $2p5\ell$ resonances is due to LS-forbidden transitions. Therefore, it is not that surprising that when LS-coupling is used, the method fails to reproduce the experimental spectrum.

The results from the AS/Breit-Pauli (see the second panel of Fig. 7.8 and R-matrix (third panel) methods agree considerably better with experiment. The R-matrix results shown here are obtained with an hydrogenic approximation for the radiative transitions. Both the AS/Breit-Pauli and the R-matrix model seem to give a total strength which agrees rather well with experiment, although some of the peaks are underestimated and others appear to be overestimated. Most of these discrepancies disappear if the hydrogenic approximation for the radiative rates used with the R-matrix method are replaced by more realistic descriptions, see Ref. [54]. However, the resonance positions still do not match with the experimental ones. The difference in peak positions is in most cases well outside the experimental uncertainties and indicate that some correlation contributions are still missing.

The results obtained with many-body perturbation theory Fig. 7.8 (last panel) combined with complex-rotation are in almost perfect agreement with experiment. The theoretical rate coefficients are slightly higher than the experimental ones, but this difference is within the experimental uncertainties. The essential difference in this calculation compared to the other calculations is the treatment of correlation to high orders.

It should be noted that the natural widths have not been taken into account in the AS spectra predicted in either of the two cases shown in the two uppermost panels in Fig. 7.8. The widths of the peaks in those theoretical spectra are only due to the instrumental resolution. A few of the lines, which are substantially broadened by their natural widths, therefore appear much sharper than in the experimental spectrum. Note also, that for the R-matrix method, contributions from both DR and RR are implicitly included in the method. For the other methods, an RR contribution has been added to the calculated DR spectra. The RR contribution was then calculated as outlined in Sec. 7.3, with $n_{max} = 16$.

The high accuracy of the many-body perturbation theory calculation is confirmed by the good agreement for the $2p_{1/2}{-}2s_{1/2}$ and $2p_{3/2}{-}2s_{1/2}$ splittings in N^{4+} with the experimental values to within a fraction of approximately $6 \cdot 10^{-5}$. The calculated splittings are 9.9766 eV and 10.0088 eV and the experimental splittings are 9.97617 eV and 10.00824 eV. Note that the inclusion of QED effects [55–57](i.e. radiative effects as self-energy and vacuum polarization) is important to get such an agreement since their contribution is one order of magnitude larger the present difference between theory and experiment.

5.2 Determination of the 2s Lamb shift in Li-like Kr ions by recombination

The treatment of QED in a many-body environment is presently a field in development. A common approach during the last decade has been to calculate the non-radiative part by many-body methods such as relativistic many-body perturbation theory (RMBPT) or relativistic configuration interaction (RCI) and then add radiative corrections in a separate calculation. Both RMBPT and RCI are able to treat relativistic electron correlation to high order and to account for the electron-electron spin interaction and retardation (Breit interaction). The dominating radiative contributions are self-energy and vacuum polarization. Accurate calculations of these contributions exist for hydrogen-like systems [55], but very approximative methods have generally been used to account for the "screening" of the nucleus by the core electrons. The approximate "screening" correction is only one effect in a hierarchy of many-body effects. The omission of higher order many-body effects on the QED contribution is the main source of theoretical uncertainty. In spite of this approximate treatment good agreement with experiments has been obtained in many cases. However, with the achievement of higher experimental accuracy [58], it has been necessary to go beyond the simple screening approximation as shown in Ref. [59]. Eventually, a rigorous QED treatment of both radiative effects and electron correlation would be desired. Until now the ground state as well as some excited states of helium [60–64] and lithium like systems [65, 66] have been treated in a strict QED manner for most of the two-photon exchange contributions, but such an approach has not been generalized to systems with more than three electrons.

Several very accurate measurements of the $2p_{1/2} - 2s_{1/2}$ Lamb shift along the lithium isoelectronic sequence have been performed [67], but most of them are in the lower Z region, where also the highest precision is achieved. Accurate experimental data are scarce in the medium and high Z ($Z > 36$) region [67] for $2p_{1/2}$-$2s_{1/2}$ energy splitting. Apart from the measurement of $2p_{1/2}$-$2s_{1/2}$ in U^{89+}, where 0.2% of QED sensitivity was achieved, only two other elements (Ag, Mo) have been measured with high enough accuracy (below 1% QED-sensitivity) to probe many-body QED effects. There is an obvious need for more data in this Z region.

Since the correlation effects are almost Z-independent, the total correlation effects and QED effects are of the same size for $n = 2$ states in the medium Z-region. Thus, theoretical approaches that treat correlation and QED effects in a more unified way can be well tested in this region. Still no such calculation exists in this region however. Here we compare the experimental results for $2p_{1/2}15l_j$, $2p_{1/2}16l_j$ and $2p_{3/2}11l_j$ resonances with the result of RMBPT calculations, corrected for QED contributions. Since Kr^{33+} is a highly charged system QED effects are rather important. The contribution to the energy positions from QED to the $2p - 2s$ splitting is $\sim$1.5 eV with an estimated error of $\sim$20 meV. Many-body effects beyond the Hartree-Fock level give on the other hand only a $\sim$0.2 eV contribution, as discussed in Ref. [50], and can be calculated to an accuracy better than 1 meV. Thus our measurement can be combined with the RMBPT calculation to extract an experimental value for the QED contribution.

Here we give a short account of the results for the $2p_{1/2}$-$2s_{1/2}$ Lamb-shift obtained by energy analysis of dielectronic recombination (DR) resonance spectrum of lithium-like Kr (for details see Ref. [50]). Fig. 7.3 shows the energy levels for Kr^{33+} and Kr^{32+} and indicates how the DR resonances are formed. The important advantage in determining the value of the $2p_{1/2} - 2s_{1/2}$ energy splitting from the DR resonances is the fact that the error in the splitting is almost completely defined by the error in the resonance positions: due to energy conservation law, the resonance energy can be written as $E_r = E_{2p_{1/2}} - E_{2s} + \Delta E$, where ΔE is the binding energy of the outer electron which can be calculated very accurately when the recombination is into a Rydberg state. This gives us the opportunity to determine the $E_{2p_{1/2}} - E_{2s_{1/2}}$ splitting (here in $\sim$ 72 eV range) with almost the same accuracy as that of the energy of the DR resonances (E_r $\sim$ 5 eV range). This was done with an accuracy of 8 meV, i.e. with 10^{-4} relative accuracy. It is remarkable that this accuracy is possible in an electron-ion scattering experiment, whereas the experimental evaluation of energy splittings in atomic systems has hitherto almost exclusively been done by spectral analysis [67]. A series of measurements to determine the $2s$ Lamb shift in Li-like U is done at the ESR cooler at GSI [68].

Table 7.1. Comparison of measurements of the $2s_{1/2} - 2p_{1/2}$ energy splitting in lithium like krypton.

	$2p_{1/2} - 2s_{1/2}$ [eV]	Error [eV]	QED Sensitivity	Experiment
Dietrich *et al.* [69]	71.194	0.106	7%	Beam Foil
Denne *et al.* [70]	71.243	0.012	0.8%	Tokamak Plasma
Hinnov *et al.* [71]	71.241	0.011	0.7%	Tokamak Plasma
Martin *et al.* [72]	71.284	0.016	1%	Beam Foil
Madzunkov *et al.* [50]	71.243	0.008	0.5%	Recombination Exp.

$^{a)}$ The QED contribution to the $2s_{1/2} - 2p_{1/2}$ energy splitting is $\sim$ 1.5 eV.

Of course, the energy calibration of the DR resonance spectra is crucial in this case. It was done by the so-called Schottky method or described as calibration method (II) in Sec. 7.4.2 above (for more details see also Ref. [50]). Our results for the $2p_{1/2}$-$2s_{1/2}$ energy splittings in lithium-like krypton are presented in Table 7.1 together with earlier determinations of this splitting. The first of these measurement of $2p_{1/2}$-$2s_{1/2}$ energy splitting was done in 1980 using a beam foil technique, however, it suffers from poor QED sensitivity [69]. Latter measurements done by Denne and Hinnov [70, 71] during 1987 and 1989 gave a much higher accuracy and they agree within the error bar with the value presented here. However, both of those results are in disagreement with the work of Martin [72]. In that paper it was argued that a possible systematic error in the measurements of Denne and Hinnov could be caused by an inaccurate determination of the Doppler shift corrections. The present calculation of $2p_{1/2}$-$2s_{1/2}$ energy splitting is in good agreement with our result within the estimated errors.

In the fourth column of the Table 7.1 we list the relative uncertainty of the determination of the QED part of the $2p_{1/2}$-$2s_{1/2}$ energy splitting. The result presented here is accurate to within 0.5% in the QED part. The estimated uncertainty in the theoretical value is 19 meV, i. e. $\approx$1% of the QED part and originates mainly from the estimated error in the screening of the radiative corrections. Improved calculations of many-body QED corrections are thus needed in order to meet the experimental precision obtained in our measurements.

5.3 Recombination of sodium-like ions; the example of Ni

As was discussed in Sec. 7.5.1 above, ab initio calculations are today able to predict the recombination spectrum for a low charge Li-like ion very precisely. Na-like system is a natural next step for testing the ability to handle more complicated atoms. Nickel is an interesting element, existing in several charge states in astrophysical plasma, and has been chosen for a first test of sodium-like ions.

In Na-like nickel the excitation energy from the $1s^22s^22p^63s$ ground state to the first excited states, $3p_j$, are around 38.7 eV ($3p_{1/2}$) and 42.5 eV ($3p_{3/2}$). This corresponds roughly to the binding energy of a $n = 10$ electron and we expect thus the recombination spectrum to show low energy $\Delta n = 0$ resonances due to $3p_j10\ell_{j'}$ doubly excited states. Resonances can also be formed if the $3s$ electron is excited to $3d$, the excitation energy is then around 95 eV. For this amount of energy to be released the initially free electron must recombine into $n = 7$. Generally we should clearly expect to see several series of resonances.

The Ni^{17+} ions were made in the electron-beam ion source CRYSIS and then injected into the storage ring CRYRING. There, after accelerating to around 8

MeV and cooling with 4513 eV electrons for 2 seconds, two types of electron-energy scans of the type described in Sec. 7.4 (see Fig. 7.5) were made. One, from 0 (cooling energy) up to 6 eV c.m. energy, and one covering the c.m. energy range from 0 up to 40 eV. The latter one should contain all resonances of the 3p series. To get the absolute energy scale of the recombination spectra calibration method (I) is applied, see Sec. 7.4.2. The uncertainty of the energy scale at 6 eV c.m. energy is estimated to 20 meV. It originates from conversion of time to voltage channel and from the drag force correction.

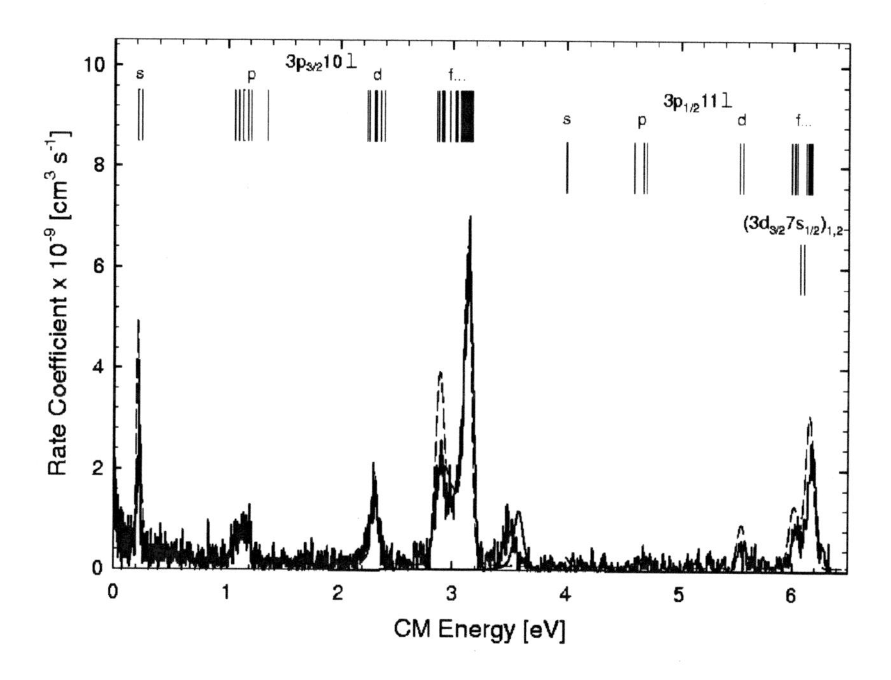

Figure 7.9. Space charge and drag force corrected spectrum averaged from the overlay of the four-part scan obtained in Fig. 7.5. The theoretical cross section was calculated and folded with electron beam temperatures for a flattened Maxwellian distribution of $kT_{trans} = 3$ meV and $kT_{long} = 0.1$ meV.

In Fig. 7.9 the experimental spectrum up to 6 eV relative energy is shown together with the calculated cross section folded with the electron beam temperatures (a flattened Maxwellian distribution corresponding to energies $kT_{trans} =$ 3 meV and $kT_{long} = 0.1$ meV). In this region three series of resonances can be seen; $3p_{3/2}10\ell_j$, $3p_{1/2}11\ell_j$ and $3d_{3/2}7\ell_j$. The identification is done through the calculation. The $3p_{3/2}10\ell_j$ is the lowest energy series while the $3p_{1/2}10l_j$

doubly excited states are all bound and do not form resonances. This is possible since the $3p$ fine structure is nearly 4 eV while the $10\ell_j$ -states spread over ~ 3 eV, see Fig. 7.9.

There is one feature in the experiment, around 3.5 eV, which does not belong to any of these series. This resonance can be identified as a $\Delta n = 1$ resonance, corresponding to one of the $(4s4p)_{J=1}$ doubly excited states. The other three $(4s4p)_J$ doubly excited states are bound with a few eV. For these symmetrically excited states the electron-electron interaction dominates over the spin-orbit interaction and the situation is most appropriately described as the ^3P-states being bound and the ^1P-state forming a resonance. A preliminary calculated result for this resonance is shown in the spectrum. The difference between the measured and the calculated position is probably due to missing higher-order electron correlation.

The agreement between the theoretical and experimental results for the resonance positions is, with the exception of the $\Delta n = 1$ resonance, within 20 meV. Also the rates agree rather well although there is a sizeable rate difference for the $3p_{3/2}10f$ resonance group at 2.9 eV, for which we have no explanation at present.

5.4 An accurate spectroscopy of the 4s - 4p energy splitting with Cu-like Pb ions

In the accumulation of heavy ions in a storage ring the limiting factor for the beam intensity is sometimes the recombination with electrons that cool the ions. A striking example is the extremely short ring-lifetime of Pb^{53+} (Cu-like). Nearby ions as Pb^{54+} (Ni-like) have almost a factor of 100 longer lifetime [73]. Dielectronic recombination resonances very closely above the ionization threshold has been regarded as a possible candidate to cause this effect. With the new super-conducting cooler at the CRYRING storage the electron temperature is so low ($\sim$1K) that electron-ion recombination resonances can be seen that are only meV or less above the ionization threshold and the Pb^{53+} spectrum reveals indeed resonances for ultra-low relative energy. The resonances which can be identified directly in the data are rather narrow. An accompanying relativistic many-body perturbation theory calculation found also broad resonances overlapping with the threshold which contribute significantly to the enhanced recombination rate.

The measurement of these resonances can also be used for critical tests of calculations: The experiment is sensitive to the $4p_{1/2} - 4s_{1/2}$ splitting in Cu-like lead at a level of 10^{-4} eV, while the QED effects (self energy and vacuum polarization) contribute with $\sim$ 2 eV to this splitting. The experiment can thus give a value for the QED contributions to the $(4s - 4p_{1/2})$ splitting which is accurate enough that higher order QED effects are clearly visible. QED calculations beyond one photon exchange and in a many particle environment is still in the development phase and need to be tested by experiments. This test requires a careful calculation of the many-body contributions to the resonance energies. The spectrum shows, however, several resonances and provides thus many check points for the calculation.

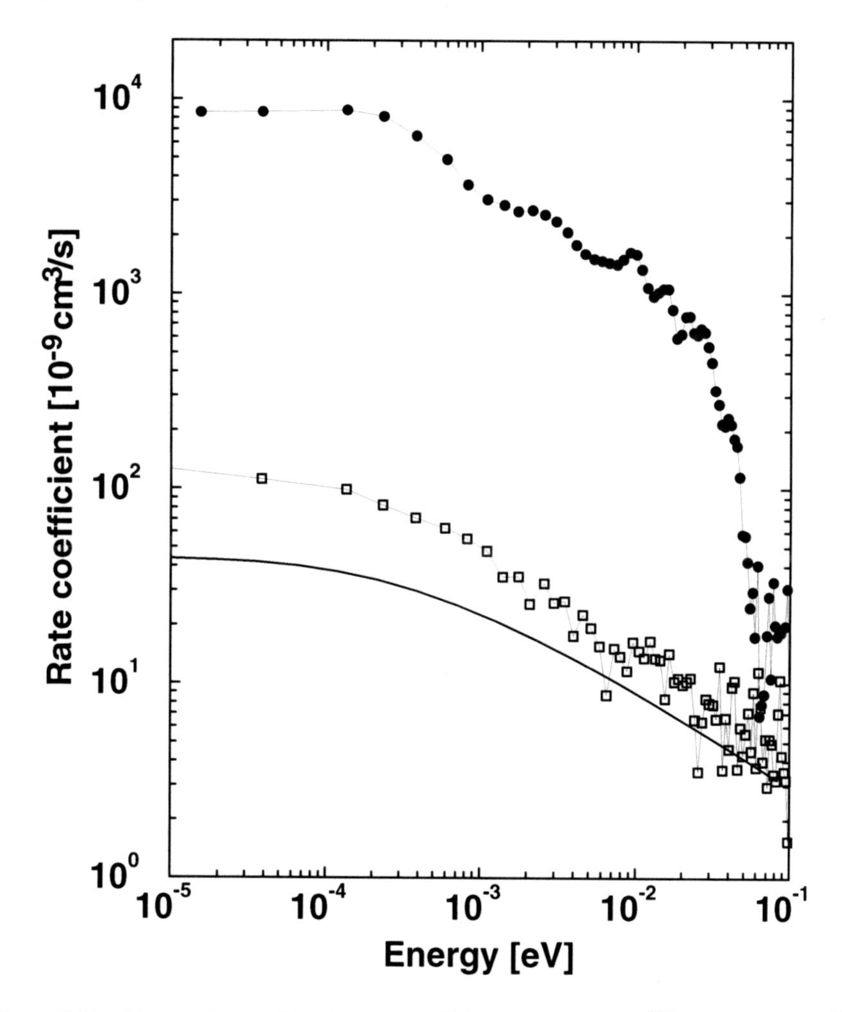

Figure 7.10. Measured recombination rate coefficients (α_{exp}) for Pb^{53+}, circles, and Pb^{54+}, squares, as a function of relative energy. The solid line shows the calculated contribution from radiative recombination for Pb^{54+}.

Fig. 7.10 shows the recombination rate as a function of relative energy for Pb^{54+} and Pb^{53+}. The Pb^{54+} spectrum must be dominated by RR since it is smooth and shows no sign of resonances. The rate for Pb^{53+} is in comparison enhanced with nearly two orders of magnitude due to dielectronic recombination resonances. Similar difference were also found for Au^{50+} and Au^{49+} [74], where also recombination spectra were recorded. Enhanced recombination rates have been detected in several storage ring experiments during the last years [75, 30, 32]. Special here are isolated resonances as close as 10^{-4} and 10^{-3} eV to threshold and the extremely high recombination rate.

In spite of its 29 electrons Pb^{53+} has a rather simple structure; one valence electron outside a closed shell core. The DR- resonances are formed in Pb^{52+} where the core is unchanged while the $4s_{1/2}$-electron is excited to $4p_{1/2}$ and the captured electron is found in a high n-state, i.e. by the following process:

$$\mathrm{Pb}^{53+}\left(4s_{1/2}\right) + e^- \rightarrow \mathrm{Pb}^{52+^{**}}\left(4p_{1/2}n\ell_j\right) \rightarrow \mathrm{Pb}^{52+^{*}} + \hbar\omega.$$

The resonances will appear for electron energies $\varepsilon_{e^-} = E\left(4p_{1/2} - 4s_{1/2}\right) + \Delta E$ where ΔE is the binding energy of the last electron, given by $\Delta E = E\left(4p_{1/2}n\ell_j\right) - E\left(4p_{1/2}\right)$. The last electron will for the present charge state have $n = 18$ or higher. The point of spectroscopy is here that if it is possible to calculate ΔE very accurately, which indeed is possible for high $n\ell$-states, the $4p_{1/2} - 4s_{1/2}$ splitting can be obtained from the energy positions of the recombination resonances. The closeness to threshold fixes the resonance positions very precisely and thus the $4p_{1/2} - 4s_{1/2}$ energy splitting can be fixed with the same precision. Our determination of the $\sim$ 118 eV large $4p_{1/2} - 4s_{1/2}$ splitting is within 1meV, i.e. below 0.001% and thus slightly more precise than the determination of the $2p_{3/2} - 2s_{1/2}$ splitting in Li-like bismuth [58].

Table 7.2 shows the accuracy with which different contributions to the Pb^{53+} $\left(4p_{1/2} - 4s_{1/2}\right)$ splitting can be calculated at present. More details can be found in [76–78], where a comparison with an earlier calculation [79] is also made. The main uncertainty in the RMBPT part of the calculation comes from the extrapolation to include contributions from partial waves with $\ell \geq 11$ and is conservatively estimated to 1meV. Higher order correlation is calculated by RMBPT in an all order formulation [36]. Table 7.2 lists also radiative contributions from Indelicato [80], evaluated in the nuclear potential (one photon exchange) as well as in a potential describing the closed shell core (screening). The result is an approximation in two ways. First, the one electron two photon diagrams are missing. Although substantial progress towards a full account of these effects has taken place recently [81] these results cannot readily be applied on a many-electron heavy ion as Pb^{53+}. An estimate of their size in Ref. [82] gave 1eV for the ground state in hydrogen-like uranium. The effects scale roughly as Z^5 and $1/n^3$, pointing to a possible 10 meV contribution to the Pb^{53+} $(4s)$ binding energy. Second, the many-body corrections to the QED contributions are only approximately accounted for. Blundell [83] has earlier found similar screening contributions. In addition he included some of the more complicated many-body corrections (called valence external exchange terms, core QED in Ref. [83]) and found ~ -40 meV contributions to the $4p_{1/2} - 4s_{1/2}$

energy splitting. The 40 meV uncertainty in the radiative contributions given in Table 7.2 is thus a rough estimate of possible additional radiative corrections and dominates completely the uncertainty of the $4p_{1/2} - 4s_{1/2}$ energy. The measured spectrum for Pb^{53+} shows a number of resonances between threshold and $\sim$ 40 meV, i.e. within a region of the same size as the uncertainty. Given this uncertainty one has now to identify the resonances. For this we have to calculate ΔE, which gives the distribution of the states relative each other and vary $E\left(4p_{1/2} - 4s_{1/2}\right)$ within the 40 meV uncertainty to match the measured spectrum as well as possible.

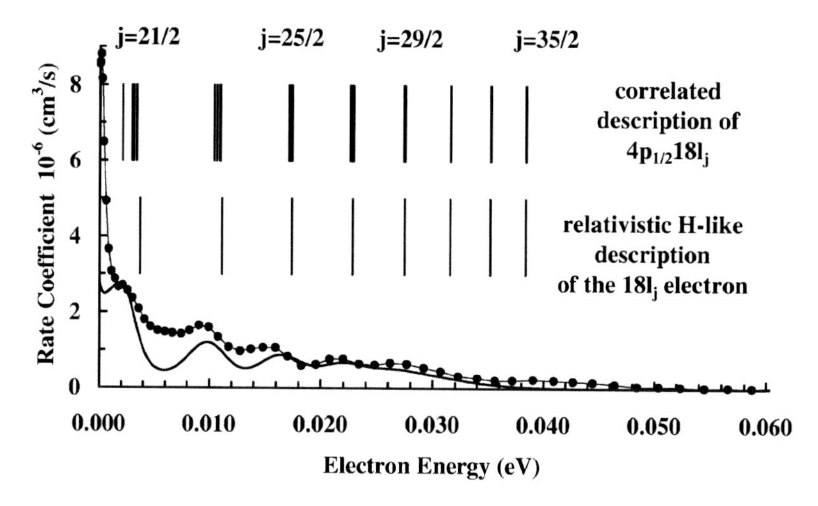

Figure 7.11. Calculated rate coefficients, solid line, are shown together with energy positions of the $4p_{1/2}18\ell_j$ resonances and α_{exp}. For each j of the outer electron there are four $\left(4p_{1/2}18\ell_j\right)_J$ states, two of each parity, spread out over $\approx$ 1 meV. The folding of the calculated cross sections with an electron velocity distribution characterized by $kT_\perp$ = 1meV and $kT_\parallel$ = 0.08meV causes a small peak shift compared to the energies of the resonant states.

Table 7.2 gives the $4p_{1/2} - 4s_{1/2}$ energy splitting to $\sim$ 118 eV. This is close to the binding energy for a $n = 18$ electron as can be estimated even from the non-relativistic hydrogen-like energy with $Z_{eff} = 53$. A relativistic, but still hydrogen-like, determination of the $18\ell_j$-states, shows that the doubly excited states due to $4p_{1/2}18\ell_j$-configurations are spread out both below and above the ionization threshold and only those including the higher $18\ell_j$-states can form resonances. The distances between the first clearly seen resonances are slightly below 10 meV and decrease for higher energy. This fits well with the distances between the $18\ell_j$-states starting from $j = 21/2$, as is seen in Fig. 7.11. The data is not compatible with the first resonances being due to configurations with $18\ell_{19/2}$ or $18\ell_{23/2}$. Already at this point the $4p_{1/2} - 4s_{1/2}$ energy splitting is determined within 10 meV, i.e. within the distance to nearby fine structure

Table 7.2. Contributions to the $4p_{1/2} - 4s_{1/2}$ splitting (eV).

	$4p_{1/2} - 4s_{1/2}$
Dirac-Fock (including nuclear size effects)	118.57088
Breit 1st order (including retardation)	1.71360
Higher order Breit corrections to the Dirac-Fock value $^{a)}$	-0.05589
Mass polarization (DF)	-0.00489
2nd order Coulomb+Breit correlation $^{b)}$	-0.1736(10)
Higher order Coulomb+Breit correlation$^{c)}$	0.00348
Total RMBPT	120.054(1)
Rad.corr. (screened self energy and vac. pol.) [80]	-2.008(40)
Sum	118.046(40)
Extracted from the present experiment	118.010(1)

$^{a)}$ also called Breit -RPA, given in Ref. [77, 78] together with the 2nd order correlation.

$^{b)}$ ℓ_{max}=10 and extrapolated from there.

$^{c)}$ ℓ_{max}=8

levels. When the interaction with the core is accounted for, through the Dirac-Fock approximation, the $18n_{21/2}$ and the $18o_{21/2}$ ($\ell = 10$ and $\ell = 11$) states split with 0.4 meV. Additional contributions of 0.1-0.2 meV arise from core polarization, but the energy is still more or less completely decided by the j-quantum number of the $n = 18$ electron. This is still the case when the full interaction between the $18\ell_j$ and the $4p_{1/2}$ electrons is accounted for. For each j of the outer electron there are four $\left(4p_{1/2}18\ell_j\right)_J$ states, two of each parity, which spread out over $\sim$ 1meV, as is seen in Fig. 7.11. The widths vary significantly. Of the states dominated by a certain $4p_{1/2}18\ell_j$ configuration one is narrow and one much broader. For $j = 21/2$ the broad states have a width of a few meV and the autoionization width of the narrow ones are $\sim$ 0.01 meV. A broad resonance arises when the $\left(4p_{1/2}18\ell_j\right)_J$ state is able to decay to $\left(4s_{1/2}\varepsilon_{\ell+1}\right)_J$ which means that no change in direction of the individual angular momenta is needed.

As stated above the identification of the resonances at $\sim$ 3 meV to be due to $4p_{1/2}18\ell_{21/2}$ doubly excited states fixes the $4p_{1/2} - 4s_{1/2}$ splitting within 10 meV. We have then varied the splitting within a few meV to obtain the best agreement with the measured spectrum. A $4p_{1/2} - 4s_{1/2}$ energy splitting of $118.010(1)$ eV is found through the comparison. This corresponds to an additional contribution from radiative corrections of $-0.036(1)$ eV compared to the calculated value in Table 7.2. The radiative contributions should thus be $-2.044(1)$ eV which is in very good agreement with the value given by Blundell, of -2.05(1) eV [83].

6. Conclusion

This review should give a flavor of the vigorous developments in electron-ion studies in recent years. Although the emphasis was here on storage ring results, and particularly on examples from the Stockholm ring, similar advances were achieved in EBIT labs and the other ion storage ring devices. One important ingredient for the advances were expanded electron beams for cooling energetic ion beams. These makes cooler rings useful instruments for highly accurate spectroscopy in few-electron atomic systems. We have shown dielectronic recombination resonances of Li-like N and Kr, Na-like Ni and Cu-like Pb that were measured with an accuracy in the energy scale in the order of a few meV. Energy positions and sizes of the cross sections are found in very good agreement with the values calculated in an isolated resonance relativistic many-body calculation. The more standard methods of calculating recombination cross sections have improved their accuracy largely. It has proven to be essential to consider the departure from LS-coupling. The data from these methods can be obtained over large impact energy ranges within short computing times. Test of many-body effects in the QED contributions to the dielectronic resonance energies can be done with presently highest accuracies. The uncertainty in the QED part originates mainly from the so called screening of the radiative corrections. Improved calculations of many-body QED corrections are thus needed in order to meet present day experimental precision in recombination measurements.

Acknowledgments

This work was supported by the Swedish Natural Science Research Council (VR).

References

[1] R. Schuch, in *Review of Fundamental Processec and Applications of Atoms and Ions*, edited by C. D. Lin (World Scientific Publ., Singapore, 1993).

[2] F. Bosch, in *Physics of Electronic and Aromic Collisions*, edited by T. Andersen *et al.* (American Institute of Physics, New York, 1993), p. 3.

[3] R. E. Marrs, P. Beiersdorfer, and D. Schneider, Physics Today, Oct. 1994, p. 27.

[4] M. Larsson, Rep. Prog. Phys. **58**, 1267 (1995).

[5] D. Bates, A. Kingston, and R. McWhirter, Proc. Roy. Soc. A **267**, 297 (1962).

[6] J. Dubau and S. Volonte, Reports on Progress in Physics **43**, 199 (1980).

[7] D. E. Osterbrock, in *Astrophys. of Gaseous Nebulae and Active Galactic Nuclei* (Univ. Science Books, Mill Valley, California, 1989).

[8] Y. Hahn, in *Atomic and Molecular Processes in Fusion Edge Plasmas*, edited by R. Janev (Plenum Publ. Corp., New York, 1995), p. 91.

[9] A. Burgess, Astrophys. J. **139**, 776 (1964).

[10] A. Burgess, Astrophys. J. **141**, 1588 (1965).

[11] W. Spies *et al.*, Phys. Rev. Lett. **69**, 2768 (1992).

[12] G. Kilgus *et al.*, Phys. Rev. **47**, 4859 (1993).

[13] D. R. DeWitt *et al.*, Phys. Rev. A **50**, 1257 (1994).

[14] D. R. DeWitt *et al.*, J. Phys. B **28**, L147 (1995).

[15] H. T. Schmidt *et al.*, Phys. Rev. Lett. **72**, 1616 (1994).

[16] S. Mannervik *et al.*, Phys. Rev. A **55**, 810 (1997).

[17] S. Mannervik *et al.*, Phys. Rev. Lett **81**, 313 (1998).

[18] R. Stensgaard, Physica Scripta T **22**, 315 (1988).

[19] K. Abrahamsson *et al.*, Nuclear Intruments and Methods B **79**, 269 (1993).

[20] Blasche, D. Bohne, B. Franzke, and H. Prange, IEEE Trans. Nucl. Sci. NS- **32**, 2657 (1985).

[21] B. Franzke, Nucl. Inst. Meth. B **24**, 18 (1987).

[22] P. Baumann *et al.*, Nucl. Inst. Meth. B **268**, 531 (1988).

[23] See e.g., CERN Accerator school, CERN 84-15 (edited by P.Bryant and S. Newman,1984), CERN 85-19 (edited by P.Bryant and S. Turner,1985), CERN 87-03 (edited by S. Turner,1987),.

[24] H. Danared *et al.*, Phys.Rev. Lett. **72**, 3775 (1994).

[25] A. Aleksandrov *et al.*, in *Proc. Workshop on Electron Cooling and New Cooling Techniques, Legnaro 1990* (World Scientific, Singapore, 1991), p. 279.

[26] R. Schuch *et al.*, Nucl. Instr. Meth. A **79**, 59 (1993).

[27] H. A. Kramers, Philos. Mag. **46**, 836 (1923).

[28] M. Stobbe, Annalen der Physik (Leipzig) **7**, 661 (1930).

[29] D. Griffin, Phys. Scripta, T **28**, 17 (1989).

[30] H. Gao *et al.*, Phys. Rev. Lett. **75**, 4381 (1995).

[31] H. Gao *et al.*, J. Phys. B **30**, L499 (1997).

[32] G. Gwinner *et al.*, Phys. Rev. Lett **84**, 4822 (2000).

[33] S. Salomonson and P. Öster, Phys. Rev. A **40**, 5559 (1989).

[34] S. Salomonson and P. Öster, Phys. Rev. A **40**, 5548 (1989).

[35] S. Salomonson and P. Öster, Phys. Rev. A **41**, 4670 (1990).

[36] E. Lindroth and J. Hvarfner, Phys. Rev. A **45**, 2771 (1991).

[37] E. Lindroth, Phys. Rev. A **49**, 4473 (1994).

[38] W. Zong *et al.*, Phys. Rev. A **56**, 386 (1997).

[39] E. Lindroth, Hyperfine Interactions **114**, 219 (1998).

[40] For an account of the early contributions to the complex rotation method see the whole No. 4 issue of Int. J. Quantum Chem. **14** (1978).

[41] Y. K. Ho, Phys. Rev. A **23**, 2137 (1981).

[42] K. T. Chung and B. F. Davis, Phys. Rev. A **26**, 3278 (1982).

[43] U. Fano, Phys. Rev. **124**, 1866 (1961).

[44] M. Tokman *et al.*, Phys. Rev. A **66** 012703 (2002)

[45] H. Danared, Physica Scripta **48**, 405 (1993).

[46] M. Fogle *et al.*, to be published.

[47] S. Asp *et al.*, Nucl. Inst. Meth. B **117**, 31 (1996).

[48] D. R. DeWitt *et al.*, Phys. Rev. A **53**, 2327 (1996).

[49] R. Schuch *et al.*, Hyperfine Interactions **99**, 317 (1996).

[50] S. Madzunkov *et al.*, Phys. Rev. A, **65**, 032505 (2002).

[51] NIST *Atomic Spectra Database.* (WWW published at http://physics.nist.gov/cgi-bin/AtData/main_asd.).

[52] N. R. Badnell, J. Phys. B **30**, 1 (1997).

[53] W. Zong *et al.*, J. Phys. B **31**, 3729 (1998).

[54] P. Glans *et al.*, Phys. Rev. A, **64**, 043609 (2001).

[55] W. R. Johnson and G. Soff, Atomic Data and Nuclear Data Tables **33**, 405 (1985).

[56] D. K. McKenzie and G. W. F. Drake, Phys. Rev. A **44**, 6973 (1991).

[57] S. A. Blundell, Phys. Rev. A **46**, 3762 (1992).

[58] P. Beiersdorfer *et al.*, Phys. Rev. Lett. **80**, 3022 (1998).

[59] K. T. Cheng, M. H. Chen, and J. Sapirstein, Phys. Rev. A **62**, 054501 (2000).

[60] S. A. Blundell, P. J. Mohr, W. R. Johnson, and J. Sapirstein, Phys. Rev. A **48**, 2615 (1993).

[61] I. Lindgren, H. Persson, S. Salomonson, and L. Labzowsky, Phys. Rev. A **51**, 1167 (1995).

[62] H. Persson, S. Salomonson, P. Sunnergren, and I. Lindgren, Phys. Rev. Lett. **76**, 204 (1996).

[63] P. J. Mohr and J. Sapirstein, Phys. Rev. A **62**, 052501 (2000).

[64] I. Lindgren, B. Åsén, S. Salomonson, and A-M. Mårtensson-Pendrill, Phys. Rev. A **64**, 062505 (2001).

[65] V. A. Yerokhin *et al.*, Phys. Rev. Lett. **85**, 4699 (2000).

[66] V. A. Yerokhin *et al.*, Phys. Rev. A **60**, 3522 (1999).

[67] P. Bosselmann *et al.*, Phys. Rev. A **59**, 1874 (1999).

[68] C. Brandau, A. Müller, *et al.* , private communication.

[69] D. D. Dietrich *et al.*, Phys. Rev. A. **22**, 1109 (1980).

[70] B. Denne and E. Hinnov, Phys. Scr. **35**, 811 (1987).

[71] E. Hinnov, TFTROperatingTeam, B. Denne, and JETOperatingTeam, Phys. Rev. A. **40**, 4357 (1989).

[72] S. Martin *et al.*, Phys. Rev. A **42**, 6570 (1990).

[73] S. Baird *et al.*, Phys. Lett. B **361**, 184 (1995).

[74] O. Uwira *et al.*, Hyperfine Interactions **108**, 149 (1997).

[75] A. Müller *et al.*, Phys. Scr. **T37**, 62 (1991).

[76] E. Lindroth *et al.*, Phys. Rev. Lett. **86**, 5027 (2001).

[77] M. Tokman *et al.*, Phys. Scr. T **92**, 406 (2001).

[78] M. Tokman *et al.*, Hyperfine Interactions **132**, 385 (2001).

[79] W. R. Johnson, S. A. Blundell, and J. Sapirstein, Physical Review A **42**, 1087 (1990).

[80] P. Indelicato, private communication.

[81] V. A. Yerokhin, Phys. Rev. Lett. **86**, 1990 (2001).

[82] S. Mallampalli and J. Sapirstein, Phys. Rev. A **57**, 1548 (1998).

[83] S. A. Blundell, Phys. Rev. A **47**, 1790 (1993).

Chapter 8

DIELECTRONIC RECOMBINATION IN EXTERNAL ELECTROMAGNETIC FIELDS

A. Müller
and S. Schippers
Institut für Kernphysik, Universität Giessen, 35392 Giessen, Germany
Alfred.Mueller@strz.uni-giessen.de, Stephan.E.Schippers@strz.uni-giessen.de

Abstract Dielectronic recombination is the most important among several possible mechanisms by which a free electron can recombine with an ion. It is characterized by resonances associated with intermediate highly excited Rydberg states. These are potentially subject to influences of the environment acting via fields, irradiation or collisions. As an example, this chapter describes the effects of external electric and magnetic fields on the cross sections for dielectronic recombination.

Keywords: dielectronic recombination, singly and multiply charged ions, free electrons, electric fields, perpendicular magnetic field components, storage rings, crossed and merged beams.

1. The Dielectronic Recombination Process

Dielectronic recombination (DR) is an elementary atomic collision process [1, 2] in which an electron is captured by an ion. It is the dominant recombination mechanism in almost all low-density plasmas and as such it is the most effective counterbalance of ionization processes induced by photons or energetic electrons. Thus, DR is most relevant to the charge-state distribution of the plasma ions which in turn determines the physical and radiative properties of the plasma [3, 4].

DR requires the presence of at least one electron in the ion core which can absorb the excess energy released when the initially free electron is captured by the ion. Assuming an initial hydrogenlike ion $\mathrm{A}^{q+}(1s)$ the capture process

F.J. Currell (ed.), The Physics of Multiply and Highly Charged Ions, Vol. 1, 269-301.

(see Fig. 8.1 right branch) can be described as

$$e + \mathrm{A}^{q+}(1s) \rightarrow [\mathrm{A}^{(q-1)+}]^{**}(nln'l'). \tag{8.1}$$

As a result of the capture and the simultaneous excitation of the core a doubly excited state $[\mathrm{A}^{(q-1)+}]^{**}(nln'l')$ is formed with n, n' representing the principal quantum numbers of the two electrons bound in the recombined ion and l, l' their angular momenta. Since this first step of the DR process requires the interaction of two electrons it was termed dielectronic capture. It can be viewed as a time-reversed Auger process, or, more generally, time-reversed autoionization. If the intermediate capture state decays by photon emission, such as

$$[\mathrm{A}^{(q-1)+}]^{**}(nln'l') \rightarrow \mathrm{A}^{(q-1)+}(1snl) + h\nu, \tag{8.2}$$

the new charge state of the ion is stabilized and the DR process is completed. It should be kept in mind that autoionization, i.e., the time-reversal of Eq. 8.1, is by far the dominant decay channel of the doubly excited intermediate state for moderately charged ions.

DR can only proceed if the energy E_e of the initially free electron plus the energy released by its binding into the ion matches the energy required for the excitation of the ion core. To be precise, in the example described by Eq. 8.1 E_e has to be equal to the difference $E_\mathrm{res} = E_\mathrm{tot}(nln'l') - E_\mathrm{tot}(1s)$ of the total energies of the intermediate capture state and the initial ion state. Since these level energies are quantized DR occurs at discrete resonance energies E_res.

DR competes with two other recombination mechanisms, radiative recombination and three-body recombination. While the latter is only relevant in very dense cold plasmas, radiative recombination (RR) (see Fig. 8.1 left branch) of any given ion including fully stripped nuclei can happen at arbitrary energies and particle densities.

$$e + \mathrm{A}^{q+}(1s) \rightarrow \mathrm{A}^{(q-1)+}(1snl) + h\nu. \tag{8.3}$$

RR is time-reversed photoionization and as such has received substantial interest even before the development of quantum mechanics [5]. As the above equations (Eqs. 8.1 - 8.3) illustrate, DR and RR cannot always be distinguished from one another, and hence, the use of different names is questionable. DR and RR are but two ways of viewing one mechanism, the so-called photorecombination. Theoretically, this calls for a unified treatment of both processes. In practice, however, and with few exceptions DR and RR can be treated as separate mechanisms. In this article, DR is viewed as a well defined resonant process that occurs independent of the RR channel.

2. Theoretical Description of Dielectronic Recombination

The scheme for DR described by Eqs. 8.1, 8.2 and by the right branch of Fig. 8.1 suggests a theoretical treatment of DR that implies two separate steps:

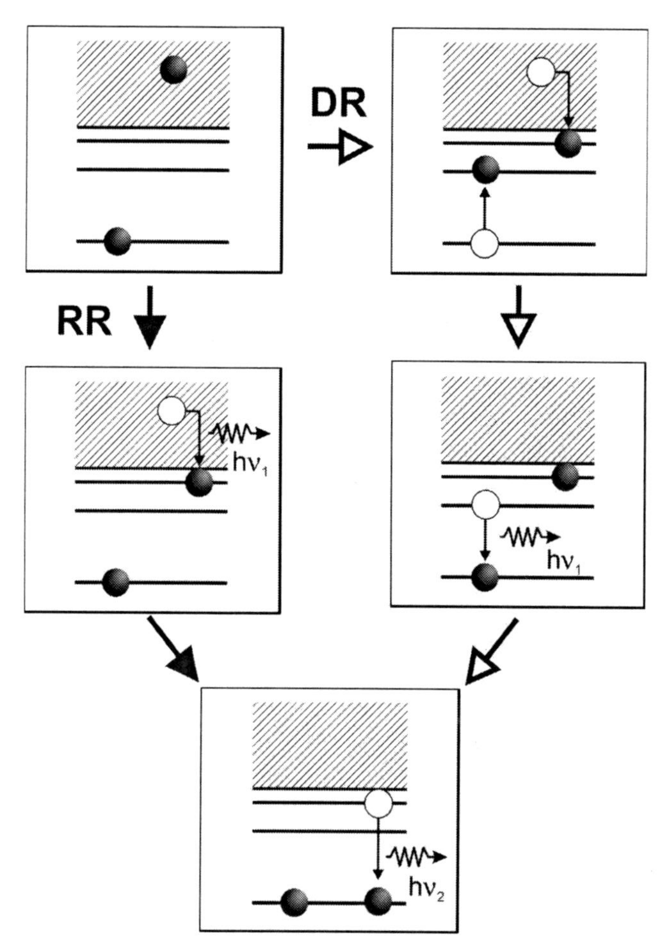

Figure 8.1. Schemes of electron-ion recombination; the generalized process is photorecombination. The two separate, but potentially non-distinguishable, channels schematically displayed here are radiative recombination (left, solid arrows) and dielectronic recombination (right, open arrows). The initial state (top left) with one bound and one free electron and the final state (bottom center) with two bound electrons and altogether two exiting photons, $h\nu_1$ and $h\nu_2$, are identical.

the dielectronic capture (DC) and the subsequent radiative stabilization. The real situation for DR of O^{7+} is shown in Fig. 8.2 which displays the level schemes of the parent O^{7+} and the recombined O^{6+} ions. As has been pointed out already above, the radiative transition from the intermediate capture state competes with autoionization which ejects the captured electron back into the continuum. Due to the principle of detailed balance, which is based on the time-reversal symmetry of nature, the cross section σ_{DC} for dielectronic capture is related to the probability of the time-reversed process, i.e., to the Auger decay

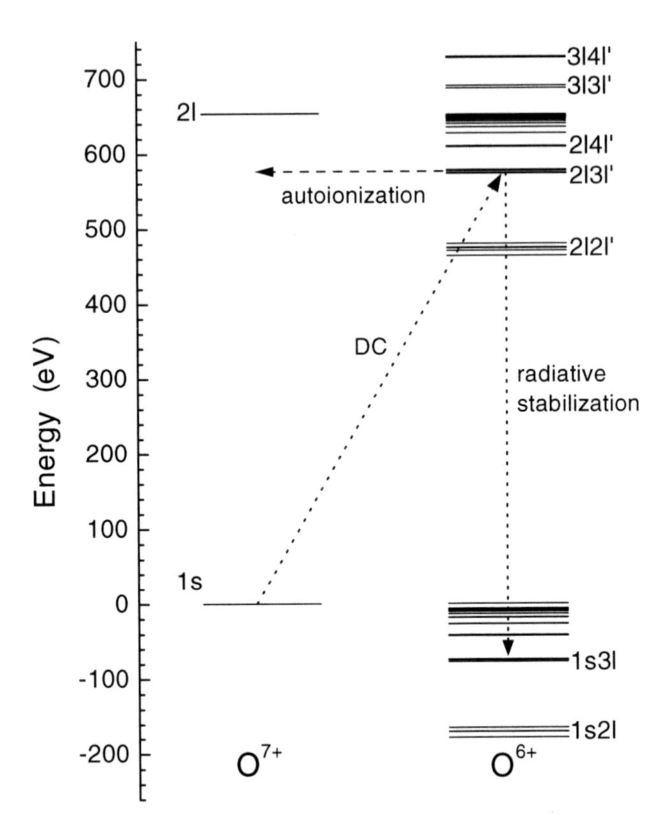

Figure 8.2. Energy level schemes relevant to dielectronic recombination of H-like O^{7+} ions; energy levels of O^{7+} and O^{6+} are displayed. A sample reaction path with dielectronic capture and subsequent radiative stabilization is indicated by arrows with dotted lines. Competing with the radiative stabilization of the intermediate capture state is the autoionization process, i.e., a transition from the resonant state ($2l3l'$ in the present example) back into the continuum of the $O^{7+} + e$ system (dashed-line arrow).

rate $A_{\rm a}$ of the capture state (described in the example by quantum numbers n, l, n', and l'). In other words, the DC cross section is proportional to the autoionization rate of the resonant doubly excited intermediate state $(nln'l')$

$$\sigma_{\rm DC} \propto A_{\rm a}. \tag{8.4}$$

The probability for the radiative stabilization that competes with the electron-emission channel of $(nln'l')$ is given by the fluorescence yield $\gamma_{\rm r} = A_{\rm r}/(A_{\rm r} + A_{\rm a})$, with the rate $A_{\rm r}$ for the radiative decay of the capture state. The two probabilities have to be multiplied to obtain the cross section for DR

$$\sigma_{\rm DR} \propto \sigma_{\rm DC}\gamma_{\rm r}. \tag{8.5}$$

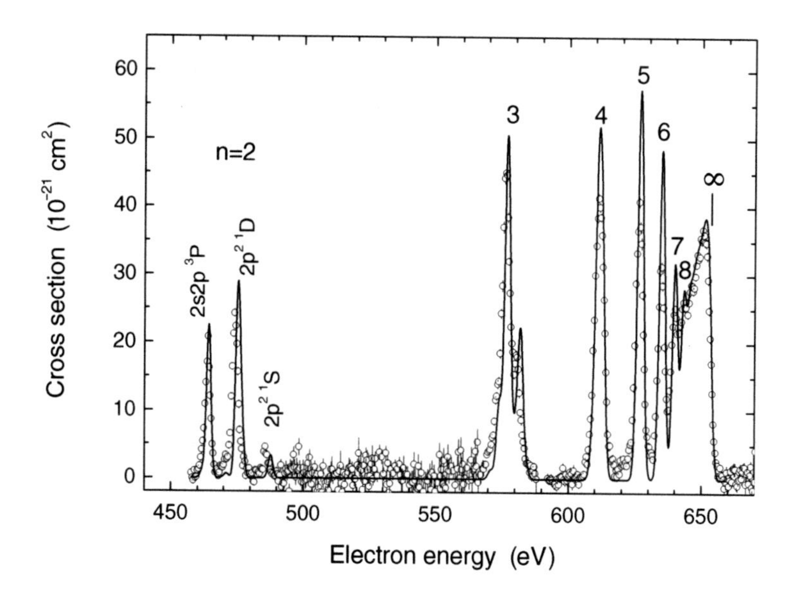

Figure 8.3. Measured (open symbols, [6]) and calculated (full line, [7]) dielectronic recombination cross sections of hydrogenlike O^{7+} ions. Individual $2lnl'$-resonances can be distinguished up to $n = 8$.

The size of the DR cross section depends on the balance between A_{r} and A_{a}. A large autoionization rate leads to a large DC cross section. However, a situation with $A_{\mathrm{a}} \gg A_{\mathrm{r}}$ results in a small fluorescence yield. Generally, in the case of a big difference between the two relevant rates, Eq. 8.5 implies that $\sigma_{\mathrm{DR}} \propto A_{<}$ with $A_{<}$ representing the smaller of the two decay rates.

In a more rigorous treatment of the problem that assumes narrow non-overlapping DR resonances the DR cross section associated with an intermediate state labelled $|d\rangle$ can be approximated by

$$\sigma_d(E_{\mathrm{e}}) = \bar{\sigma}_d L_d(E_{\mathrm{e}}) \tag{8.6}$$

with the electron-ion center-of-mass (c.m.) frame energy E_{e}, the Lorentzian line shape $L_d(E)$ normalized to $\int L_d(E)dE = 1$, and the resonance strength, i.e. the energy-integrated DR cross section

$$\begin{aligned} \bar{\sigma}_d &= 4.95 \times 10^{-30}\mathrm{cm}^2\mathrm{eV}^2\mathrm{s} \\ &\times \frac{1}{E_{\mathrm{res}}} \frac{g_d}{2g_i} \frac{A_{\mathrm{a}}(d \to i) \sum_f A_{\mathrm{r}}(d \to f)}{\sum_k A_{\mathrm{a}}(d \to k) + \sum_{f'} A_{\mathrm{r}}(d \to f')} \end{aligned} \tag{8.7}$$

where E_{res} is the resonance energy, g_i and g_d are the statistical weights of the initial ionic core $|i\rangle$ and the doubly excited intermediate state $|d\rangle$, $A_{\mathrm{a}}(d \to i)$ denotes the rate for an autoionizing transition from $|d\rangle$ to $|i\rangle$ and $A_{\mathrm{r}}(d \to f)$

the rates for radiative transitions from $|d\rangle$ to states $|f\rangle$ below the first ionization limit. The summation indices k and f' run over all states which can either be reached from $|d\rangle$ by autoionization or by radiative transitions, respectively. Thus, the denominator of Eq. 8.7 is equal to $2\pi\Gamma/h$ with the total width Γ of the intermediate capture state $|d\rangle$. As an example for the above considerations Fig. 8.3 shows experimental [6] and theoretical results [7] for DR of hydrogenlike O^{7+} ions. The calculated cross sections basically result from the application of Eq. 8.7. The resonance energies follow from the level scheme displayed in Fig. 8.2. Fine structure in the $(2l2l')$ and $(2l3l')$ configurations has been resolved. Some individual states such as the strong $2p^2\ ^2D_1$ resonance are indicated.

3. Field Effects - the Physical Picture

Soon after the establishment of DR as an important process governing the charge state balance of ions in the solar corona by Burgess [8], it has been realized by Burgess and Summers [9], and Jacobs, Davies and Kepple [10], that DR cross sections should be sensitive to external electric fields particularly when the incident electron is captured into high Rydberg states. Such fields are present in plasma environments because of the thermal motion of charged particles in the vicinity of the recombining ion-electron pair. The resulting microfields can enhance the DR cross section contributions of those high Rydberg states. The reason is in the electric-field dependence of the autoionization rates of states involving a Rydberg electron. These rates strongly depend on the angular momentum quantum number l and the principal quantum number n of the Rydberg electron.

Clearly, the physical process behind the ejection of an electron from a doubly excited state is a collision of the two active electrons by which one of them is knocked down to a lower orbital and the other gains the energy necessary to leave the ion. The probability for such a collision depends on the frequency of encounters of the two electrons which reside in different quantum states. A Rydberg electron on a circular orbit hardly ever meets its counterpart residing deep inside the ion, and this is all the more true the larger the diameter of that orbit. A Rydberg electron with a highly eccentric, elliptical orbit and low angular momentum l, however, may meet the core electron with a vastly enhanced probability. Quantitatively, the autoionization rates are found to be proportional to the inverse cube of the principal quantum number n. The dependence on l is more complicated. It has been approximated by high powers of $1/l$ or by an exponentially decaying function. For the present discussion we use the

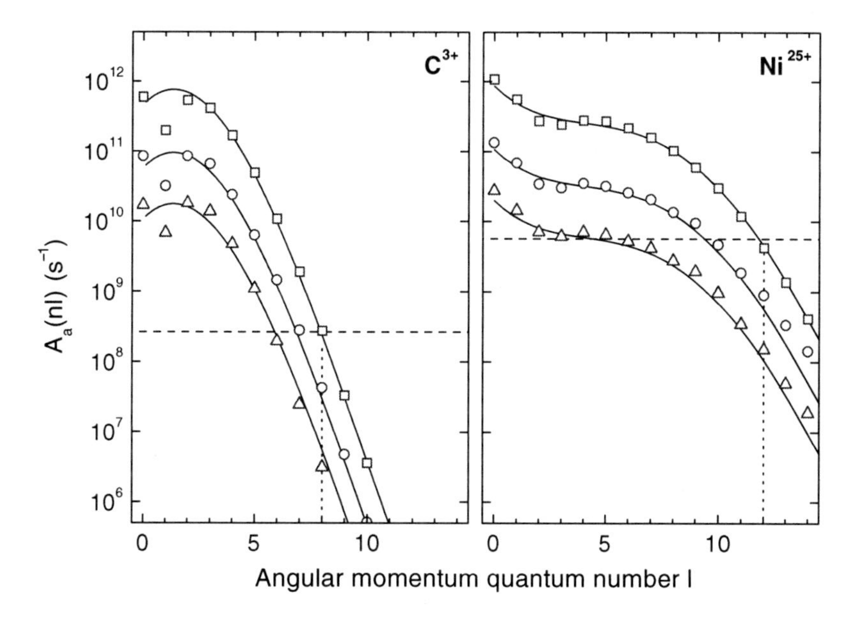

Figure 8.4. Calculated decay rates for $C^{3+}(2pnl)$ and $Ni^{25+}(2pnl)$ states versus the orbital momentum quantum number l for three different principal quantum numbers $n = 20$ (squares), $n = 40$ (circles) and $n = 70$ (triangles). The solid lines are fits to the calculated $n = 20$-rates using the representation of Eq. 8.8 scaled to $n = 40$ and $n = 70$ by employing the n^{-3}-scaling. Also shown are the $2p \to 2s$ radiative transition rates of the core electron (horizontal dashed lines). The dotted vertical lines indicate the cut-off values $l_{\rm cut}$ for $n = 20$.

following representation for the field-free autoionization rates:

$$A_{\rm a}(nl) = \frac{\tilde{A}_{\rm a}}{n^3} \exp\left(- \sum_{i=1}^{i_{\rm max}} d_i \, l^i \right), \tag{8.8}$$

with parameters $\tilde{A}_{\rm a}$, $i_{\rm max}$ and d_i to be adjusted to the specific case.

Generally, the stabilization of the doubly excited intermediate state $|d\rangle$ is accomplished by a radiative transition of the excited core electron back to its original orbit. The energy of the emitted photon is influenced by the presence of the second (spectator) electron. This effect gives rise to the so-called dielectronic satellites which are observed in the line spectrum of a plasma. The radiative rate for the core transition is practically independent of the quantum numbers of the Rydberg companion electron. An example for n- and l-dependences of autoionization rates and of the $2p \to 2s$ radiative rate of the core electron is shown in Fig. 8.4 for doubly excited $C^{2+}(2pnl)$ and $Ni^{24+}(2pnl)$ ions formed by DR of Li-like $C^{3+}(1s^2 2s)$ and $Ni^{3+}(1s^2 2s)$, respectively. The rates that have been calculated with the AUTOSTRUCTURE atomic structure computer code [11] behave as described above.

Concentrating on one $C^{3+}(1s^2 2pnl)$ Rydberg state, for example on $n = 20$, the figure shows the overwhelming excess of A_a over A_r as long as l is not too large. Above $l_{cut} = 8$ the radiative rate becomes greater than the autoionization rate and $\sigma_{DR}(l)$ takes a dive proportional to $\exp[-d_1 l - d_2 l^2 - \ldots]$. The number of states involved in DR proceeding via a configuration $(2pnl)$ with $n = 20$ is $N_{max} = \xi_p \times 2n^2 = 4800$ with the multiplicity $\xi_p = 6$ of the $2p$ sublevel and considering the electron spins. Out of these 4800 states only states with $l \leq l_{cut}$ effectively contribute to DR. For states with higher angular momenta the DR cross sections drop to zero very fast. With $l_{cut} = 8$ the number of contributing states is $N = 12 \times \sum_{l=0}^{l_{cut}}(2l + 1) = 12(l_{cut} + 1)^2 = 972$. For higher values of n the ratio N/N_{max} becomes smaller and smaller.

When DR occurs in the presence of an electric field, the spherical symmetry of the doubly excited state is broken and l is no longer a good quantum number. In other words, the Stark effect mixes states which, in the field-free case, were characterized by different l quantum numbers. As a result, the autoionization rates that depend so strongly on l are changed. States which previously had a low l quantum number and high autoionization rates are now mixed with high-l states and thus their autoionization rate is reduced. Vice versa, high-l states with initially small A_a are mixed with low-l states and by that gain in A_a. At the same time the radiative rate of the excited core electron remains constant. The effect of these changes is, that l_{cut} is pushed to higher values. Consequently, more states become accessible and contribute to the total DR cross section and, thus, σ_{DR} is enhanced.

More precisely, the Rydberg states in a not too strong electric field can be described in linear Stark representation with the parabolic quantum numbers n_1 and n_2 and the orientation quantum number m, which satisfy the relations $n = n_1 + n_2 + |m| + 1$ and $k = n_2 - n_1$. The wave function of the Rydberg electron transforms from the field-free Φ_{nlm} to

$$\Psi_{nkm} = \sum_{l=|m|}^{n-1} C_{k,m}^{n,l} \Phi_{nlm} \tag{8.9}$$

where the expansion coefficients are the Clebsch-Gordan coefficients

$$C_{k,m}^{n,l} = \left\langle \begin{array}{cc|c} (n-1)/2 & (n-1)/2 & l \\ (m-k)/2 & (m+k)/2 & m \end{array} \right\rangle \tag{8.10}$$

for the transformation to Stark states. Accordingly, the autoionization rates, $A_a = A_a^{n,l,m}$, are transformed and assume the values for the new Stark states [12]

$$A_a^{n,k,m} = \sum_{l=|m|}^{n-1} \left| C_{k,m}^{n,l} \right|^2 A_a^{n,l,m}. \tag{8.11}$$

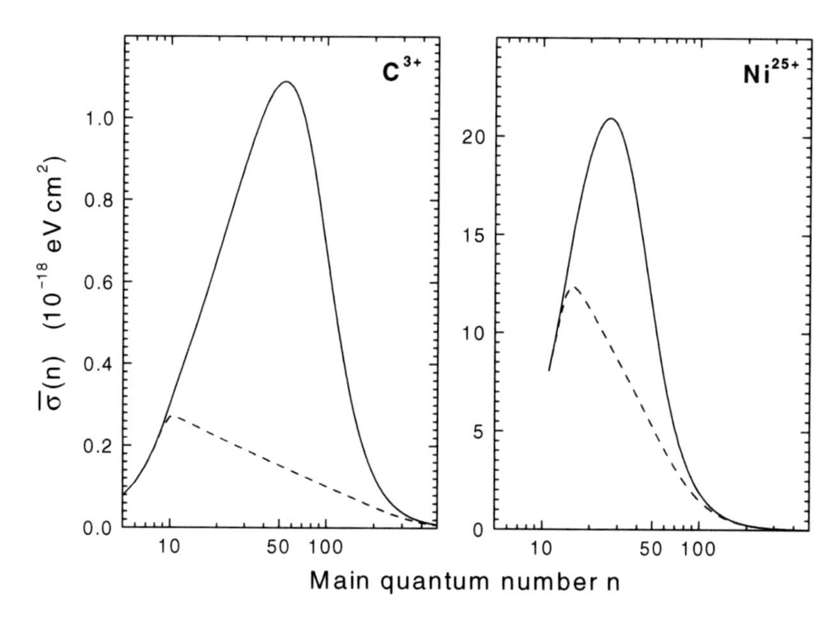

Figure 8.5. Collision strengths, i.e., energy-integrated DR resonance contributions $\bar{\sigma}(n)$, calculated for C^{3+} and Ni^{25+} by using the rates from Fig. 8.4 and Eqs. 8.7 and 8.11. The dashed lines show the field-free DR data. The solid lines are the results for full Stark mixing. Note the logarithmic scale for the n-axis which may distort the perception of total resonance strength, i.e., the sum of $\bar{\sigma}(n)$ over all n.

While l-states are mixed by an electric field, the associated orientation quantum numbers m are still conserved. Mixing occurs only among states which can be characterized by $|lm\rangle$ with arbitrary l but fixed m. This condition limits the gain of the number of states which, by pure Stark mixing in an electric field, can be made to contribute to DR. Ultimately, the number of contributing states can thus be enhanced by $\Delta N = \xi_p \times 2 \times (2l_{\text{cut}} + 1)(n - 1 - l_{\text{cut}})$. For $n = 20$ and $l_{\text{cut}} = 8$ $\Delta N = 2244$ for the example of DR of C^{3+} producing $C^{2+}(2pnl)$. The situation is shown schematically in Fig. 8.6. The states belonging to a given n-manifold $|nlm\rangle$ are represented by circles in the m-l plain. The electron-spin multiplicity is not separately addressed in the figure.

Using simple representations of A_{r} such as $A_{\text{r}} = Z^4 \times 3 \times 10^8\,\text{s}^{-1}$ and of $A_{\text{a}}^{n,l,m}$ such as $A_{\text{a}}^{n,l,m} = 6 \times 10^{15}\,\text{s}^{-1} \times n^{-3} \times \exp[-0.25 \times l^2]$ and plugging these into Eqs. 8.7 and 8.11 one can easily demonstrate the global dependences of the cross sections for DR on n, the atomic number Z along an isoelectronic sequence and the effect of Stark mixing. Sample calculations are shown in the paper on DR of Mg^+ by Müller *et al* [13]. For the two examples discussed above, C^{3+} and Ni^{25+}, the scaled rates displayed in Fig. 8.4 were transformed using Eq. 8.11 and then the resonance strengths were calculated from Eq. 8.7 for the

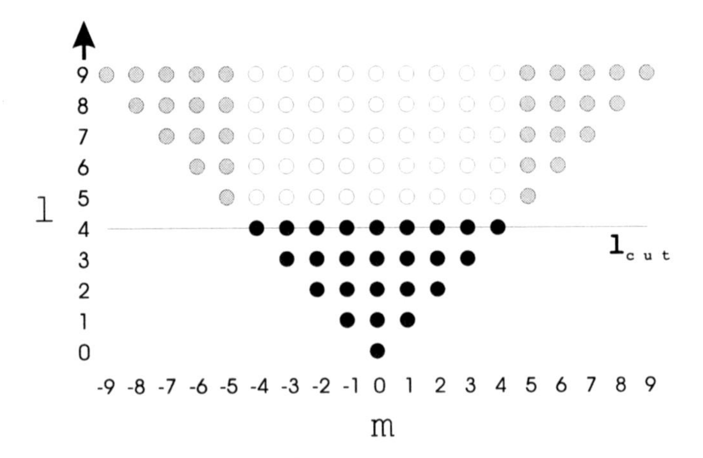

Figure 8.6. Counting of states $|nlm\rangle$ participating in DR for a given n-manifold of intermediate capture states. The electron-spin multiplicity is not considered, nor the statistical weight of the orbit of the excited core electron. For each n there are states with orbital momentum quantum numbers l which range from 0 to $n-1$. The associated orientation quantum numbers m range from $-l$ to $+l$. States with $l \leq l_{\rm cut}$ (full circles) contribute to DR (see Fig. 8.4 and discussion in the text) as long as there are no external fields. Pure electric fields make additional states (open circles) available to DR by mixing l-states along columns with a given m. When the axial symmetry is broken by an additional magnetic field with components perpendicular to the electric field, then also different m-states can be mixed and thus, additional states (shaded circles) can potentially contribute to DR.

field-free case and for full field mixing. The results are shown in Fig. 8.5. The DR enhancement by an electric field is clearly visible as well as the reduction of this effect when going from the low Z ion, C^{3+}, to the higher Z ion, Ni^{25+}.

A quantitative theoretical description of DR in an electric field was made tractable for example by the DRFEUD computer package [14] which diagonalizes the Hamiltonian containing the field term $H = eEz$ with the electron charge e and an electric field E in z-direction. By interaction of theory with experiment and in search of the reasons for everpresent deficiencies in the theoretical description of measured cross sections, it became finally obvious that external magnetic fields also have to be considered [15]. It was known from previous work that very strong purely magnetic fields can lead to effects on DR [16]. The Hamiltonian for an electron (mass $m_{\rm e}$) in a uniform magnetic field B pointing in the direction of the z-axis can be written as

$$H = BL_z e/(2m_{\rm e}) + B^2(x^2 + y^2)e^2/(8m_{\rm e}). \qquad (8.12)$$

The first contribution, the paramagnetic term, contains the z-component of the electron angular momentum operator $\mathbf{L}$. It only causes a shift of level energies and has no other influence on DR. The second expression, the diamagnetic term, leads to a mixing of states, however, the fields have to be very strong for that with $B > 10$ T [16]. New insight came when the presence of crossed electric and magnetic field components was considered. Assuming an electric field in y-direction and a magnetic field in z-direction the field Hamiltonian becomes

$$H = eEy + BL_z e/(2m_\mathrm{e}) + B^2(x^2 + y^2)e^2/(8m_\mathrm{e}). \tag{8.13}$$

At small fields, the diamagnetic term can be ignored. The crossed fields break spherical and cylindrical symmetries and now, neither l nor m are good quantum numbers anymore. As a consequence, the mixing of states gives access also to the rest of the states in the $|nlm\rangle$ manifold and all $2n^2$ states can potentially contribute to DR. In the example discussed above, another 1584 states get in the business (the states represented by shaded circles in Fig. 8.6 with accounting for spin multiplicity and statistical weight of the excited core).

With the last symmetry gone, the calculational effort to diagonalize the resulting Hamiltonian becomes enormous [17]. Even with the vastly improved computing power, as compared to the first calculations with DRFEUD about 15 years ago, the new calculations for crossed electric and magnetic fields including intermediate coupling are limited to relatively low-n states. Otherwise the matrices become too big. Comparison with experiments is therefore considerably hampered at this point.

4. Dielectronic Recombination in External Fields: Experimental Aspects

After the discussion of the previous section the question arises whether field effects can be studied in experiments. Clearly, DR is a process that involves free electrons and free ions. For a meaningful measurement one has to provide conditions such that the collision energies of the electrons and ions are well defined, the electron and ion densities are known, the interaction volume can be measured, and the ions are in a given electronic state. The most direct method to study electron-ion collisions under these conditions is to use well defined, interacting beams of electrons and ions. The beams have to be parallel in themselves but can collide under angles between 0 and 180 degrees. Measurements usually employ 90-degree interaction, i.e., crossed beams, or 0-degree interaction, i.e., one of the beams is merged within the other (merged-beam technique).

4.1 Applying Electric and Magnetic Fields to Electron-Ion Collisions

Introducing strong magnetic fields to a crossed-beams experiment will distort at least one of the two beams, e.g. the ion beam when the electron beam is guided by the magnetic field lines along the beam axis. In a merged-beam experiment, co-axial magnetic fields are not harmful since they maintain the direction of the

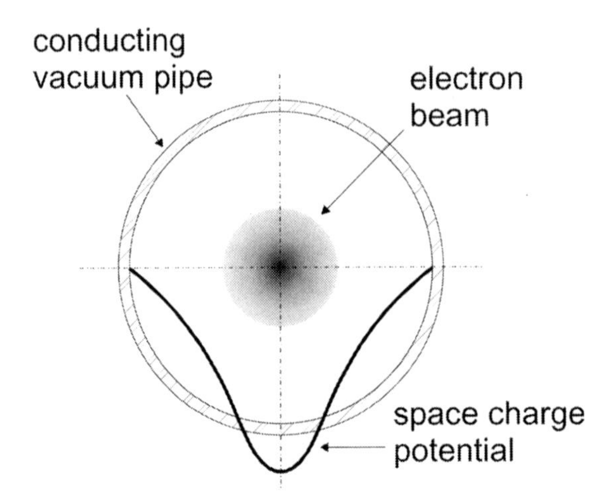

Figure 8.7. Schematic of the space charge potential $V_{\mathrm{sc}}(r)$ (solid curve) of a cylindrical electron beam within a conducting beam pipe. The space charge potential is zero at the inner surface of the beam pipe and reaches its minimum on the axis of the electron beam where, in turn, the electric field is zero.

electron and ion beams. However, electric fields in the collision region applied for example by a capacitor would influence again both beams by accelerating the charged particles. This would not only cause deflections of the beams but also introduce energy spreads across the collision region. A relatively small electric field of 100 V/cm within an interaction volume of 2 mm diameter would cause an energy spread of 20 eV compared to usually much less than 1 eV in a typical low-energy electron beam.

Even without applied electro-magnetic fields, beams of charged particles produce their own electric fields by the space charge (Fig. 8.7). Imagine a 1 keV electron beam of 1 cm diameter with 100 mA of electrical current. The electron density within this beam is assumed to be constant with $n_{\mathrm{e}} \approx 4 \times 10^{8}\ \mathrm{cm}^{-3}$. The resulting electrical charge density produces an electric field

$$\vec{E}_{\mathrm{sc}}(\vec{r}) = -\vec{\nabla} V_{\mathrm{sc}}(r) = -\frac{n_{\mathrm{e}} e}{2\epsilon_0}\vec{r} \tag{8.14}$$

that increases linearly with the distance r from the beam center and reaches its maximum, $|\vec{E}_{\mathrm{sc}}| = 190$ V/cm, at the electron beam edge. Here, ϵ_0 is the permittivity of vacuum. Only very dilute beams with electrical currents in the μA range can be produced "quasi field-free" at least as far as practical purposes are concerned. Intense beams of electrons (and all other charged particles as well) always produce electric space charge fields. Slow positive ions which are generated by electron-impact ionization of the residual gas are attracted by the negative potential and can compensate a usually unknown fraction of the electron space charge.

The discussion in the preceding paragraphs shows that the application and the control of external fields in experiments using intense ion and electron beams is a challenge. Employing external electrical fields in the laboratory is almost forbidden in an electron-ion collision experiment, and this includes the space-charge fields. Electron beam intensities therefore always have to be limited. Strong magnetic fields can only be applied parallel to the axis of a merged-beam experiment. Magnetic fields transverse to one of the two beams in any arrangement have to be limited. The same is true for any possible electric field. The only way out of some of these limitations of a colliding-beams experiment is to make use of motional (or Lorentz) electric fields $\vec{E} = \vec{v}_{\rm i} \times \vec{B}$ experienced by an ion in its rest frame [18] when the ion moves with velocity $\vec{v}_{\rm i}$ in a magnetic field $\vec{B}$. If $|\vec{v}_{\rm i}| = v_{\rm i} = \beta c = 0.1c$, i.e., 10% of the speed of light, and the perpendicular magnetic field $B = |\vec{B}|$ is only 5×10^{-5} T, i.e., of the magnitude of the earth magnetic field, then the motional electric field already amounts to $E = 15$ V/cm. Such fields may already have dramatic effects on DR cross sections (see for example Fig. 8.12).

The number of measurements on electron-ion collisions with well-known applied fields is quite limited. Work has been restricted to two classes of experimental arrangements: (1) low-energy crossed beams where a relatively substantial magnetic field up to 20 mT was applied along the electron beam axis giving rise to Lorentz electric fields of the order of 10 to 30 V/cm seen by ions moving with about 0.1 $\%$ of the speed of light; (2) accelerator-based merged-beam experiments where strong coaxial magnetic fields up to almost 0.2 T and relatively small perpendicular components of less than 3 mT could be applied so that the Lorentz electric fields reached up to 1400 V/cm at $\beta \approx 0.15$.

A different class of experiments in which laser-excited barium atoms in autoionizing states are observed in a scenario resembling the DR process, real electric and magnetic fields of the size up to 28 V/cm and 25 mT, respectively, were applied to the decaying neutral atoms.

4.2 Field Ionization of Rydberg States

External electromagnetic fields influence DR mainly through their effects on the decay properties of high Rydberg states. Such states, however, are also subject to the possible destruction by field ionization which limits the size of applicable fields. Overlaying the Coulomb field of the nucleus in a hydrogenlike ion with a uniform electric field $\vec{E}$ produces a saddle in the resulting total potential, i.e., a barrier over which electrons can escape from initially bound states with high principal quantum numbers (see Fig. 8.8). From this concept a critical principal quantum number $n_{\rm c}$ can be derived above which all Rydberg

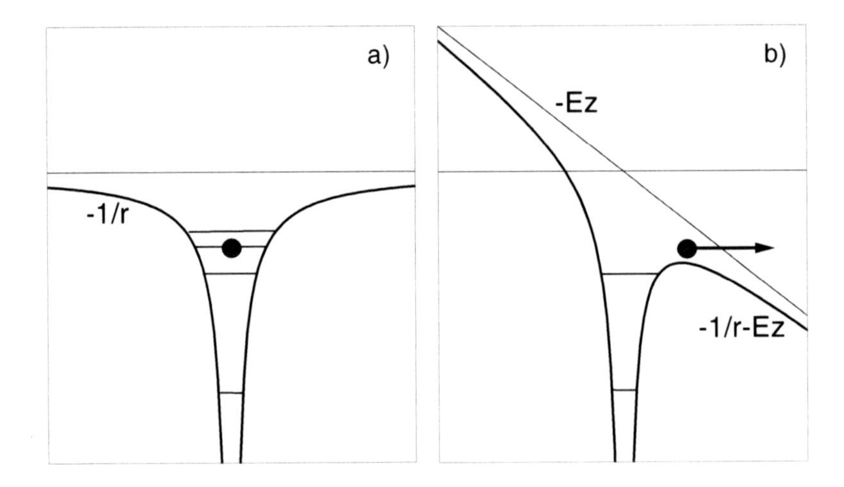

Figure 8.8. Schematic of an electron in a hydrogenic atom bound by the Coulomb potential of the nucleus (a) and released when an electric field E in z direction is switched on (b).

states are field ionized. By a simple calculation not considering the change of the electron binding energy in the presence of an electric field nor the possibility of quantum tunneling through a potential barrier one obtains for a hydrogenic system $n_{\rm c} = (Z^3 E_0/16E)^{1/4}$ with the atomic number Z and the electric field in atomic units ($E_0 = 5.142 \times 10^9 \mathrm{V/cm}$). A more elaborate but still simplifying approach [19] leads to

$$n_{\rm c} = (Z^3 E_0/9E)^{1/4}. \tag{8.15}$$

For example, the latter equation yields $n_{\rm c} = 49$ for a neutral Rydberg atom (with core charge $1e$) in a field of 100 V/cm.

In low-energy interacting-beams experiments the separation of the recombined particles from the parent ion beam requires electric fields of the order of only 10 to 50 V/cm. This leaves a wide range of principal quantum numbers for Rydberg states that survive such fields. In the high-energy merged-beam experiments, however, the fast ions have to pass strong magnetic dipole fields for charge-state separation. A typical Lorentz field in such a magnet (with an ion velocity $v_{\rm i} = 0.1c$ and a magnetic dipole field $B = 1$ T) is $E = 300$ kV/cm, which causes field stripping of hydrogenlike Rydberg atoms with $n \geq n_{\rm c} = 7$.

Reality is yet more complicated. Field ionization is not only dependent on the principal quantum number. It is different for different Stark states within one n-manifold and even for different orientations characterized by the m quantum number. Hence, there is not a sharp cut-off number $n_{\rm c}$ but rather a distribution that depends on the population of the different Stark states. Moreover, ions in Rydberg states with $n > n_{\rm c}$ populated by DR in the collision region can radiatively decay to a state with $n' < n_{\rm c}$ before they arrive at the charge-state analyzing magnet and thus survive the strong dipole field. Sorting out these intricacies complicates the interpretation of experiments studying field effects on DR. In fact, these difficulties presently prevent us from a detailed quantitative comparison of theory and experiment: due to limitations in the available computing power theory can presently provide calculations only for few n values of a given ion where n may not be too high (due to the multiplicity $2n^2$ for each Rydberg electron) while the experiment can presently not deliver measured DR results for a single well characterized n-manifold of states.

4.3 Measurement of DR Cross Sections and Rate Coefficients

The DR process is experimentally accessible by detecting at least one of its reaction products, for example an ion which has changed its charge state by -1 (see Eq. 8.2). In order to distinguish DR (Eqs. 8.1 and 8.2) from RR (Eq. 8.3) the resonance character of DR can be exploited. For the determination of an absolute cross section mono-energetic beams of electrons and ions in a well defined geometry are especially suited, although DR cross sections have also been derived from experiments with trapped ions [20].

In a crossed-beams experiment with beam diameters of typically about 1 mm the reaction volume is rather small. This leads to comparatively low signal count rates but gives access to the reaction products over a solid-angle range of nearly 4π. A small reaction volume is mandatory if time-of-flight detection techniques are to be used with good time resolution.

Generally, the reaction rate of colliding electrons and ions with densities $n_{\rm e}$ and $n_{\rm i}$ and velocities $\vec{v}_{\rm e}$ and $\vec{v}_{\rm i}$ in a volume τ for a given cross section $\sigma(v)$ and relative velocity between the particles $v_{\rm rel} = |\vec{v}_{\rm e} - \vec{v}_{\rm i}|$ is given by

$$R = \int_\tau \int_v n_{\rm e}(\vec{r})\, n_{\rm i}(\vec{r})\, \sigma(v)\, v\, f(\vec{v}, v_{\rm rel})\, d^3\vec{v}\, d^3\vec{r}. \tag{8.16}$$

For the deduction of cross sections from reaction rates R the spatial density distributions of the particles and the distribution function $f(\vec{v}, v_{\rm rel})$ of the relative velocities have to be determined. Eq. 8.16 implies that f is independent of the location $\vec{r}$. Therefore, the integration over velocity space can be carried out independently and produces the rate coefficient

$$\alpha(E_{\rm rel}) = \langle v\sigma \rangle = \int v\, \sigma(v) f(\vec{v}, v_{\rm rel}(E_{\rm rel}))\, {\rm d}^3\vec{v}. \tag{8.17}$$

In non-relativistic collisions the relative energy is given by

$$E_{\rm rel} = \frac{\mu}{2} v_{\rm rel}^2, \tag{8.18}$$

where $\mu = m_{\rm e} m_{\rm i}/(m_{\rm e} + m_{\rm i})$ is the reduced mass of the colliding particles, electrons with mass $m_{\rm e}$ and ions with mass $m_{\rm i}$. In experiments with mono-energetic particle beams $f(\vec{v}, v_{\rm rel})$ is a narrow distribution of $v = |\vec{v}|$ around $v_{\rm rel}$. Therefore, with not too small $v_{\rm rel}$, an experimental (apparent) cross section can be derived from a measurement using the expression $\langle v\sigma\rangle / v_{\rm rel}$.

In Eq. 8.16 the particle densities $n_{\rm e}$ and $n_{\rm i}$ are related to the electron and ion flux densities $J_{\rm e}$ and $J_{\rm i}$ via the continuity equation $J_{\rm e,i} = n_{\rm e,i} v_{\rm e,i}$, respectively. Accordingly, Eq. 8.16 can be rewritten as

$$R = \frac{\langle v\sigma\rangle}{v_{\rm e} v_{\rm i}} \Omega \qquad (8.19)$$

with the form factor Ω accounting for the spatial overlap of the two beams

$$\Omega = \iiint J_{\rm e}(x, y, z)\; J_{\rm i}(x, y, z)\, dx\, dy\, dz. \qquad (8.20)$$

In a crossed-beams experiment with zero-divergence beams located in the x-y plain the form factor can be simplified since there is no density dependence of the flux densities then in the direction of each beam. The integrations over coordinates x and y can be readily performed producing a new expression for the beam overlap now explicitly containing the angle θ between the two beams

$$\Omega = \frac{1}{q e^2 \sin\theta} \int i_{\rm e}(z)\; i_{\rm i}(z)\, dz. \qquad (8.21)$$

The charge state q of the ions enters Eq. 8.21 by the replacement of particle fluxes by electrical currents which can be determined directly by a measurement. Experimentally the integral can be evaluated by measuring differential currents $i_{\rm e}(z)$ and $i_{\rm i}(z)$ in each beam transmitted through a narrow horizontal slit at a number of positions z_k and replacing the integral by a sum over k. The essential device for this measurement is indicated by the 'beam probe' displayed in Fig. 8.11.

It is apparent that collision volumes in crossed beams are small and, hence, also the observable reaction rates are small in most of these experiments. If the primary concern is to achieve a count rate as high as possible the merged-beam arrangement is the method of choice. In the case of the electron-ion interaction studies of interest in this article the electron beam overlaps the ion beam over a distance of typically 0.5–2.5 m. Consequently the interaction volume is large. The merging of the electron beam with the ion beam (and the demerging) is achieved by electric or magnetic fields or a combination of the two. In order to prevent especially the electron beam from being blown up by its own space charge (see Sec. 8.4.1) it is magnetically guided over its entire path by a magnetic field with the field vector $\vec{B}$ being aligned parallel to the desired electron beam direction.

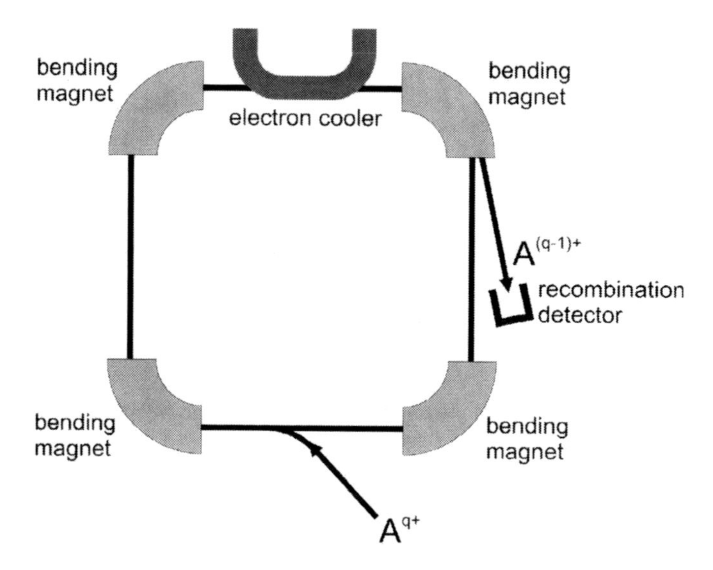

Figure 8.9. Schematic of a heavy ion storage ring set-up used for electron-ion recombination experiments. The electron cooler usually serving for ion-beam quality improvement is employed here as an electron target. The ring bending magnets are employed for the separation of recombined ions from the circulating parent ion beam.

The merged-beam method was recently reviewed [21]. It has been most successfully implemented at heavy-ion storage rings [22, 23] where high-quality ion beams can be prepared with the aid of the electron-cooling technique. Electron cooling [24] is the transfer by elastic collisions of the internal kinetic energy of the ensemble of ions to a heat bath made up of cold electrons. It leads to a small diameter and a small momentum spread of the ion beam, i.e., to a smaller volume in phase space occupied by the ion beam. Consequently, more ions can be stored in a given volume and thus higher ion currents be achieved when the ion beam is cooled in between successive injections of ions into the storage ring.

Electron cooling is most effective when both the electron and ion ensembles move with identical mean velocities. The corresponding laboratory electron energy is termed "cooling energy" and can be calculated from the ion laboratory energy E_{i} as $E_{\mathrm{cool}} = (m_{\mathrm{e}}/m_{\mathrm{i}})E_{\mathrm{i}}$ where m_{e} and m_{i} are the electron mass and the ion mass, respectively. At cooling the relative energy between electron and ion beam is $E_{\mathrm{rel}} = 0$. Nonzero relative energies are achieved by setting the electron laboratory energy E_{lab} to values different from E_{cool} yielding, in a non-relativistic approximation,

$$E_{\mathrm{rel}} = \left(\sqrt{E_{\mathrm{lab}}} - \sqrt{E_{\mathrm{cool}}}\right)^2. \tag{8.22}$$

For a relativistically correct expression see e.g. [25].

Figure 8.10. Schematic of an electron-ion merged-beam arrangement as it is used in heavy ion storage rings; a relatively wide beam (diameters up to 5 cm) of cold electrons is guided by a co-axial magnetic field produced by the indicated solenoid. The beam of fast cooled ions has a diameter of typically several mm and is well centered within the electron beam such as to minimize space charge fields in the collision region. This is accomplished by using additional magnetic coils indicated by the lines on top of the solenoid that serve for the fine alignment of the two interacting beams. The coils are used to produce small transverse magnetic field components in any desired azimuthal direction so that the resulting total magnetic field and hence also the electron beam point in the direction of the ion beam with angular deviations as low as 0.01^o. The same coils are used to produce Lorentz electric fields in the electron-ion collision region in order to study field effects on DR.

A typical arrangement for an electron-ion collision experiment at a heavy-ion storage ring is shown in Fig. 8.9. In a recombination experiment the electron cooler after having served its purpose to help prepare an intense, well defined ion beam circulating in the storage ring, is now used as an electron target from which the ions may capture an electron. After such a recombination event the ion will no longer travel on a closed orbit in the storage ring, since it is less deflected by the first bending magnet behind the cooler. After having left the bending magnet at a position different from that of the circulating ions, recombined ions are easily collected and counted with standard single-particle detection techniques. The generation of motional electric fields needed for the study of the field dependence of DR is explained in the caption of Fig. 8.10. It should be noted that the motional electric field is perpendicular to the magnetic guiding field inside the electron cooler.

As in all scattering experiments the quantity measured in a storage ring merged-beam recombination experiment is primarily the rate coefficient $\alpha(E_{\mathrm{rel}})$ (see Eq. 8.17). Since the electron cooler produces a beam of uniform electron density and since the ion beam is totally immersed in the electron beam, the form factor Ω can be simply determined by pulling the electron flux density J_{e}

out of the integral in Eq. 8.20 so that one obtains

$$\Omega = \frac{j_e I_i L}{q e^2} \tag{8.23}$$

with the electrical current density $j_e = e n_e v_e$ of the electrons, the length L of the interaction region and the effective electrical current I_i of the stored ions. Considering the relativistic length contraction and time dilatation one obtains

$$\alpha(E_{rel}) = \frac{R_{exp}\, v_i\, eq\, \gamma^2}{n_e\, I_i\, L\, \eta}. \tag{8.24}$$

The recombination rate $R = R_{exp}/\eta$ is obtained from the measured counting rate R_{exp} by correcting for the detection efficiency η. The electron density in the laboratory is n_e and γ is the relativistic factor for the transformation between the laboratory frame and the ion rest frame. Eq. 8.17 can be used to compare results from merged-beam measurements with theoretical cross sections σ. For such a comparison also the apparent cross section defined as $\langle v\sigma \rangle / v_{rel}$ may be used as has been done e.g. in Fig. 8.3.

In an electron cooler the electron velocity distribution can be characterized by two temperatures $T_\parallel$ and $T_\perp$ parallel and perpendicular to the electron beam direction. It reads

$$f(\vec{v}, v_{rel}) = \left(\frac{m_e}{2\pi k_B}\right)^{3/2} \frac{1}{T_\parallel^{1/2} T_\perp} \exp\left(-\frac{m_e (v_\parallel - v_{rel})^2}{2 k_B T_\parallel} - \frac{m_e v_\perp^2}{2 k_B T_\perp}\right) \tag{8.25}$$

with $v_\parallel$ and $v_\perp$ denoting the components of $\vec{v}$ perpendicular and parallel to the electron beam direction, respectively, and k_B being the Boltzmann constant. With $T_\parallel = T_\perp$ Eq. 8.25 represents an isotropic Maxwellian electron velocity distribution. In an electron cooler, however, $T_\parallel \ll T_\perp$ holds. Typical values are $k_B T_\parallel = 0.1$ meV and $k_B T_\perp = 10$ meV. Therefore, Eq. 8.25 with $T_\parallel < T_\perp$ has been termed "flattened" Maxwellian. The width of this distribution determines the experimental electron energy spread

$$\Delta E_{rel}(\text{FWHM}) = \left[(\ln(2)\, k_B T_\perp)^2 + 16 \ln(2)\, k_B T_\parallel\, E_{rel}\right]^{1/2}. \tag{8.26}$$

According to Eq. 8.26 the experimental energy spread increases when going to higher relative energies. This consequence of the merged-beam kinematics eventually limits the capability of resolving individual high-n DR resonances. The resolution of such resonances would be desirable for the study of the effect of external fields on a single n-manifold of Rydberg states (see discussion below).

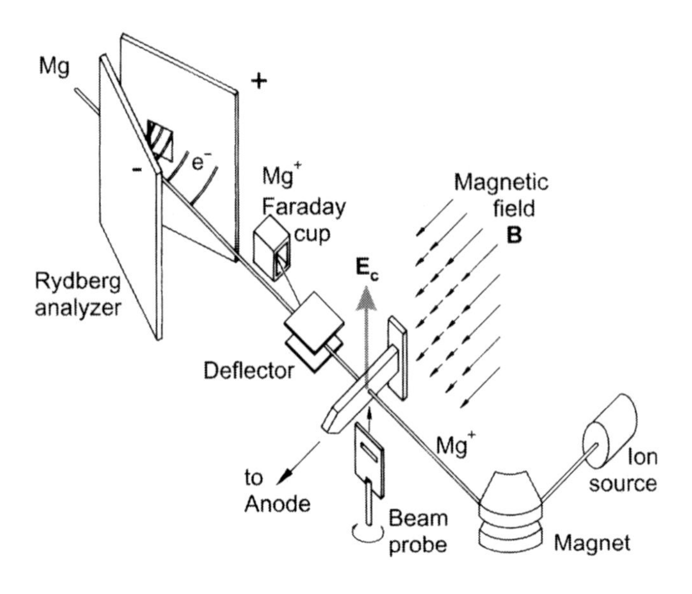

Figure 8.11. Schematic of an electron-ion crossed-beams recombination experiment using controlled and known magnetic $\vec{B}$ and motional electric $\vec{E}_c$ fields in the collision region. The rotatable beam probe serves for measuring the beam overlap.

5. Overview of Experiments Studying Field Effects on DR

5.1 Crossed-Beams Experiments

Results from early recombination experiments were obtained without considering the possible presence and influence of external fields (see reviews by Dunn [26], and Dittner and Datz [27]). Equally, theoretical calculations did not include influences of static electric fields although earlier work had shown the effect of plasma microfields on DR [10]. Experimental data exceeded the theoretical calculations by large factors which soon led to the inclusion of effects of static electric fields in the theory [28]. Experimental data could be brought in agreement with the calculations under the assumption that external electric fields had been present in the interaction region. The first experiment where external fields were applied under well controlled conditions (see Fig. 8.11) was performed by Müller *et al* [29], who investigated DR in the presence of known external fields of singly charged Mg^+ ions. They observed an enhancement beyond theoretical zero-field predictions (see Fig. 8.12) by about an order of magnitude and found an increase of the measured n-dependent DR cross section by factors up to about 3 depending on n when increasing the motional $\vec{v} \times \vec{B}$ electric field from 7.2 to 23.5 V/cm. The agreement of these results with theoretical predictions that included effects of electric fields [1, 28, 30]

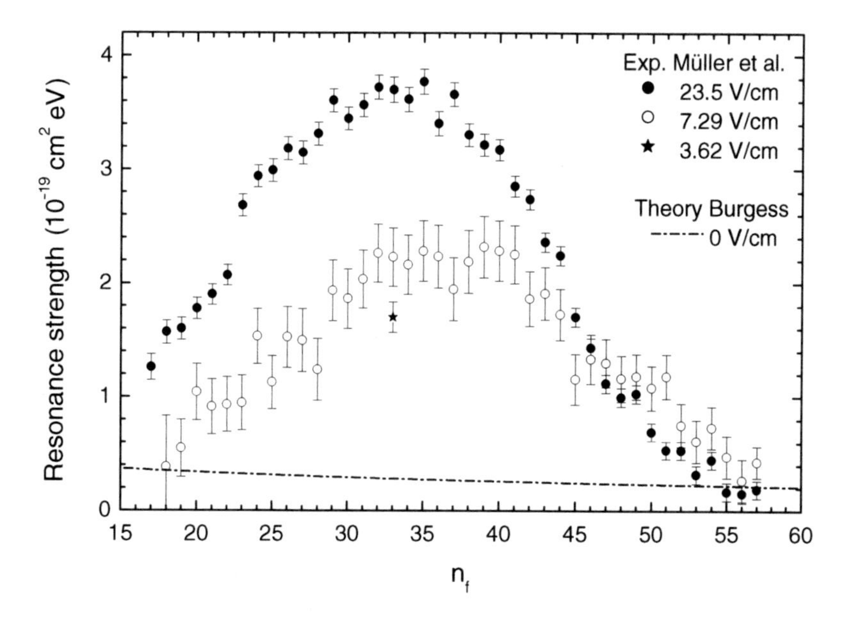

Figure 8.12. Collision strengths for resonance contributions to dielectronic recombination of Mg^+ ions. The experimental data were obtained with the arrangement displayed in Fig. 8.11 [13, 29] for three motional electric fields as indicated. The dash-dotted line represents a theoretical calculation by Burgess [13] for zero applied field in the collision region. No effects of field ionization are included in the calculation. The horizontal axis is meant to represent the Rydberg quantum number n. In the experiment n was determined by employing a Rydberg detector (see Fig. 8.11) based on field ionization. The experimental quantity n_f is a function of the electric field E applied to the Rydberg analyzer and for this display the relation between n_f and E, $n_f = (E_0/16E)^{1/4}$, was used (see also Eq. 8.15).

was at the 20 % level. Further experiments with multiply charged C^{3+} ions also revealed drastic DR rate enhancements by electric fields [31, 32]; however, the large uncertainties of these measurements left ambiguities. It should be noted that the Mg^+ experiment of Müller *et al* [13, 29] is unique because it directly determined the Rydberg state distribution of the recombined ions. In the storage-ring experiments discussed below only the total recombination cross sections have been measured so far.

A serious problem with using multiply charged ions in a DR experiment at low ion energies is related to the electron capture of the ions from the residual gas in the apparatus. Such electron transfer processes give rise to strong background that masks the signal from electron-ion recombination. The effect is less severe for singly charged ions. The background-producing cross section σ_c for electron capture from a residual gas particle has a very strong energy dependence. Above a specific energy of roughly 25 keV/u σ_c drops with an ion velocity dependence that reaches a v^{-11} proportionality. Therefore, the use of accelerated ion beams in a merged-beam experiment provides many advantages and is apparently the only viable scheme to study DR of highly charged ions. In particular, the use of heavy-ion storage rings has become the method of choice to study DR at low relative energies in the electron-ion c.m. frame.

5.2 Accelerator-Based Merged-Beam Experiments

The first DR experiment studying field effects with highly charged ions at a storage ring was carried out with Si^{11+} ions by Bartsch *et al* [33]. It produced results with an unprecedented accuracy, enabling a detailed comparison with theory that included pure electric fields at that time. The observed enhancement of the measured DR rate reached about a factor of 3 when the field was increased from 0 V/cm to 183 V/cm. This magnitude of the effect and the overall behavior of the field dependent cross section was in fair agreement with the theory, however, considering the quality and detail of the experimental results, this agreement was not fully satisfactory. This finding stimulated theoretical investigations of the role of the additional magnetic field which is always present in storage ring DR experiments, since it is needed to guide and confine the electron beam within the electron cooler.

In a model calculation Robicheaux *et al* [34], found that in a configuration of crossed E and B fields indeed the magnetic field influences the field-induced rate enhancement through the mixing of m levels. More detailed calculations confirmed these results. We note here again that in theoretical calculations by Huber and Bottcher [16], no influence of a pure magnetic field of at least up to 5 T on DR was found. It is indeed the combination of electric and magnetic fields with non-vanishing mutually perpendicular components that produces additional effects.

Inspired by these predictions storage ring DR experiments with controlled and variable electric and magnetic fields were performed using Li-like Cl^{14+} ions ([35], Figs. 8.13 and 8.14) and Ti^{19+} ions [36] and a distinct effect of the magnetic field strength on the magnitude of the DR rate enhancement was clearly discovered. The electric field effect decreased monotonically with the B field increasing from 30 mT to 80 mT. A decrease of the electric field enhancement by a crossed magnetic field is also predicted by the model calculation of Robicheaux, Pindzola, and Griffin [34] for magnetic fields larger than approximately 20 mT where, due to a dominance of the magnetic over the electric interaction energy, the l-mixing weakens and consequently the number of states participating in DR decreases.

Most of the storage ring measurements on DR in fields (DRF) have been carried out on Li-like ions. This has two reasons. First, low-energy DR resonances with their comparatively large autoionization rates are the most likely to be subject to field effects and low-energy DR measurements can exploit the

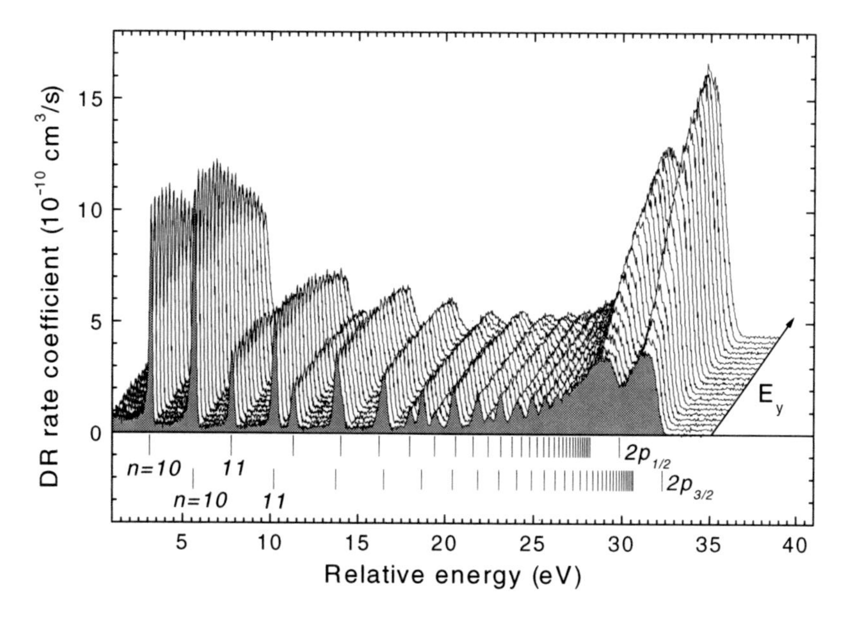

Figure 8.13. Dielectronic recombination rate coefficients of Li-like Cl^{14+} [35] at different motional electric fields ranging from 0 to 380 V/cm. Level energies $E(n)$ of the $2p_{1/2}nl$ and $2p_{3/2}nl$ resonances are indicated. They are calculated using the Rydberg formula $E(n) = \Delta E_j - q^2\, 13.6\,\text{eV}/n^2$ with $q = 14$ the parent ion charge state and ΔE_j, $(j = 1/2, 3/2)$, the $2s \to 2p_j$ core excitation energies.

special strength of the storage ring merged-beam technique. Second, the ratio of ion charge q over ion mass m_i is quite large for Li-like ions. It would only be higher for He-like and H-like ions which do not support sizable DR resonances at very low electron-ion collision energies. High values for q/m_i at a given ion velocity result in lower rigidity of the ion beam, i.e., the dipole magnets of the ring can be set to lower magnetic fields. One consequence of this is a reduced Lorentz field experienced by the recombined ions and hence survival and detection of Rydberg (DR product) states with higher n quantum number. High-n Rydberg states are the best candidates for studying electric field effects.

The aim of the most recent experimental investigations of DRF at heavy ion storage rings with Li-like Ni^{25+} [25], Ne^{7+} [37] and O^{5+} [38] was to extend previous studies to ions in a wider range of atomic numbers Z. Strong field effects require a drastic l-dependence of field-free autoionization rates. As exemplified in Fig. 8.4, higher-Z ions have a flatter l-dependence than low-Z ions. Hence the field effect on DR is expected to decrease for more highly charged ions. Moreover, when A_r exceeds the highest autoionization rate within a n-manifold of Rydberg states the mixing of l-states becomes irrelevant for the DR cross section. Due to the strong Z-dependence of A_r this limit is reached at

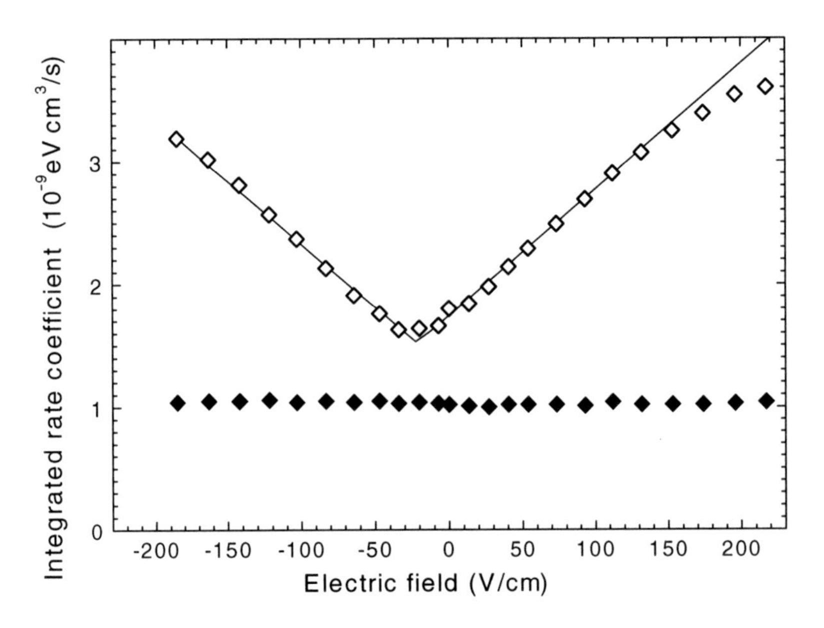

Figure 8.14. Integrated Cl^{14+} DR rate coefficients as a function of the motional electric field strength [35]. The low energy integration range (full symbols) and the high energy integration range (open symbols) have been 2–15 eV and 24.6–35 eV, respectively. The low energy range comprises $2p_{1/2}nl$ resonances with $10 \leq n \leq 14$ and $2p_{3/2}nl$ resonances with $10 \leq n \leq 13$ (see Fig. 8.13). These are not influenced by the electric field as opposed to the $2p_{1/2}nl$ resonances with $n \geq 23$ and $2p_{3/2}nl$ resonances with $n \geq 19$ contained in the high energy range. The full line is a linear fit to the high-energy integrals.

lower n for heavier isoelectronic ions. Results obtained previously by Müller *et al* [13], using a simple model calculation and DRFEUD calculations by Griffin and Pindzola [39], already predicted this behavior. A specific aspect of going to high Z is that the separation in energy of Rydberg resonances in a given range of n increases with Z^2 and therefore different Rydberg states can be resolved more easily if the ion charge state is high. This has been demonstrated by Shi *et al* [40], who resolved $2p_{1/2}nl$ DR resonances up to $n = 45$ in the recombination spectrum of Li-like Bi^{80+} ions

The improved capability for resolving high-n levels in highly charged ions is counteracted by the decreasing influence of fields on DR for high-Z ions. Moreover, efficient mixing of l-states in a given n-manifold that is well resolved at zero external electric field requires strong fields which deteriorate the experimental resolution and which produce a fan of Stark states that ultimately overlaps with the Stark states associated with the next lower or next higher Rydberg quantum number n. Thus, it has not been possible to observe and measure field effects on one isolated Rydberg n-manifold.

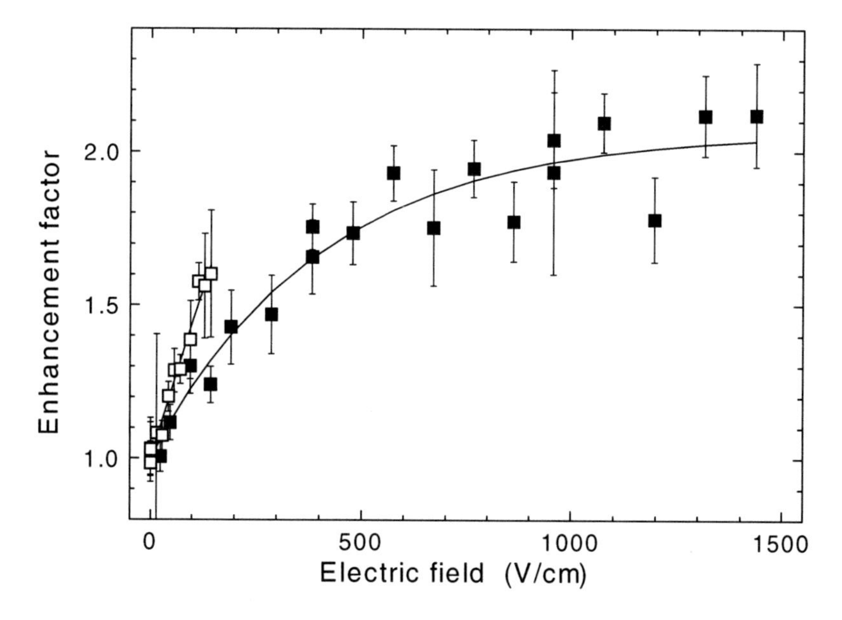

Figure 8.15. Measured enhancement of the DR cross section of Ne^{7+} ions by external electric fields and crossed magnetic fields [37]. The enhancement factor is defined as a normalized integrated recombination rate coefficient shown in Fig. 8.14. Clearly the enhancement factor as a function of the electric field rises more steeply in a lower magnetic field (30 mT, open symbols) than in a higher magnetic field (180 mT, closed symbols). At the higher magnetic field, where higher electric fields could be applied, the saturation of the field effect is observed. The full lines are fits through the data points to guide the eye.

For lower-Z ions the field effects are potentially more pronounced and therefore also lighter Li-like ions were studied. With Ne^{7+} ions experimental conditions were applied such as to demonstrate saturation of field enhancement of the DR cross section (Fig. 8.15). The principle of such saturation is quite obvious: once the l-states are sufficiently mixed so that all states corresponding to the black and the white circles in Fig. 8.6 can contribute to the DR process, no further enhancement will be possible. Only by changes of the magnetic field more states can be made available.

The latest attempt to study field effects on DR for a narrow bandwidth of Rydberg states was made with O^{5+} ions. In that experiment the field effects were measured with different energies of the O^{5+} parent ion beam. The philosophy of changing the ion energy is to vary the Lorentz field met by the ions in the storage ring dipole magnets. The effect is two-fold. Reduced energy means reduced ion velocity and, hence the motional fields seen by the ions in a given magnetic field decrease. Moreover, at lower energy the ion beam becomes less rigid and therefore the ring magnets have to be lowered in their fields such as to keep the parent ion beam on its closed orbit. Different Lorentz fields seen by the recombined Rydberg ions mean different field-ionization cut-off numbers (cf. Eq. 8.15). Thus, the measurement comprises different ranges of (unresolved) Rydberg n-states. In a simple approach, one can subtract from one another measured cross sections obtained at different energies (and hence with different cut-off quantum numbers $n_c^{(1)}$ and $n_c^{(2)}$) and relate the difference with the bandwidth of n-states with $n_c^{(1)} \leq n \leq n_c^{(2)}$. This is an important step towards the desired quantitative comparison of experiment with theory which, at this time, can only provide results on DR in crossed electric and magnetic fields for a very limited range of n-states as mentioned already several times above.

The simple approach to obtain n-differential DR results is similar to the method applied by Müller *et al* [13, 29], in their pioneering experiment on Mg^+ ions. It had been shown there already that the simple picture of a sharp cut-off n_c of Rydberg states is not realistic. It is rather necessary to determine the survival and detection probabilities of Rydberg states in the specific experiment. This would in principle require the knowledge of the population of all Stark states of recombined ions along the flight path from the collision region to the entrance of the ring dipole magnet that separates the recombined product from the parent ions which continue circulating the ring. Experimental determination of such population distributions is not feasible. However, the Rydberg states under consideration can be reliably treated in a hydrogenic approach. Thus, the mixing, the radiative decay, and the field ionization of the Rydberg states can be modeled.

Such model calculations have been carried out in the spirit of the earlier treatment by Müller *et al* [29], now also for highly charged ions where the radiative stabilization ($A_r \propto Z^4$) along the flight path between the locations of Rydberg-ion production (by DR) and field ionization becomes substantial. The extended model for the detection probability of Rydberg ions has recently been described by Schippers *et al* [41]. It has been applied to the case of O^{5+} ions and the associated studies at different ion beam energies [38]. The n-distribution following the simple approach of Eq. 8.15 would be a constant in the range from $n_c^{(1)}$ to $n_c^{(2)}$ resulting from the variation of the beam energy. In reality, a slightly broadened bell-shaped probability distribution is found by the model. From the experiment with O^{5+} ions [38] one can therefore indeed get n-differential information on the contribution and the field effect on DR involving narrow n-bins of Rydberg states.

5.3 Laser-Excitation of Atoms Providing a Model System for Investigating Field Effects on DR

A completely different experimental approach has been chosen by Gallagher and coworkers to study field effects on DR [42, 43]. They studied recombination

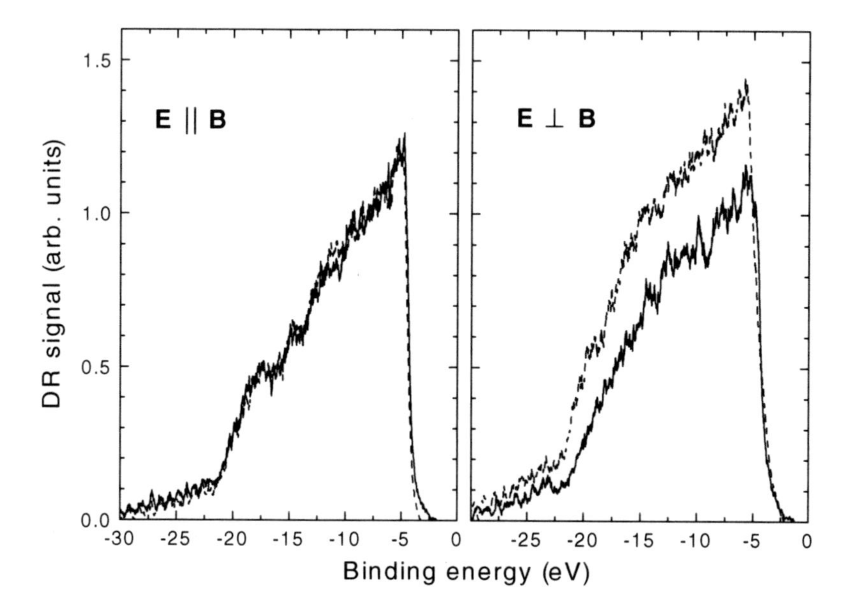

Figure 8.16. Measured recombination signal from laser ionized Ba^+ ions [43] in external electric and magnetic fields (dashed lines) compared to measurements in a pure magnetic field (full lines) for two field configurations, $\vec{E}||\vec{B}$ (left panel) and $\vec{E}\perp\vec{B}$ (right panel). The abscissa is the binding energy of the Rydberg electron in the intermediate DR state. The binding energy is approximately equal to $-13.6\,\mathrm{eV}\,/\,n^2$ and thus smoothly increases with the principal Rydberg quantum number n.

of Ba^+ ions from a continuum of finite bandwidth which they had prepared by laser excitation of neutral Ba atoms. This system resembles a DR process and provides a model case for detailed studies of field effects within this special system. The method is not suitable for the determination of DR cross sections and rates for free electrons colliding with free ions as are the other experiments discussed so far.

For a given electric field strength of 0.5 V/cm, they find that the recombination rate is increasingly enhanced by crossed magnetic fields up to about 20 mT. However, there is no effect of the magnetic field when it is directed parallel to the electric field vector (see Fig. 8.16. For the m-mixing to occur the crossed E and B arrangement is essential. In the case of parallel B and E fields m remains a good quantum number and no influence of the magnetic field is expected. The group has studied effects of pure static electric fields [42, 44], crossed static electric and magnetic fields [43] and microwave fields [45] on the DR process.

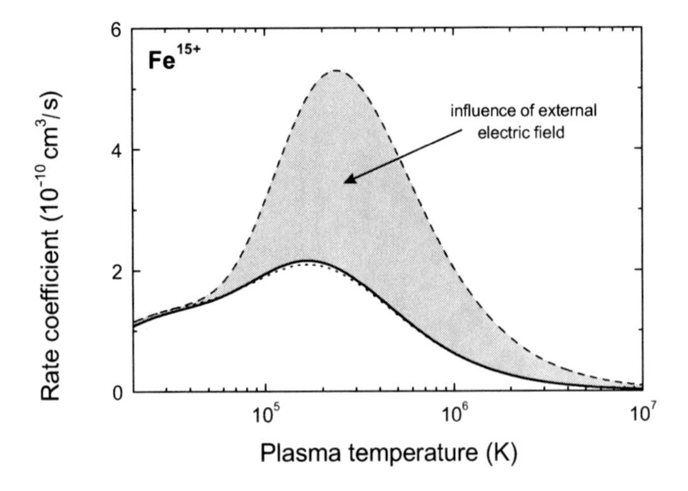

Figure 8.17. Influence of an external electric field on Maxwellian-averaged plasma rate coefficients for DR of Fe^{15+} ions with electrons. The rate coefficients are for DR processes involving $3s \rightarrow 3l$ and $3s \rightarrow 4l$ core excitations in the energy range 0 to 40 eV only. The solid line was calculated [46] by convolution of experimental results obtained at the Heidelberg heavy-ion storage ring TSR [47]. These data were taken with at most little external field (< 10 V/cm) in the collision region. The theoretical (dotted and dashed) curves were obtained by convoluting cross sections published by Griffin and Pindzola [39], for zero field and for a field of 10 kV/cm, respectively, where for the latter saturation of the field enhancement was expected. The calculations were limited to Rydberg states up to n=100. The experimental data are truncated by a field-ionization cutoff at about the same principal quantum number.

6. Summary

Dielectronic recombination via high Rydberg states is influenced by static electric fields. Relatively small fields already can enhance cross sections for dielectronic recombination by as much as an order of magnitude. This is particularly true for low-charge state ions. Such field effects are potentially important in plasmas, whether of natural or technical origin. The quantity describing the probability of collisions in a plasma is a rate coefficient that is obtained from Eq. 8.25 and 8.17 with a constant temperature $T = T_\perp = T_\parallel$ and with the relative velocity of the electron and ion ensembles $v_{rel} = 0$. An example for the magnitude of electric field effects on such a plasma rate coefficient is shown in figure 8.17 for dielectronic recombination of a moderately highly charged ion, Fe^{15+}. Depending on the plasma temperature the rate coefficient may be enhanced by up to a factor of three in a sufficiently strong electric field.

Pure magnetic fields have no effect on dielectronic recombination unless their strength exceeds approximately 10 T. Magnetic field components, however, perpendicular to an electric field have an additional influence on the magnitude of the cross section. Time-varying fields, such as the plasma microfields produced by high-density charged particles, or electromagnetic radiation can also strongly influence dielectronic recombination. A grand goal of further experimental and theoretical work is the test of calculations for measured field effects of single Rydberg states and the extension of this work to complete Rydberg series when both electric and perpendicular magnetic fields are present in the electron-ion collision region.

Acknowledgments

We are grateful for the fruitful and enjoyable collaboration on DR in fields with T. Bartsch, S. Böhm, W. Shi (Giessen, Germany), G. Gwinner, D. Schwalm, A. Wolf (Heidelberg, Germany), H. Danared, R. Schuch (Stockholm, Sweden), G. H. Dunn (Boulder, Colorado, USA), M. S. Pindzola (Auburn, Alabama, USA), D. C. Griffin (Winter Park, Florida, USA), and numerous other colleagues who helped to collect the material and to gain the present understanding of DR in fields summarized in this article.

References

[1] Y. Hahn and K. J. LaGattuta, *Dielectronic recombination and related resonance processes* Phys. Rep. **166**, 195–268 (1988).

[2] Y. Hahn, *Electron-ion recombination processes — an overview* Rep. Prog. Phys. **60**, 691–759 (1997).

[3] D. R. Bates and A. Dalgarno, Electronic recombination. in: *Atomic and Molecular Processes*, ed. D. R. Bates, chapter 7, pages 245–271. Academic Press, New York (1962).

[4] J. Dubau and S. Volonté, *Dielectronic recombination and its applications in astronomy* Rep. Prog. Phys. **43**, 199–252 (1980)

[5] H. A. Bethe and E. E. Salpeter, *Quantum Mechanics of One- and Two-Electron Atoms.* Springer, Berlin, Heidelberg, New York (1957)

[6] G. Kilgus, J. Berger, P. Blatt, M. Grieser, D. Habs, B. Hochadel, E. Jaeschke, D. Krämer, R. Neumann, G. Neureither, W. Ott, D. Schwalm, M. Steck, R. Stokstad, E. Szmola, A. Wolf, R. Schuch, A. Müller, and M. Wagner, *Dielectronic recombination of hydrogenlike oxygen in a heavy-ion storage ring* Phys. Rev. Lett. **64**, 737–740 (1990)

[7] M. S. Pindzola, N. R. Badnell and D. C. Griffin, *Dielectronic recombination cross sections for H-like ions* Phys. Rev. A **42**, 282–285 (1990)

[8] A. Burgess, *Dielectronic recombination and the temperature of the solar corona* Astrophys. J. **139**, 776–780 (1964)

[9] A. Burgess and H. P. Summers, *The effect of electron and radiation density on dielectronic recombination* Astrophys. J. **157**, 1007–1021 (1969)

[10] V. L. Jacobs, J. Davies and P. C. Kepple, *Enhancement of dielectronic recombination by plasma electric microfields* Phys. Rev. Lett. **37**, 1390–1393 (1976)

[11] N. R. Badnell, *Dielectronic recombination of* Fe^{22+} *and* Fe^{21+} J. Phys. B **19**, 3827–3835 (1986) http://amdpp.phys.strath.ac.uk/autos/

[12] V. L. Jacobs and J. Davis, *Properties of Rydberg autoionizing states in electric fields* Phys. Rev. A **19**, 776–786 (1979)

[13] A. Müller, D. S. Belić, B. D. DePaola, N. Djurić, G. H. Dunn, D. W. Mueller, and C. Timmer, *Experimental measurements of field effects on dielectronic recombination cross sections and Rydberg product-state distributions* Phys. Rev. A **36**, 599–613 (1987)

[14] D. C. Griffin, M. S. Pindzola, and C. Bottcher, *Distorted-wave calculations of dielectronic recombination as a function of electric field strength* Phys. Rev. A **33**, 3124–3132 (1986)

[15] F. Robicheaux and M. S. Pindzola, *Enhanced dielectronic recombination in crossed electric and magnetic fields* Phys. Rev. Lett. **79**, 2237–2240 (1997)

[16] W. A. Huber and C. Bottcher, *Dielectronic recombination in a magnetic field* J. Phys. B **13**, L399 (1980)

[17] D. C. Griffin, F. Robicheaux, and M. S. Pindzola, *Theoretical calculations of dielectronic recombination in crossed electric and magnetic fields* Phys. Rev. A **57**, 2708–2717 (1998)

[18] J. D. Jackson, *Classical Electrodynamics* John Wiley & Sons, New York, 2nd edition (1975).

[19] T. F. Gallagher, *Rydberg Atoms* Number 3 in Cambridge Monographs on Atomic, Molecular and Chemical Physics. Cambridge University Press, Cambridge, U.K. (1994)

[20] D. A. Knapp, The use of electron beam ion traps in the study of highly charged ions in: *Physics with multiply charged ions*, ed. D. Liesen, pages 143–169. Plenum Press, New York (1995)

[21] R. A. Phaneuf, C. C. Havener, G. H. Dunn, and A. Müller, *Merged-beams experiments in atomic and molecular physics* Rep. Prog. Phys. **62**, 1143–1180 (1999)

[22] R. Schuch, Cooler storage rings: New tools for atomic physics In: *Review of Fundamental Processes and Applications of Atoms and Ions*, ed. C. D. Lin, pages 169–212. World Scientific, Singapore (1993)

[23] A. Müller and A. Wolf, Heavy ion storage rings In: *Accelerator-based atomic physics techniques and applications*, ed. J. C. Austin and S. M. Shafroth, pages 147–182. AIP Press, Woodbury (1997)

[24] H. Poth, *Electron cooling: theory, experiment, application* Phys. Rep. **196**, 135–297 (1990)

[25] S. Schippers, T. Bartsch, C. Brandau, A. Müller, G. Gwinner, G. Wissler, M. Beutelspacher, M. Grieser, A. Wolf, and R. A. Phaneuf, *Dielectronic recombination of lithiumlike* Ni^{25+} *ions — high resolution rate coefficients and influence of external crossed E and B fields* Phys. Rev. A **62**, 022708 (2000)

[26] G. H. Dunn, Early dielectronic recombination measurements: Singly charged ions. In: *Recombination of Atomic Ions* ed. W. G. Graham, W. Fritsch, Y. Hahn, and J. A. Tanis, volume 296 of *NATO ASI Series B: Physics*, pages 115–131. Plenum Press, New York (1992)

[27] P. F. Dittner and S. Datz, Early measurements of dielectronic recombination: Multiply charged ions. In: *Recombination of Atomic Ions* ed. W. G. Graham, W. Fritsch, Y. Hahn, and J. A. Tanis, volume 296 of *NATO ASI Series B: Physics*, pages 133–141. Plenum Press, New York (1992)

[28] K. LaGattuta and Y. Hahn, *Effect of extrinsic electric fields upon dielectronic recombination:* Mg^{1+} Phys. Rev. Lett. **51**, 558–561 (1983)

[29] A. Müller, D. S. Belić, B. D. DePaola, N. Djurić, G. H. Dunn, D. W. Mueller, and C. Timmer, *Field effects on the Rydberg product-state distribution from dielectronic recombination* Phys. Rev. Lett. **56**, 127–130 (1986)

[30] C. Bottcher, D. C. Griffin, and M. S. Pindzola, *Dielectronic recombination of* Mg^{+} *in the presence of electric fields* Phys. Rev. A **34**, 860–865 (1986)

[31] A. R. Young, L. D. Gardner, D. W. Savin, G. P. Lafyatis, A. Chutjian, S. Bliman and J. L. Kohl, *Measurement of* C^{3+} *dielectronic recombination in a known external field* Phys. Rev. A **49**, 357–362 (1994)

[32] D. W. Savin, L. D. Gardner, D. B. Reisenfeld, A. R. Young and J. L. Kohl, *Absolute measurement of dielectronic recombination for* C^{3+} *in a known external field* Phys. Rev. A **53**, 280–289 (1996)

[33] T. Bartsch, A. Müller, W. Spies, J. Linkemann, H. Danared, D. R. DeWitt, H. Gao, W. Zong, R. Schuch, A. Wolf, G. H. Dunn, M. S. Pindzola, and D. C. Griffin, *Field enhanced dielectronic recombination of* Si^{11+} *ions* Phys. Rev. Lett. **79**, 2233–2236 (1997)

[34] F.Robicheaux, M. S. Pindzola, and D. C. Griffin, *Effect of interacting resonances on dielectronic recombination in static fields* Phys. Rev. Lett. **80**, 1402–1405 (1998)

[35] T. Bartsch, S. Schippers, A. Müller, C. Brandau, G. Gwinner, A. A. Saghiri, M. Beutelspacher, M. Grieser, D. Schwalm, A. Wolf, H. Danared, and G. H. Dunn, *Experimental evidence for magnetic field effects*

on dielectronic recombination via high Rydberg states Phys. Rev. Lett. **82**, 3779–3781 (1999)

[36] T. Bartsch, S. Schippers, M. Beutelspacher, S. Böhm, M. Grieser, G. Gwinner, A. A. Saghiri, G. Saathoff, R. Schuch, D. Schwalm, A. Wolf, and A. Müller, *Enhanced dielectronic recombination of lithium-like* Ti^{19+} *ions in external* $E \times B$ *fields* J. Phys. B **33**, L453–L460 (2000)

[37] S. Böhm, S. Schippers, W. Shi, A. Müller, N. Djurić, G. H. Dunn, W. Zong, B. Jelenković, H. Danared, N. Eklöw, P. Glans, and R. Schuch, *Influence of electromagnetic fields on the dielectronic recombination of* Ne^{7+} *ions* Phys. Rev. A, submitted for publication (2002).

[38] S. Böhm, S. Schippers, W. Shi, A. Müller, N. Eklöw, R. Schuch, H. Danared, and N. R. Badnell, *Measurement of the field-induced dielectronic recombination-rate enhancement of* O^{5+} *ions differential in the Rydberg quantum number* n Phys. Rev. A **65**, 052728 (2002)

[39] D. C. Griffin and M. S. Pindzola, *Dielectronic recombination of highly ionized iron* Phys. Rev. A **35**, 2821–2831 (1987)

[40] W. Shi, T. Bartsch, S. Böhm, C. Böhme, F. Bosch, C. Brandau, B. Franzke, N. Grün, A. Hoffknecht, S. Kieslich, H. Knopp, C. Kozhuharov, A. Krämer, A. Müller, F. Nolden, W. Scheid, S. Schippers, Z. Stachura, M. Steck, T. Steih, T. Stöhlker, and T. Winkler, *Dielectronic recombination and spectroscopy of the heaviest Li-like ions* Phys. Scr., **T92** 191-199 (2001)

[41] S. Schippers, A. Müller, G. Gwinner, J. Linkemann, A. A. Saghiri, and A. Wolf, *Storage ring measurement of the* C IV *recombination rate coefficient* Astrophys. J., **555** 1027-1037 (2001)

[42] L. Ko, V. Klimenko, and T. F. Gallagher, *Enhancement of dielectronic recombination by an electric field* Phys. Rev. A **59**, 2126–2133 (1999)

[43] V. Klimenko, L. Ko, and T. F. Gallagher, *Magnetic field enhancement of dielectronic recombination from a continuum of finite bandwidth* Phys. Rev. Lett. **83**, 3808–3811 (1999)

[44] J. G. Story, B. J. Lyons, and T. F. Gallagher, *Electric-field enhancement of dielectronic recombination from a continuum of finite bandwidth* Phys. Rev. A **51**, 2156–2161 (1995)

[45] V. Klimenko and T. F. Gallagher, *Resonant enhancement of dielectronic recombination by a microwave field* Phys. Rev. Lett. **85**, 3357–3360 (2000)

[46] A. Müller, *Plasma rate coefficients for highly charged ion-electron collisions: New experimental access via ion storage rings* Int. J. Mass Spec. **192**, 9–22 (1999)

[47] J. Linkemann, J. Kenntner, A. Müller, A. Wolf, D. Habs, D. Schwalm,, W. Spies, O. Uwira, A. Frank, A. Liedtke, G. Hofmann, E. Salzborn, N. R. Badnell, and M. S. Pindzola, *Electron impact ionization and dielectronic recombination of sodium–like iron ions* Nucl. Instrum. Methods **98**, 154–157 (1995)

Chapter 9

ELECTRON ION SCATTERING USING AN R-MATRIX APPROACH

M.P.Scott and P.G.Burke

Department of Applied Mathematics and Theoretical Physics
The Queen's University of Belfast
Belfast BT7 1NN, U K
m.p.scott@qub.ac.uk

Abstract This chapter presents the general theory of low energy electron scattering by atoms and ions. Extensions of the theory to include relativistic effects for heavy atoms and ions and to treat resonances are also presented. The R-matrix approach for the solution of this problem is then introduced and extensions of this approach to treat electron scattering at intermediate energies is discussed. Finally a number of applications are included to illustrate the theory.

Keywords: electron scattering, excitation, inelastic, ionization, R-matrix, resonances, ion.

1. Introduction

The study of electron scattering by atomic ions is of fundamental importance in many application areas. For example, accurate scattering data is required for diagnostics of astrophysical and laboratory plasmas; in fusion research; in the development of ion lasers and in the study of the solar corona and chromosphere. The theoretical study of the electron scattering process provides confirmation of existing experimental data; a greater understanding of the mechanisms involved in the scattering process and a powerful predictive tool to study electron scattering where experiments may not be possible. In many applications a large quantity of highly accurate data is often required, particularly at low incident electron energies. The need for a highly reliable, efficient and general theoretical method to provide these data for a range of ions is therefore evident. In this chapter we will concentrate on one of the most successful theoretical methods for low energy electron–atom/ion scattering – the R-matrix method.

F.J. Currell (ed.), The Physics of Multiply and Highly Charged Ions, Vol. 1, 303-332.

This chapter is arranged as follows. In section 1 we present the general theory of electron–atom/ion scattering, deriving the basic scattering equations for low energy scattering. We then consider the inclusion of relativistic effects for scattering from heavy atoms and ions. Section 1 concludes with a discussion of resonances and quantum defect theory. In section 2 we present the R-matrix method for electron–atom/ion scattering at low energies and discuss how the method can be extended to describe accurately the scattering process at intermediate energies. Finally in section 3 we present some key applications of the R-matrix method to illustrate the approach.

2. Introduction to Electron–Atom/Ion Scattering Theory

In this section we present the general theory for electron scattering by complex atoms and ions. We concentrate first on scattering of low energy electrons by light atoms and ions where we can ignore relativistic effects. We will then discuss the inclusion of relativistic effects for scattering from heavy atoms and ions and conclude the section with a discussion on the theory of resonances and quantum defect theory. By 'low energy' we usually mean that the velocity of the scattering electron is of the same order as that of the target electrons which play an active part in the scattering process. The term 'intermediate energy' usually implies that the scattering electron has sufficient energy to enable one of the target electrons to be ionized. This region can extend to about two to three times the ionization threshold.

2.1 Basic Scattering Equations

We consider the process

$$e^- + A_i \rightarrow e^- + A_j \tag{9.1}$$

where A_i and A_j are the initial and final states of the target atom or ion. The time independent Schrödinger equation describing this scattering process for a target atom or ion with N electrons and nuclear charge Z is given by

$$H_{N+1}\Psi = E\Psi \quad . \tag{9.2}$$

E is the total energy of the $(N+1)$-electron system and H_{N+1} is the non-relativistic $(N+1)$-electron Hamiltonian operator given by

$$H_{N+1} = \sum_{i=1}^{N+1} \left(-\frac{1}{2}\nabla_i^2 - \frac{Z}{r_i} \right) + \sum_{i>j=1}^{N+1} \frac{1}{r_{ij}} \tag{9.3}$$

In the above we have assumed the use of atomic units. The quantity $r_{ij} = |\mathbf{r}_i - \mathbf{r}_j|$ where $\mathbf{r}_i$ and $\mathbf{r}_j$ are the vector coordinates of the ith and jth electrons.

The target nucleus is taken as the origin of the coordinate system and is assumed to have infinite mass.

In order to describe the N-electron target atom or ion, we introduce the target eigenstates Φ_i and the corresponding eigenenergies E_i through the equation

$$\langle \Phi_i | H_N | \Phi_j \rangle = E_i \delta_{ij} \tag{9.4}$$

where H_N is the N-electron target Hamiltonian defined by equation (9.3) with N+1 replaced by N. We will assume that it is possible to obtain accurate target eigenstates and will return to their determination later. The solution of equation (9.2) corresponding to the process defined by equation (9.1) when the target is an atom then has the asymptotic form

$$\Psi_i \underset{r\to\infty}{\sim} \Phi_i \chi_{\frac{1}{2}m_i} e^{ik_i z} + \sum_j \Phi_j \chi_{\frac{1}{2}m_j} f_{ji}(\theta, \phi) \frac{e^{ik_j r}}{r} \tag{9.5}$$

where $\chi_{\frac{1}{2}m_i}$ and $\chi_{\frac{1}{2}m_j}$ are the spin eigenfunctions of the incident and scattered electrons, where the direction of spin quantization is usually taken to be the incident electron beam direction, and $f_{ji}(\theta, \phi)$ is the scattering amplitude, the spherical polar coordinates of the scattered electron being denoted by r, θ and ϕ. Also the wave numbers k_i and k_j of the incident and scattered electrons in equation (9.5) are related to the total energy E of the electron plus atom system by

$$E = E_i + \frac{1}{2}k_i^2 = E_j + \frac{1}{2}k_j^2. \tag{9.6}$$

The outgoing wave term in equation (9.5) contains contributions from all target states that are energetically allowed, that is, for which $k_j^2 \geq 0$. The remaining target states, for which $k_j^2 < 0$, can only occur virtually in the scattering process. While they do not occur explicitly in the asymptotic form, these states can play an important role in the scattering process, giving rise, for example, to intermediate resonances. If the energy lies above the ionization threshold, then the expansion in equation (9.5) also includes target continuum states. We will discuss this possibility further in section 2 when we consider electron scattering at intermediate energies. When the target is an ion, the exponents in equation (9.5) must be modified by including logarithmic phase factors caused by the long-range Coulomb distortion of the scattered electron by the ion. The differential cross section for a transition from an initial state $|i\rangle \equiv |\mathbf{k}_i, \Phi_i, \chi_{\frac{1}{2}m_i}\rangle$ to a final state $|j\rangle \equiv |\mathbf{k}_j, \Phi_j, \chi_{\frac{1}{2}m_j}\rangle$ can now be obtained by considering the incident and scattered fluxes in equation (9.5). We obtain

$$\frac{d\sigma_{ji}}{d\Omega} = \frac{k_j}{k_i} |f_{ji}(\theta, \phi)|^2. \tag{9.7}$$

The total cross section is obtained by averaging over the initial spin states, summing over the final spin states and integrating over all scattering angles.

In order to solve the Schrödinger equation (9.2) at low scattering energies, most approximation methods start from a partial wave expansion of the total wave function of the form

$$\begin{aligned}\Psi_i^\Gamma(\mathbf{X}_{N+1}) &= \mathcal{A}\sum_{j=1}^{n}\tilde{\Phi}_j^\Gamma(\mathbf{x}_1,\ldots,\mathbf{x}_N;\hat{\mathbf{r}}_{N+1}\sigma_{N+1})r_{N+1}^{-1}F_{ji}^\Gamma(r_{N+1})\\ &+ \sum_{j=1}^{m}\chi_j^\Gamma(\mathbf{X}_{N+1})b_{ji}^\Gamma \end{aligned} \tag{9.8}$$

where $\mathbf{X}_{N+1} \equiv \mathbf{x}_1,\ldots,\mathbf{x}_{N+1}$ represents the space and spin coordinates of all $N+1$ electrons, $\mathbf{x}_i \equiv \mathbf{r}_i\sigma_i$ represents the space and spin coordinates of the ith electron and $\mathcal{A}$ is the operator that antisymmetrizes the first expansion with respect to exchange of all pairs of electrons in accordance with the Pauli exclusion principle. The channel functions $\tilde{\Phi}_j^\Gamma$ in equation (9.8) are obtained by coupling the orbital and spin angular momenta of the target states Φ_j with those of the scattered electron to form eigenstates of the total orbital and spin angular momenta L and S, their z components M_L and M_S and the parity π, where the quantum numbers

$$\Gamma \equiv LSM_LM_S\pi \tag{9.9}$$

are conserved in the scattering process, since relativistic effects are being neglected. The χ_j^Γ in the second expansion in equation (9.8), are square integrable correlation functions which allow for additional correlation effects not adequately represented by the first expansion. These correlation functions are usually constructed from the same orbital basis that is used to construct the target states Φ_j . The second expansion also includes terms that ensure completeness if, as is usual, the functions F_{ji}^Γ, which represent the scattered electron, are orthogonalized to the orbitals used to construct the Φ_j.

We now substitute expansion (9.8) into the Schrödinger equation (9.2), project onto the channel functions $\tilde{\Phi}_j^\Gamma$ and onto the square integrable functions χ_j^Γ and eliminate the expansion coefficients b_{ji}^Γ between these equations. This yields n coupled integrodifferential equations satisfied by the radial functions F_{ji}^Γ,

$$\begin{aligned}&\left(\frac{d^2}{dr^2} - \frac{\ell_j(\ell_j+1)}{r^2} + \frac{2Z}{r} + k_j^2\right)F_{ji}^\Gamma\\ &= 2\sum_{l=1}^{n}\{V_{jl}^\Gamma(r)F_{li}^\Gamma(r) + \int_0^\infty\left[W_{jl}^\Gamma(r,r') + X_{jl}^\Gamma(r,r')\right]F_{li}^\Gamma(r')dr'\},\\ &\qquad j = 1,\ldots,n \ ,\end{aligned} \tag{9.10}$$

where ℓ_j is the orbital angular momentum of the scattered electron and V_{jl}^{Γ}, W_{jl}^{Γ} and X_{jl}^{Γ} are the local direct, non-local exchange and non-local correlation potentials. If we omit the correlation terms in expansion (9.8) then equations (9.10) are referred to as the close coupling equations. These equations were first derived by Percival and Seaton[1] for e$^-$–H scattering but because of their complexity for multi-electron atoms and ions they are now set up and solved by computer.

The scattering amplitudes and cross sections can be obtained by solving equations (9.10), for each set of conserved quantum numbers Γ, subject to the following asymptotic boundary conditions

$$\begin{aligned} F_{ji}^{\Gamma}(r) &\underset{r\to\infty}{\sim} k_j^{-1/2}(\sin\theta_j\delta_{ji} + \cos\theta_j K_{ji}^{\Gamma}), \quad (k_j^2 \geq 0) \\ F_{ji}^{\Gamma}(r) &\underset{r\to\infty}{\sim} 0, \quad (k_j^2 < 0) \end{aligned} \tag{9.11}$$

where

$$\theta_j = k_j r - \frac{1}{2}\ell_j\pi + \frac{z}{k_j}\ln 2k_j r + \sigma_j \tag{9.12}$$

and where $z = Z - N$ is the residual charge on the ion and $\sigma_j = \arg\Gamma(\ell_j + 1 - iz/k_j)$ is the Coulomb phase shift. The S-matrix $\mathbf{S}^{\Gamma}$ and the T-matrix $\mathbf{T}^{\Gamma}$ are related to the K-matrix $\mathbf{K}^{\Gamma}$, defined by equation (9.11), by the matrix equations

$$\mathbf{S}^{\Gamma} = \frac{\mathbf{I} + \mathbf{iK^{\Gamma}}}{\mathbf{I} - \mathbf{iK^{\Gamma}}}, \quad \mathbf{T}^{\Gamma} = \mathbf{S}^{\Gamma} - \mathbf{I} = \frac{\mathbf{2iK^{\Gamma}}}{\mathbf{I} - \mathbf{iK^{\Gamma}}} \tag{9.13}$$

where the dimensions of the matrices in these equations are $n_a \times n_a$ where n_a is the number of open channels $(k_j^2 \geq 0)$ at total energy E under consideration, for the given Γ. The hermiticity and time reversal invariance of the Hamiltonian ensures that $\mathbf{K}^{\Gamma}$ is real and symmetric. It then follows from equation (9.13), that $\mathbf{S}^{\Gamma}$ is unitary and symmetric.

The scattering amplitude defined by equation (9.5) can be expressed in terms of the T-matrix elements. For a neutral atomic target

$$\begin{aligned} f_{ji}(\theta,\phi) &= i\left(\frac{\pi}{k_i k_j}\right)^{\frac{1}{2}} \sum_{LS\pi\ell_i\ell_j} i^{\ell_i-\ell_j}(2\ell_i+1)^{\frac{1}{2}} \\ &\quad \times(L_i M_{L_i}\ell_i 0|LM_L)(S_i M_{S_i}\tfrac{1}{2}m_i|SM_S) \\ &\quad \times(L_j M_{L_j}\ell_j m_{\ell_j}|LM_L)(S_j M_{S_j}\tfrac{1}{2}m_j|SM_S) \\ &\quad \times T_{ji}^{\Gamma} Y_{\ell_j m_{\ell_j}}(\theta,\phi) \end{aligned} \tag{9.14}$$

which describes a transition from an initial state $\alpha_i L_i S_i M_{L_i} M_{S_i} m_i$ to a final state $\alpha_j L_j S_j M_{L_j} M_{S_j} m_j$, where α_i and α_j represent any additional quantum numbers required to completely define the initial and final states. Also in equation (9.14) the quantities $(abcd|ef)$ are Clebsch-Gordan coefficients and $Y_{\ell_j m_{\ell_j}}(\theta, \phi)$ are spherical harmonics defined by Rose[2]. The corresponding total cross section is obtained by averaging the differential cross section over the initial magnetic quantum numbers, summing over the final magnetic quantum numbers and integrating over all scattering angles. We obtain

$$\sigma_{\mathrm{Tot}}(i \to j) = \frac{\pi}{2k_i^2(2L_i+1)(2S_i+1)} \sum_{LS\pi\ell_i\ell_j} (2L+1)(2S+1)|T_{ji}^{\Gamma}|^2 \quad (9.15)$$

in units of a_0^2, which describes a transition from an initial target state $\alpha_i L_i S_i$ to a final target state $\alpha_j L_j S_j$. We now define the dimensionless quantity called the collision strength by

$$\Omega(i,j) = k_i^2(2L_i+1)(2S_i+1)\sigma_{\mathrm{Tot}}(i \to j). \quad (9.16)$$

The collision strength is symmetric with respect to interchange of the initial and final states denoted by i and j. For electron scattering by an ion, the above expression for the scattering amplitude must be modified by the inclusion of the Coulomb scattering amplitude when the initial and final states are identical.

In an ionized plasma, it is necessary to average the electron scattering cross section over a Maxwell distribution of electron velocities. We introduce the collisional transition probability $q(i \to j)N_e$ where

$$q(i \to j) = \int_0^\infty \sigma_{\mathrm{Tot}}(i \to j) v_i f(v_i, T_e) dv_i \quad (9.17)$$

and where $f(v_i, T_e)$ is the Maxwell velocity distribution function, v_i is the velocity of the incident electron, N_e is the electron density and T_e is the electron temperature of the plasma. We also introduce the effective collision strength

$$\Upsilon(j,i) = \int_0^\infty \Omega(j,i) \exp\left[-\frac{\epsilon_j}{kT_e}\right] d\left[\frac{\epsilon_j}{kT_e}\right], \quad (9.18)$$

where $\epsilon_j = \frac{1}{2}mv_j^2$ and k is Boltzmann's constant. It can be shown that the probability of de-excitation in $\mathrm{cm}^3\ \mathrm{sec}^{-1}$ is related to the effective collision strength by

$$q(j \to i) = \frac{8.63 \times 10^{-6}\Upsilon(j,i)}{\omega_j T_e^{\frac{1}{2}}}, \quad E_j \geq E_i \quad (9.19)$$

where ω_j is the statistical weight of level j and T_e is in degrees Kelvin. It is usually sufficient to determine $\Upsilon(j,i)$ rather than $\Omega(j,i)$ for the transitions of interest.

For incident electron energies insufficient to excite the atom or ion, only elastic scattering is possible. For low energy elastic scattering by a neutral atom such as helium in a ^{1}S ground state, the expression for the scattering amplitude given by equation (9.14) simplifies to

$$f(\theta) = \frac{1}{2ik}\sum_{\ell=0}^{\infty}(2\ell+1)(e^{2i\delta_\ell}-1)P_\ell(\cos\theta) \tag{9.20}$$

where ℓ, which now equals $L = \ell_i = \ell_j$, is the angular momentum of the scattered electron, k is its wave number, $P_\ell(\cos\theta)$ is a Legendre polynomial and the phase shift δ_ℓ can be expressed in terms of the K-matrix by

$$\tan\delta_\ell = K_{11}^{\Gamma} \tag{9.21}$$

where $\mathbf{K}^{\Gamma}$ now has only one element since there is only one open channel. The corresponding expression for the total cross section is

$$\sigma_{\mathrm{Tot}} = \frac{4\pi}{k^2}\sum_{\ell=0}^{\infty}(2\ell+1)\sin^2\delta_\ell. \tag{9.22}$$

Overviews of early electron scattering calculations have been given by Burke and Smith[3] and by Mott and Massey[4]. Currently, most of the theoretical cross section data is calculated by solving the coupled integrodifferential equations (9.10) using general computer program packages. These include computer codes based on the convergent close coupling method, the R-matrix method, the J-matrix method, the linear algebraic equations method, the noniterative integral equations method and the Kohn and the Schwinger variational methods. These methods and the associated computer program packages have been reviewed for example by Burke and Eissner[5], Henry[6], Burke and Berrington [7] and Bartschat[8].

2.2 Relativistic Effects for Heavy Atoms and Ions

As the nuclear charge Z of the target increases, relativistic effects become important even for low energy electron scattering. There are two ways in which these effects play a role. First, there is relativistic distortion of the wave function describing the scattered electron by the strong Coulomb potential near the nucleus and second, there is a change in the charge distribution of the target due to the use of relativistic target wave functions.

For atoms and ions with small Z, the K-matrices can first be calculated in LS coupling, neglecting relativistic effects. The K-matrices are then recoupled to yield transitions between the fine-structure levels of the target. This is best achieved by introducing a pair-coupling scheme

$$\mathbf{L}_{\mathrm{i}} + \mathbf{S}_{\mathrm{i}} = \mathbf{J}_{\mathrm{i}}, \quad \mathbf{J}_{\mathrm{i}} + \boldsymbol{\ell}_i = \mathbf{K}_{\mathrm{i}}, \quad \mathbf{K}_{\mathrm{i}} + \mathbf{s}_{\mathrm{i}} = \mathbf{J}, \tag{9.23}$$

where $\mathbf{L_i}$ and $\mathbf{S_i}$ are the orbital and spin angular momenta of the target, $\mathbf{J_i}$ is the total angular momentum of the target, $\boldsymbol{\ell}_i$ and $\mathbf{s_i}$ are the orbital and spin angular momenta of the scattered electron and $\mathbf{J}$ is the total angular momentum which, with the parity π, is conserved. The required recoupling coefficient $\langle((L_iS_i)J_i,\ell_i)K_is_i;JM_J|(L_i\ell_i)L,(S_is_i)S;JM_J\rangle$ can be expressed in terms of the product of two Racah coefficients[5]. A computer program based on this approach has been developed by Saraph[9].

For atoms and ions with higher Z values relativistic effects must be included explicitly in both the target and scattering Hamiltonians. One approach for achieving this is to use the Breit-Pauli Hamiltonian[10]

$$H_{N+1}^{\rm BP} = H_{N+1}^{\rm nr} + H_{N+1}^{\rm rel}, \tag{9.24}$$

where $H_{N+1}^{\rm nr}$ is defined by equation (9.3) and $H_{N+1}^{\rm rel}$ consists of the one- and two-body relativistic terms. The one-body terms are

$$\begin{aligned} H_{N+1}^{\rm mass} &= -\frac{1}{8}\alpha^2\sum_{i=1}^{N+1}\nabla_i^4, \quad \text{mass-correction term,} \\ H_{N+1}^{\rm D_1} &= -\frac{1}{8}\alpha^2 Z\sum_{i=1}^{N+1}\nabla_i^2\left(\frac{1}{r_i}\right), \quad \text{Darwin term,} \\ H_{N+1}^{\rm SO} &= \frac{1}{2}\alpha^2 Z\sum_{i=1}^{N+1}\frac{1}{r_i}\frac{\partial V}{\partial r_i}(\boldsymbol{\ell}_i\cdot\mathbf{s_i}), \quad \text{spin-orbit term.} \end{aligned} \tag{9.25}$$

Relativistic effects using this approach have been incorporated in an R-matrix program for electron scattering by a general atom or ion described by Berrington *et al.*[11]

For high Z atoms and ions the most satisfactory approach is to adopt the Dirac Hamiltonian defined by

$$H_{N+1}^{\rm D} = \sum_{i=1}^{N+1}\left(c\boldsymbol{\alpha}.\mathbf{p} + \beta' c^2 - \frac{Z}{r_i}\right) + \sum_{i>j=1}^{N+1}\frac{1}{r_{ij}}, \tag{9.26}$$

where $\beta' = \beta - 1$ and $\boldsymbol{\alpha}$ and β are the usual Dirac matrices. The expansion of the total wave function for a particular set of conserved quantum numbers J, M_J and π takes the general form of equation (9.8) where now j-j coupling is adopted. The target bound orbitals and the scattered electron orbitals are represented by Dirac spinors with the form

$$\phi(\mathbf{r},\sigma) = \frac{1}{r}\begin{pmatrix} p_a(r)\eta_{\kappa m}(\hat{\mathbf{r}},\sigma) \\ iq_a(r)\eta_{-\kappa m}(\hat{\mathbf{r}},\sigma) \end{pmatrix} \tag{9.27}$$

for the bound orbitals and

$$F(\mathbf{r},\sigma) = \frac{1}{r}\begin{pmatrix} p_c(r)\eta_{\kappa m}(\hat{\mathbf{r}},\sigma) \\ iq_c(r)\eta_{-\kappa m}(\hat{\mathbf{r}},\sigma) \end{pmatrix} \tag{9.28}$$

for the continuum orbitals, where $p(r)$ and $q(r)$ are the large and small radial components respectively. The spherical spinors $\eta_{\kappa m}$ in equation (9.27) and (9.28) are defined by

$$\eta_{\kappa m}(\hat{\mathbf{r}},\sigma) = \sum_{m_\ell m_s} (\ell m_\ell \frac{1}{2} m_s | jm) Y_{\ell m_\ell}(\theta,\phi)\chi_{\frac{1}{2}m_s}(\sigma) \tag{9.29}$$

where $\kappa = j + \frac{1}{2}$ when $\ell = j + \frac{1}{2}$ and $\kappa = -j - \frac{1}{2}$ when $\ell = j - \frac{1}{2}$. We can derive first-order coupled integrodifferential equations for the functions $p_c(r)$ and $q_c(r)$ describing the motion of the scattered electron in a similar way to equation (9.10). The K-matrix and T-matrix can be determined from the asymptotic form of the solution of these equations. The total cross section is then given by

$$\sigma_{\rm Tot}(i \to j) = \frac{\pi}{2k_i^2(2J_i+1)} \sum_{J\pi j_i \ell_i j_j \ell_j} (2J+1)|T_{ji}^{J\pi}|^2 \tag{9.30}$$

in units of a_0^2, where $j_i\ell_i$ and $j_j\ell_j$ are the initial and final total and orbital angular momenta of the scattered electron. The corresponding collision strength is

$$\Omega(i,j) = k_i^2(2J_i+1)\sigma_{\rm Tot}(i \to j) \tag{9.31}$$

and the effective collision strength is defined as before by equation (9.18). A Dirac R-matrix program has been written by Norrington and Grant[12].

2.3 Resonances and Quantum Defect Theory

The formation of resonance structures plays an important role in low energy electron scattering by atoms and ions, particularly in electron–ion scattering where scattering cross sections are dominated by series of Rydberg resonances converging onto excited state thresholds of the ion. General resonance theories have been developed by Feshbach[13, 14] and Fano[15] and their role in electron scattering by atoms, ions and molecules has been discussed by Burke[16] and by Schulz[17, 18].

There are two types of resonances, namely open-channel or shape resonances and closed channel or Feshbach resonances. Shape resonances occur when the effective potential between the electron and the target has a characteristic shape with an inner attractive well and an outer repulsive barrier, the latter usually caused by the centrifugal repulsion. Feshbach resonances occur for scattered electron energies just below excitation thresholds where the scattered electron is captured temporarily into a bound state with the target in an excited state. We will briefly consider the main features of the theory.

Following Feshbach, we introduce projection operators P and Q, where P projects onto the set of open channels in expansion (9.8), assumed finite in number, and Q projects onto the remaining closed channels, where we restrict our consideration to the space corresponding to a particular set of conserved quantum numbers Γ. Hence

$$P^2 = P, \quad Q^2 = Q, \quad P + Q = 1 \quad . \tag{9.32}$$

The Schrödinger equation (9.2) can then be written as

$$\begin{aligned} P(H - E)(P + Q)\Psi &= 0 \\ Q(H - E)(P + Q)\Psi &= 0, \end{aligned} \tag{9.33}$$

where for convenience we have omitted the subscript $N+1$ on H. Substituting for $Q\Psi$ from the second of these equations into the first yields the equation

$$P(H + V_{\text{opt}} - E)P\Psi = 0, \tag{9.34}$$

where the optical potential V_{opt} defined here by

$$V_{\text{opt}} = -PHQ\frac{1}{Q(H - E)Q}QHP, \tag{9.35}$$

allows for propagation in the Q channel space.

We now introduce the eigenstates ϕ_i and eigenvalues ϵ_i of the operator QHQ by

$$QHQ\phi_i = \epsilon_i\phi_i. \tag{9.36}$$

Hence we can write

$$V_{\text{opt}} = -\sum_i \int \frac{PHQ|\phi_i\rangle\langle\phi_i|QHP}{\epsilon_i - E} d\epsilon_i \tag{9.37}$$

where the summation in this equation goes over the bound eigenstates and the integral goes over the continuum eigenstates of QHQ. If the energy E is in the neighbourhood of an isolated eigenvalue ϵ_i, we can rewrite equation (9.34) as

$$P\left(H - \sum_{j \neq i} \int \frac{PHQ|\phi_j\rangle\langle\phi_j|QHP}{\epsilon_j - E} d\epsilon_j - E\right) P\Psi = \frac{PHQ|\phi_i\rangle\langle\phi_i|QHP}{\epsilon_i - E} P\Psi \tag{9.38}$$

where the rapidly varying term in the optical potential has been separated out on the right-hand-side of this equation. Equation (9.38) can be solved by introducing the Green's function G_0 and the solution $\mathbf{\Psi}_0^+$ corresponding to the operator on the left-hand-side of this equation, where $\mathbf{\Psi}_0^+$ is normalized to a delta function in energy and satisfies boundary conditions corresponding to in-going waves in the incident channel and outgoing waves in all channels. The pole term on the right-hand-side of equation (9.38) gives rise to a resonance whose position in the complex energy plane is

$$W_i = \epsilon_i + \Delta_i - \frac{1}{2}i\Gamma_i = E_i - \frac{1}{2}i\Gamma_i \tag{9.39}$$

where the resonance shift

$$\Delta_i = \langle\phi_i|QHPG_0PHQ|\phi_i\rangle \tag{9.40}$$

and the resonance width

$$\Gamma_i = 2\pi|\langle\phi_i|QHP|\mathbf{\Psi}_0^+\rangle|^2. \tag{9.41}$$

In the neighbourhood of the resonance energy E_i, the S-matrix varies rapidly with energy and can be written as

$$\mathbf{S} = \mathbf{S}_0^{\frac{1}{2}}\left(\mathbf{I} - i\Gamma\frac{\gamma_i \times \gamma_i}{E - E_i + \frac{1}{2}i\Gamma_i}\right)\mathbf{S}_0^{\frac{1}{2}}, \tag{9.42}$$

where $\mathbf{S}_0$ is the slowly varying non-resonant background S-matrix associated with $\mathbf{\Psi}_0^+$ and the partial widths γ_i are defined by

$$\Gamma_i^{\frac{1}{2}}\gamma_i\mathbf{S}_0^{\frac{1}{2}} = \langle\phi_i|QHP|\mathbf{\Psi}_0^+\rangle. \tag{9.43}$$

The above theory enables the analysis of individual resonances. However, in electron–ion scattering we are usually concerned with whole series of resonances which are associated with particular excited states of the ion. Multi-channel quantum defect theory (MQDT), developed by Seaton[19, 20], is well suited for this situation. MQDT uses general analytic properties of the Coulomb wave functions to relate the S-matrices and the K-matrices above and below the threshold to which the resonance series converge. In particular, Seaton showed that the K-matrix defined by equation (9.11) can be written below this threshold as

$$\mathbf{K}_{\rm op}^{\Gamma} = \mathbf{K}_{\rm oo}^{\Gamma} - \mathbf{K}_{\rm oc}^{\Gamma}\left(\mathbf{K}_{\rm cc}^{\Gamma} + \tan\pi\boldsymbol{\nu}_c\right)^{-1}\mathbf{K}_{\rm co}^{\Gamma} \tag{9.44}$$

where the elements of the sub-matrices $\mathbf{K}_{\rm oo}^{\Gamma}$, $\mathbf{K}_{\rm oc}^{\Gamma}$, $\mathbf{K}_{\rm co}^{\Gamma}$ and $\mathbf{K}_{\rm cc}^{\Gamma}$ are defined by partitioning the K-matrix into channels which are open and channels which are closed below this threshold as follows

$$\mathbf{K}^{\Gamma} = \begin{pmatrix} \mathbf{K}_{\rm oo}^{\Gamma} & \mathbf{K}_{\rm oc}^{\Gamma} \\ \mathbf{K}_{\rm co}^{\Gamma} & \mathbf{K}_{\rm cc}^{\Gamma} \end{pmatrix}. \tag{9.45}$$

The K-matrix elements on the right-hand-side of equation (9.44) are determined by analytical continuation from the K-matrix elements calculated above this threshold. Also in equation (9.44), ν_i is defined in the closed channels by

$$k_i^2 = -\frac{z^2}{\nu_i^2} \ . \tag{9.46}$$

The factor $\tan \pi \boldsymbol{\nu}_c$ causes the last term in equation (9.44) to contain an infinity of poles converging to the threshold of interest from below, giving rise to resonance series. Equation (9.44) thus enables cross sections and resonance structure to be predicted in an energy region where some channels are closed, by calculating the K-matrix at a few energies where these channels are open and then extrapolating the K-matrix to the energy region where the channels are closed.

3. Introduction to R-matrix Theory

R-matrix theory was first introduced into nuclear physics in 1947 by Wigner and Eisenbud[22] to assist in the interpretation of resonance structures which had been observed in nuclear reactions. These resonances could be thought of as a temporary compound state of the colliding nuclei and the problem solved using many-body nuclear structure methods. The region of configuration space in which this reaction occurred was enveloped by a sphere of radius $r = a$ and referred to as the 'internal region'. Outside this region, where the colliding nuclei were further apart, the problem reduced to a much simpler two-body problem. This region outside the sphere was referred to as the 'external region'. To connect the solution to the scattering problem in these two regions, Wigner and Eisenbud proposed that on the boundary of the sphere the wavefunction in both regions should be related to its derivative at $r = a$ through the same matrix entity which was referred to as the R-matrix.

The atomic and molecular R-matrix method was developed in the early seventies when it was appreciated that collision processes involving the scattering of low energy electrons, with the possible formation of intermediate resonance states, could be described using an R-matrix type theory (see e.g. Burke and Seaton[23], Burke *et al.*[24]). In the atomic and molcular R-matrix method configuration space describing the electron–atom (ion, molecule) complex is divided into two regions - the 'internal' and 'external' regions, using a sphere of radius $r = a$ which has been chosen to completely envelop the electronic orbitals of the target (see Figure 9.1). In the 'internal' region the interaction between the scattering electron and the target electrons is strong and electron exchange is also possible. Atomic (or molecular) structure techniques can be used to solve the scattering problem in this region. In the 'external' region electron exchange is no longer possible and the scattering electron moves in some long-range 'average' potential of the target. The problem thus simplifies considerably. The solutions in the 'internal' and 'external' regions are again connected by relating the wavefunction in each region to its derivative on the boundary through the R-matrix.

While we will concentrate on electron scattering by atoms and ions, the R-matrix theory which we will develop in this chapter is immediately applicable to atomic/ionic photionization discussed by Wilson and Bell elsewhere in this book, as well as to electron molecule scattering and multiphoton processes.

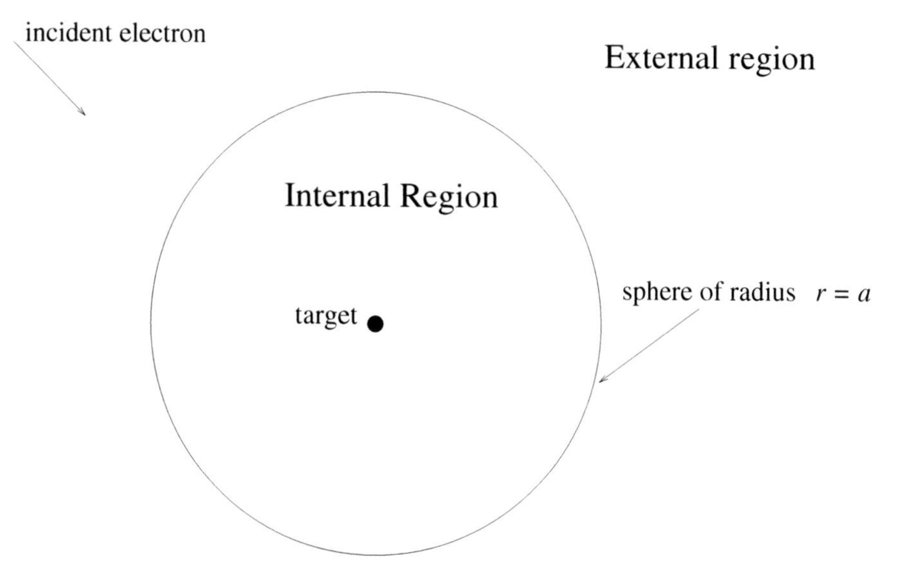

Figure 9.1. Partitioning of configuration space in R-matrix theory.

3.1 Target Eigenstates

Before considering the $(N+1)$-electron scattering problem we need to consider methods for constructing the target atomic or ionic eigenstates in equation (9.4). The target eigenstates Φ_i are usually written as a configuration interaction expansion in terms of a set of basis functions ϕ_j as

$$\Phi_i(\mathbf{x}_1, \ldots, \mathbf{x}_N) = \sum_j \phi_j(\mathbf{x}_1, \ldots, \mathbf{x}_N)c_{ij} \tag{9.47}$$

where the ϕ_j are constructed from a set of orthonormal one-electron orbitals which can be either bound physical orbitals or pseudo-orbitals, included to represent electron correlation effects or to represent the target continuum. Each one-electron orbital can be written in the form

$$U_{nlm_l}(\mathbf{r}, m_s) = \frac{1}{r}P_{nl}(r)Y_{lm_l}(\theta, \phi)\chi_{\frac{1}{2}m_s} \ , \tag{9.48}$$

where the radial functions $P_{n\ell}(r)$ satisfy the orthogonality relation

$$\int_0^\infty P_{n\ell}(r)P_{n'\ell}(r)dr = \delta_{nn'} \quad . \tag{9.49}$$

Each ϕ_j in equation (9.47) is antisymmetric with respect to interchange of the space and spin coordinates of any two electrons. The expansion coefficients c_{ij} in equation (9.47) are obtained by diagonalizing the target Hamiltonian in equation (9.4).

3.2 Internal Region

3.2.1 The wavefunction expansion. In the internal region, which is defined by a sphere of radius $r = a$ completely enveloping the electronic orbitals of the target, the scattering electron is indistinguishable from the N target electrons and the $(N+1)$-electron complex behaves much like a bound state. It is therefore appropriate to use a configuration interaction expansion for the total $(N+1)$-electron wavefunction. We expand the wavefunction $\Psi_E^\Gamma(\mathbf{X}_{N+1})$ at energy E in the internal region in terms of an orthonormal set of R-matrix basis states, ψ_k as

$$\Psi_E^\Gamma = \sum_k A_{Ek}\psi_k \tag{9.50}$$

where the A_{Ek} are energy dependent coefficients, E being the total energy of the $(N+1)$-electron system. The R-matrix basis states, which depend on the set of quantum numbers Γ defined by equation (9.9), are independent of the total energy E. These basis states are expanded in the form

$$\begin{aligned} \psi_k(\mathbf{x}_1,\ldots,\mathbf{x}_N) &= \mathcal{A}\sum_{ij}\tilde{\Phi}_i(\mathbf{x}_1,\ldots,\mathbf{x}_N;\hat{\mathbf{r}}_{N+1}\sigma_{N+1})r_{N+1}^{-1}u_{ij}(r_{N+1})c_{ijk} \\ &+ \sum_j \chi_j(\mathbf{x}_1,\ldots,\mathbf{x}_{N+1})d_{jk} \end{aligned} \tag{9.51}$$

where, for clarity, we have omitted the Γ superscripts. The coefficients c_{ijk} and d_{jk} are obtained by diagonalizing the $(N+1)$-electron Hamiltonian matrix in the internal region, i.e.

$$(\psi_k\,|H_{N+1}|\,\psi_{k'}) = E_k\delta_{kk'} \quad . \tag{9.52}$$

The round brakets indicate that the integration over the radial coordinates is limited to the region $r \leq a$. As in equation (9.8), the χ_j functions in the second expansion in equation (9.51), are square integrable correlation functions which allow for additional correlation effects not adequately represented by the first expansion. These correlation functions are constructed from the same orbital basis that is used to construct the target states Φ_i .

3.2.2 Continuum orbitals. The u_{ij} functions in equation (9.51) form a set of basis orbitals which are used to describe the radial motion of the scattered electron. They are solutions of the zeroth order differential equation

$$\left(\frac{d^2}{dr^2} - \frac{\ell_i(\ell_i+1)}{r^2} + V_0(r) + k_{ij}^2\right) u_{ij}(r) = \sum_{n=\ell_i+1}^{n_{\max}(\ell_i)} \lambda_{ijn} P_{n\ell_i}(r) \quad (9.53)$$

subject to the boundary conditions

$$\begin{aligned} u_{ij}(0) &= 0 \\ \frac{a}{u_{ij}(a)} \frac{du_{ij}(r)}{dr}\bigg|_{r=a} &= b_i \ . \end{aligned} \quad (9.54)$$

The summation index n runs over all one-electron physical orbital functions which are included in the expansion of the atomic or ionic target states for each orbital angular momentum ℓ_i. The λ_{ijn} are Lagrange multipliers which are chosen to ensure that the $u_{ij}(r)$ are orthogonal to the $P_{n\ell}(r)$ functions, that is

$$(u_{ij}\,|P_{n\ell_i}) = \int_0^a u_{ij}(r)P_{n\ell_i}(r)dr = 0 \quad \text{for all} \ \ n,j \ . \quad (9.55)$$

It follows from equations (9.53) and (9.54) that the $u_{ij}(r)$ are orthogonal and can be normalized according to:

$$\left(u_{ij}\middle|\,u_{ij'}\right) = \delta_{jj'} \ . \quad (9.56)$$

It then follows that the orbitals

$$P_{n_{\min}(\ell_i)\ell_i}(r), \ldots, P_{n_{\max}(\ell_i)\ell_i}(r), u_{i1}(r), u_{i2}(r), \ldots \quad (9.57)$$

form a complete set of functions over the range $r = 0$ to $r = a$. The potential $V_0(r)$ in equation (9.53) is a zeroth order potential which behaves as $2Z/r$ near the target nucleus. While it is usually chosen to be the average static potential of the target atom or ion, the choice of $V_0(r)$ is not crucial. Finally, the value of the constant b_i in equation (9.54) is arbitrary, but is usually chosen as zero.

If pseudo-orbitals are included in the construction of the target states in equation (9.50), these orbitals are not retained on the right-hand-side of equation (9.53), but instead the continuum basis orbitals u_{ij} are orthogonalized to these orbitals using the Schmidt method.

3.2.3 Derivation of the R-matrix. The R-matrix is determined by solving the equation

$$H_{N+1}\Psi_E = E\Psi_E \quad (9.58)$$

in the internal region. Since H_{N+1} is not hermitian in this region due to the surface terms at $r = a$ which arise from the kinetic energy operators $-\frac{1}{2}\nabla_i^2$ in equation (9.3), we introduce a Bloch operator[25]

$$L_{N+1} = \sum_{i=1}^{N+1} \frac{1}{2}\,\delta(r_i - a)\left(\frac{\partial}{\partial r_i} - \frac{b-1}{r_i}\right) \tag{9.59}$$

where b is the arbitrary constant in equation (9.54) which is taken to be independent of the channel. It follows that $H_{N+1} + L_{N+1}$ is hermitian. We then rewrite equation (9.58) as

$$(H_{N+1} + L_{N+1} - E)\,\Psi_E = L_{N+1}\Psi_E \tag{9.60}$$

which can be formally solved yielding

$$\Psi_E = (H_{N+1} + L_{N+1} - E)^{-1}\,L_{N+1}\Psi_E \quad . \tag{9.61}$$

We now represent the operator on the right-hand-side of this equation in terms of the basis ψ_k defined by the equations (9.51) and (9.52), where we remember that our choice of b in equation (9.54) implies that $L_{N+1}\psi_k = 0$. Equation (9.61) then becomes

$$|\Psi_E\rangle = \sum_k |\psi_k\rangle \frac{1}{E_k - E}\langle\psi_k|L_{N+1}|\Psi_E\rangle \quad . \tag{9.62}$$

Projecting this equation onto the channel functions $\tilde{\Phi}_i$ and evaluating on the boundary of the internal region $r = a$ then yields

$$F_i(a) = \sum_j R_{ij}\left(a\frac{dF_i}{dr} - bF_j\right)\Bigg|_{r=a} \tag{9.63}$$

where we have introduced the R-matrix by

$$R_{ij}(E) = \frac{1}{2a}\sum_k \frac{w_{ik}w_{jk}}{E_k - E} \;, \tag{9.64}$$

the reduced radial wave functions F_i by

$$r^{-1}F_i(r) = (\tilde{\Phi}_i \mid \Psi_E) \tag{9.65}$$

and the surface amplitudes w_{ik} by

$$a^{-1}w_{ik} = (\tilde{\Phi}_i \mid \psi_k)_{r=a} \quad . \tag{9.66}$$

We note that the integration in equations (9.65) and (9.66) is carried out over all electronic space and spin coordinates except the radial coordinate of the scattered electron. The surface amplitudes w_{ik} and the R-matrix poles E_k are obtained from the eigenvalues and eigenvectors of the $(N+1)$-electron Hamiltonian matrix diagonalized in the basis ψ_k. We see that the R-matrix is thus obtained for all energies by this single diagonalization for each set of conserved quantum numbers Γ.

3.2.4 Buttle correction. In practice, the summation in equation (9.64) needs to be truncated. However, since the continuum basis (9.53) satifies homogeneous boundary conditions (9.54), the neglected levels make a significant contribution to the diagonal elements of the R-matrix where they add coherently. We can correct for these omitted terms by using a procedure proposed by Buttle[26]. We solve the differential equation

$$\left(\frac{d^2}{dr^2} - \frac{\ell_i(\ell_i+1)}{r^2} + V_0(r) + k_i^2\right) u_i^0(r) = \sum_{n=\ell_i+1}^{n_{\max}(\ell_i)} \lambda_{ijn} P_{n\ell_i}(r) \ . \quad (9.67)$$

This is the same as equation (9.53) but we solve it at channel energies k_i^2 without imposing the boundary conditions given by equation (9.54). If we retain $\mathcal{N}$ terms in each channel in equation (9.64) then the correction, R_{ii}^c, which should be added to the diagonal elements of the R-matrix is given by

$$\begin{aligned} R_{ii}^c &\approx \frac{1}{a} \sum_{j=\mathcal{N}+1}^{\infty} \frac{u_{ij}(a)^2}{k_{ij}^2 - k_i^2} \\ &= \left(\frac{a}{u_i^0(a)} \left.\frac{du_i^0}{dr}\right|_{r=a} - b\right)^{-1} - \frac{1}{a} \sum_{j=1}^{\mathcal{N}} \frac{u_{ij}(a)^2}{k_{ij}^2 - k_i^2} \ . \end{aligned} \quad (9.68)$$

The channel energy, k_i^2, in equation (9.67) is given by equation (9.6) which can be rewritten as

$$k_i^2 = 2(E - E_i) \ . \quad (9.69)$$

In the low energy region outside the energy range spanned by the eigenvalues k_{ij}^2, $j = \mathcal{N}+1, \ldots, \infty$, R_{ii}^c is a continuous, monotonic function of k_i^2. It is then possible to evaluate R_{ii}^c at a small number of channel energies and to fit to a convenient smooth functional form.

3.3 External region

If the radius of the sphere defining the internal region is chosen to completely envelop the electronic orbitals of the target, then for $r \geq a$ exchange effects can be neglected and the scattered electron can be considered distinct from the target electrons. The wavefunction Ψ_E can then be expanded in the form

$$\Psi_E(\mathbf{x}_1,\ldots,\mathbf{x}_{N+1}) = \sum_i \tilde{\Phi}_i(\mathbf{x}_1,\ldots,\mathbf{x}_N;\hat{\mathbf{r}}_{N+1}\sigma_{N+1})r_{N+1}^{-1}F_i(r_{N+1}) \tag{9.70}$$

where we again omit the Γ superscripts for clarity. Substituting equation (9.70) into equation (9.2) and projecting onto the channel functions $\tilde{\Phi}_i$, we obtain the following set of coupled differential equations which are satisfied by the reduced radial function $F_i(r)$

$$\left(\frac{d^2}{dr^2} - \frac{\ell_i(\ell_i+1)}{r^2} + \frac{2Z}{r} + k_i^2\right) F_i(r) = 2\sum_{j=1}^{n} V_{ij}(r)F_j(r) \qquad i = 1,\ldots,n \ \ (r \geq a)\ . \tag{9.71}$$

n is the number of channels retained in expansion (9.51) or (9.70), ℓ_i and k_i^2 are the channel angular momenta and energies and the direct potential matrix is given by

$$V_{ij}(r_{N+1}) = \left\langle \tilde{\Phi}_i \left| \sum_{k=1}^{N} \frac{1}{r_{kN+1}} \right| \tilde{\Phi}_j \right\rangle \ . \tag{9.72}$$

The integration in equation (9.72) is carried out over all coordinates except the radial coordinate of the scattered electron. We now use the expansion

$$\sum_{k=1}^{N} \frac{1}{r_{kN+1}} \underset{r_{N+1}>r_k}{=} \sum_{\lambda=0}^{\infty} r_{N+1}^{-\lambda-1} \sum_{k=1}^{N} r_k^{\lambda} P_\lambda(\cos\theta_{kN+1}) \tag{9.73}$$

where $\cos\theta_{kN+1} = \hat{\mathbf{r}}_k.\hat{\mathbf{r}}_{N+1}$. Defining the long range potential coefficients a_{ij}^{λ} by

$$a_{ij}^{\lambda} = \left\langle \tilde{\Phi}_i \left| \sum_{k=1}^{N} r_k^{\lambda} P_\lambda(\cos\theta_{kN+1}) \right| \tilde{\Phi}_j \right\rangle , \tag{9.74}$$

equation (9.71) can be rewritten as

$$\left(\frac{d^2}{dr^2} - \frac{\ell_i(\ell_i+1)}{r^2} + \frac{2(Z-N)}{r} + k_i^2\right) F_i(r) = 2\sum_{\lambda=1}^{\infty}\sum_{j=1}^{n} a_{ij}^{\lambda} r^{-\lambda-1} F_j(r) \qquad i = 1,\ldots,n \ \ (r \geq a) \tag{9.75}$$

The asymptotic form of the solution defines the K-matrix through the equations

$$\left.\begin{array}{ll} F_{ij} \underset{r\to\infty}{\sim} k_i^{-1/2}(\sin\theta_i\delta_{ij} + \cos\theta_i K_{ij}) & k_i^2 > 0 \\ F_{ij} \underset{r\to\infty}{\sim} \exp(-\phi_i)\delta_{ij} & k_i^2 < 0 \end{array}\right\} \quad j = 1,\ldots,n_a. \tag{9.76}$$

The second index j has been introduced to distinguish the n_a linearly independent solutions of equation (9.75), where n_a is the number of open channels and

$$\begin{aligned}\theta_i &= k_i r - \frac{1}{2}\ell_i \pi - \eta_i \ln 2k_i r + \arg\Gamma(\ell_i + 1 + i\eta_i) \\ \eta_i &= -\frac{(Z-N)}{k_i} \\ \phi_i &= |k_i| r - \frac{(Z-N)}{|k_i|}\ln(2|k_i|r) \ .\end{aligned} \tag{9.77}$$

Having specified the scattering wavefunction in both the internal and external regions we are now able to link these two regions to complete the solution. We can rewrite the reduced radial wavefunction of the scattered electron on the boundary of the internal region (given by equation (9.65)) in matrix form as

$$\mathbf{F} = a\mathbf{R}\dot{\mathbf{F}} - b\mathbf{R}\mathbf{F} \qquad (r = a) \ . \tag{9.78}$$

To relate the $n \times n$ dimensional R-matrix with the $n_a \times n_a$ dimensional K-matrix defined by equation (9.76) we introduce $n + n_a$ linearly independent solutions, $s_{ij}(r)$ and $c_{ij}(r)$, of equation (9.75) satisfying the boundary conditions

$$\begin{aligned} s_{ij}(r) &\underset{r\to\infty}{\sim} k_i^{-\frac{1}{2}} \sin\theta_i \delta_{ij} + O(r^{-1}) && j = 1, \ldots, n_a \\ c_{ij}(r) &\underset{r\to\infty}{\sim} k_i^{-\frac{1}{2}} \cos\theta_i \delta_{ij} + O(r^{-1}) && j = 1, \ldots, n_a \\ c_{ij}(r) &\underset{r\to\infty}{\sim} \exp(-\phi_i)\delta_{ij} + O(r^{-1}\exp(-\phi_i)) && j = n_a + 1, \ldots, n \end{aligned} \tag{9.79}$$

for $i = 1, \ldots, n$. We expand the reduced radial wavefunction of the scattered electron in the external region as a linear combination of these solutions. In matrix form this becomes

$$\mathbf{F} = \mathbf{s} + \mathbf{c}\mathbf{K_T} \qquad (r \geq a) \tag{9.80}$$

where $\mathbf{K_T}$ has dimension $n \times n_a$. Differentiating this gives

$$\dot{\mathbf{F}} = \dot{\mathbf{s}} + \dot{\mathbf{c}}\mathbf{K_T} \qquad (r \geq a) \ . \tag{9.81}$$

We now substitute this into equation (9.78), matching the solutions for the internal and external regions on the boundary $r = a$ and eliminating $\mathbf{F}$ and $\dot{\mathbf{F}}$. This gives the matrix equation

$$\mathbf{s} + \mathbf{c}\mathbf{K_T} = a\mathbf{R}(\dot{\mathbf{s}} + \dot{\mathbf{c}}\mathbf{K_T}) - b\mathbf{R}(\mathbf{s} + \mathbf{c}\mathbf{K_T}) \ . \tag{9.82}$$

We define the matrices $\mathbf{A}$ and $\mathbf{B}$ as follows

$$\begin{aligned}\mathbf{A} &= a\mathbf{R}(\dot{\mathbf{s}} - \frac{b}{a}\mathbf{s}) - \mathbf{s} \\ \mathbf{B} &= \mathbf{c} - a\mathbf{R}(\dot{\mathbf{c}} - \frac{b}{a}\mathbf{c}) \ .\end{aligned} \tag{9.83}$$

The matrix $\mathbf{K_T}$ is then given by

$$\mathbf{K_T} = \mathbf{B}^{-1}\mathbf{A} \ , \tag{9.84}$$

where the upper n_a rows correspond to the $n_a \times n_a$ dimensional K-matrix, $\mathbf{K}$. The $n_a \times n_a$ dimensional S-matrix is related to the K-matrix by

$$\mathbf{S} = \frac{\mathbf{1} + i\mathbf{K}}{\mathbf{1} - i\mathbf{K}} \ . \tag{9.85}$$

Finally, the contribution to the cross section for the transition from an initial state i to a final state j is given by

$$\sigma_{i \to j} = \frac{\pi}{2k_i^2} \sum_{\ell_i \ell_j} \frac{(2L+1)(2S+1)}{(2L_i+1)(2S_i+1)} \left| S_{ij} - \delta_{ij} \right|^2 \ . \tag{9.86}$$

While the R-matrix in the internal region is obtained at all energies by a single diagonalization of the Hamiltonian matrix in equation (9.52), the solutions s_{ij} and c_{ij} must be calculated and the matching equation (9.84) solved at each energy E where the cross section is required. However, since exchange effects are negligible in the external region, the calculation of the K-matrix once the R-matrix has been evaluated is only a small part of the the total computational effort. It is this feature of the R-matrix method which makes the approach very efficient for resolving resonance structure dominating low energy electron–ion scattering cross section data.

3.4 Intermediate energy R-matrix theory

In recent years R-matrix theory has been developed to enable accurate treatment of electron–atom/ion scattering at intermediate electron scattering energies. The intermediate energy regime is usually thought of as covering incident electron energies from the ionization threshold to about three to four times this value. Hence the incident electron has sufficient energy to ionize one of the electrons of the target atom or ion. This means that any accurate theoretical treatment must be able to include the effects of the infinite number of continuum states of the ionized target plus ejected electron as well as the infinite number of bound states of the target atom or ion lying below the ionization threshold. Other theoretical approaches to intermediate electron–atom/ion scattering include the use of optical potentials to take account of the loss of flux into the infinity of open channels in the continuum (for example, Byron and Joachain[27], Bransden and Coleman[28], Bransden *et al.*[29] and Bray *et al.*[30]); the inclusion of higher order terms in the Born approximation (for a review of perturbation methods in electron–atom/ion scattering see Walters[31]); the application of time-dependent methods to directly solve the time-dependent Schrödinger equation[32, 33] and the extension of the close-coupling expansion used in many low-energy scattering approaches. These close coupling approaches include the very successful convergent close coupling (CCC) method of Bray and Stelbovics[34]; the pseudo-state close coupling approach of Callaway and Wooten[35, 36]; the intermediate energy R-matrix method (IERM) of Burke *et al.*[37] and the R-matrix with pseudo-states (RMPS) approach of Bartschat *et al.*[38].

The IERM and RMPS methods are extensions of the low energy R-matrix method discussed above. In these theories the close-coupling expansion is extended to include, not only target eigenstates of interest, but also a number of well constructed pseudo-states to represent in some average way the high-lying Rydberg states and continuum states of the target that cannot be included explicitly. In the RMPS approach the pseudo-states are constructed from the same orbital basis as in equation (9.48) which has been augmented by additional pseudo-orbitals. These pseudo-orbitals are obtained by taking the minimum linear combination of Sturmian-type orbitals of the form $r^i e^{-\alpha r}$ which are orthogonal to the physical orbitals of the same orbital angular momentum. This RMPS basis is useful for studying electron impact excitation and ionization over a wide range of incident electron energies. In the IERM method the motion of both the valence electron of the target atom or ion and the scattering electron are described in terms of the one-electron numerical continuuum orbitals given by equation (9.53). This gives rise to a very densely packed pseudo-state basis which is more appropriate in the study of near-threshold electron impact ionization. As an example, the IERM target state basis used in the study of electron impact ionization of atomic hydrogen is given in Figure (9.2). The basis consists of up to $n = 7$ physical states for each target angular momentum $\ell = 1, 2$ and 3 augmented by 18 additional pseudo-states per angular momentum.

4. Illustrative Examples

4.1 Electron Impact Excitation of He^+

We consider first electron scattering from He^+, the simplest ionic system. Figure (9.3) gives the total cross section for 1s-2s excitation by electron impact in the energy region between the $n = 2$ and $n = 3$ thresholds of He^+. A number of different theoretical approaches are illustrated which include the 15-state R-matrix calculation of Aggarwal *et al.*[39]; the 2-dimensional R-matrix propagator results of Dunseath *et al.*[40]; CCC data of Bray *et al.*[41] and the hybrid multichannel algebraic variational calculation of Morgan[42]. Experimental data of Dance *et al.*[43] and Dolder and Peart[45] are also presented. Resonance structure due to the different $3snl$, $3pn'\ell'$ and $3dn''\ell''$ Rydberg series converging onto the $n = 3$ thresholds is clearly visible in the three theoretical calculations between 45.0 eV and 48.0 eV. The target state expansion in the

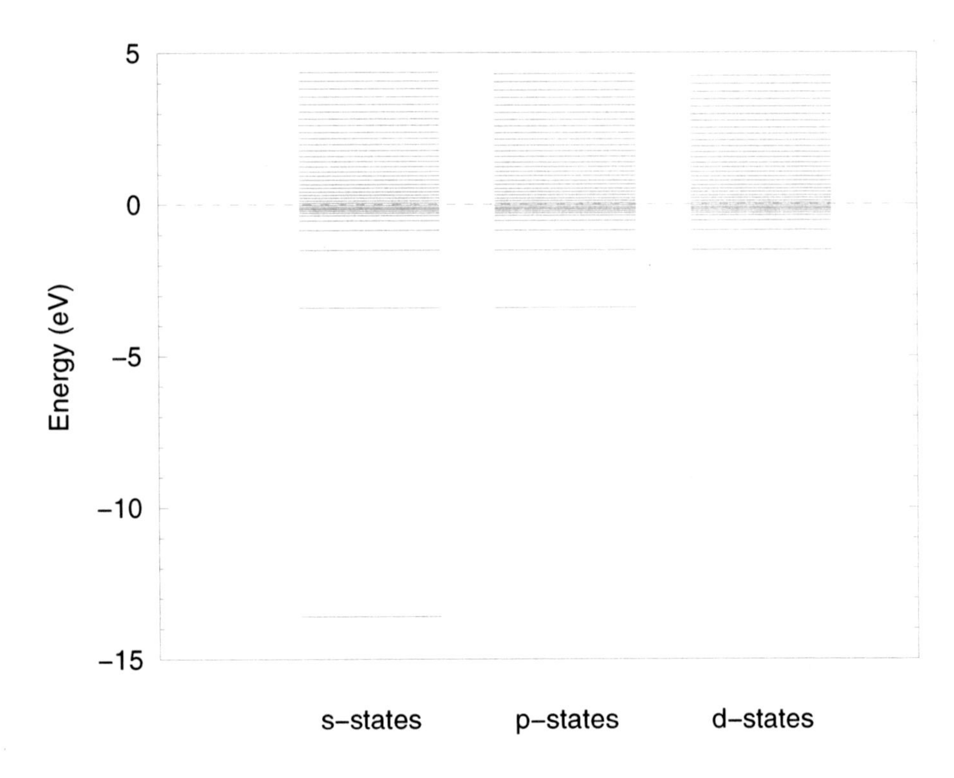

Figure 9.2. Distribution of IERM basis target states for electron impact ionization of atomic hydrogen near threshold.

R-matrix calculation of Aggarwal *et al.*included all the physical states of He^+ for $n \leq 5$, but did not contain any pseudo-states to represent the continuum states of the target ion. In the 2-dimensional R-matrix propagator method the (r_1, r_2) space is subdivided into a number of sectors, which allows a larger region of configuration space to be accurately treated. The Schrodinger equation is solved using a basis-state expansion in each sector and elementary R-matrices relating the radial functions to their derivatives on the boundaries of each sector are evaluated. These are then used to propagate a global R-matrix, $\mathcal{R}$, across the internal region. On the boundary of the internal region, the R-matrix is transformed into the close-coupling representation (9.70) using the target states defined by diagonalising the appropriate one-electron Hamiltonian in the global internal region. In the calculation of Dunseath *et al.*the target states used included all the physical states of He^+ for $n \leq 5$ together with an additional 15 pseudo-states per angular momentum to represent the continuum states of the target ion. In this way, the 2-dimensional R-matrix propagator calculation can be regarded as an extension of the 15-state R-matrix calculation of Aggarwal *et al.*to include the effect of the higher lying bound states and the target contin-

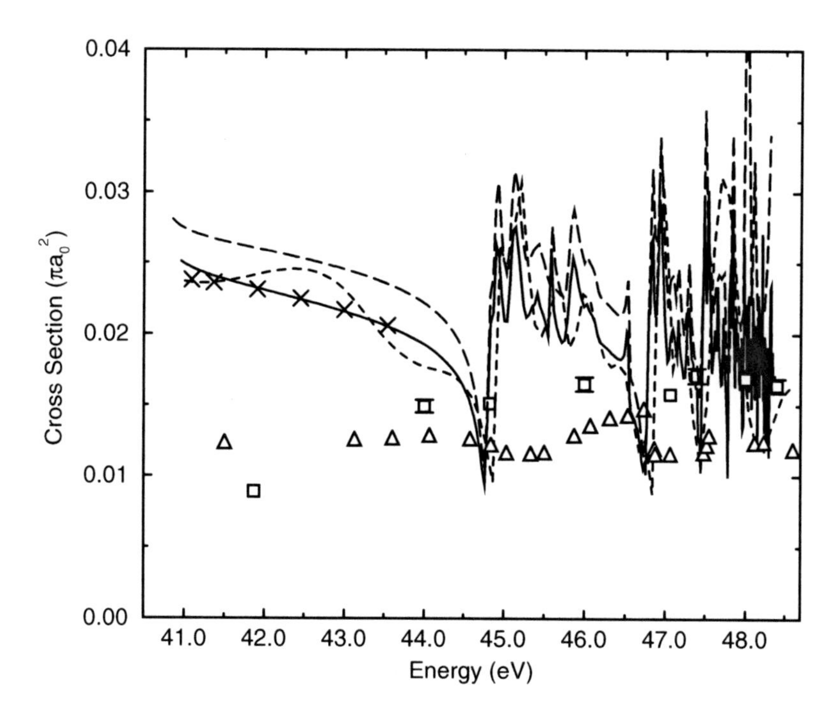

Figure 9.3. Total cross section for 1s-2s excitation at incident electron energies between the $n = 2$ (40.8175 eV) and $n = 3$ (48.3763 eV) thresholds. Theoretical values : ——, 2D R-matrix propagator (Dunseath *et al*[40]); — — —, 15-state R-matrix (Aggarwal *et al*[39]); – – –, CCC (Bray *et al*[41]); ×, hybrid multichannel algebraic variational calculation (Morgan[42]). Experimental data: □, Dance *et al*[43]; △, Dolder and Peart[45] . (From ref.[40], copyright IOP Publishing.)

uum. Comparing results from these two calculations we see that, even at low energies, it is important to include these effects. We note that the results of Dunseath *et al.*are in excellent agreement with the hybrid multichannel algebraic variational calculation of Morgan and that, apart the energy region below 44.0 eV, the CCC calculation is also in good agreement with the 2-dimensional R-matrix propagator data. Both the CCC method and the approach of Morgan include pseudo-states to allow for the effect of the higher lying bound states and the target continuum. These calculations are also in excellent agreement close to threshold with earlier close coupling plus correlation calculations by Burke and Taylor[44], indicating that there is probably a normalization problem in the experimental measurements.

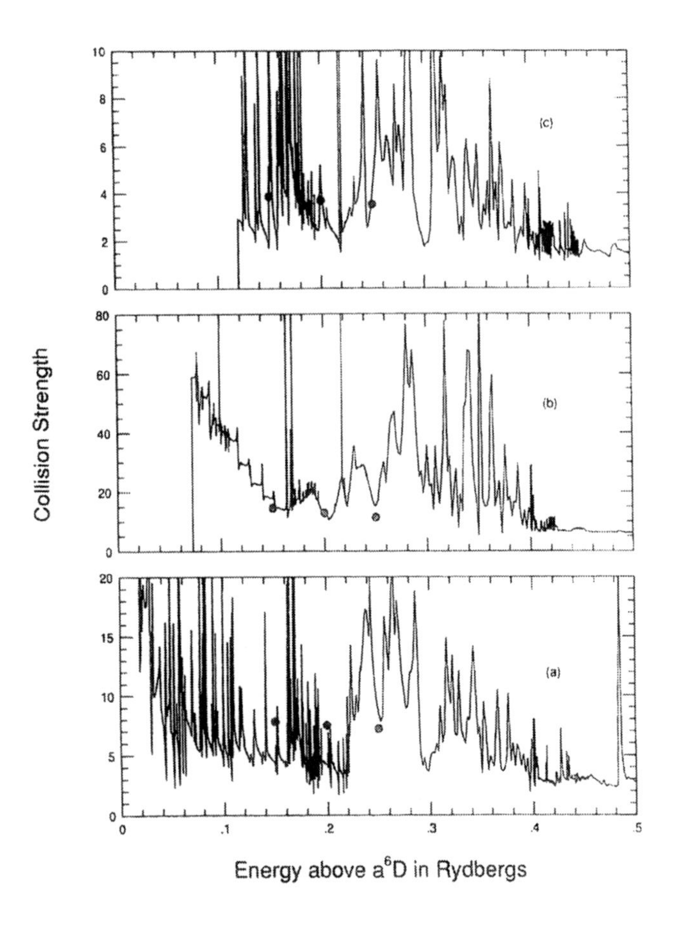

Figure 9.4. e^-–Fe^+ collision strengths for excitation of the a^6D ground state to the metastable states: (a) a^4F, (b) a^4D and (c) a^4P. ——, 38-state R-matrix calculation[46]; •, distorted wave calculation[47]. (From ref.[46], copyright IOP Publishing.)

4.2 Electron Impact Excitation of Fe^+

Accurate knowledge of electron–ion scattering cross sections is required in the analyses of observations of atomic physics processes in many laboratory and astrophysical plasmas. An ion of importance in these application is Fe^+, and in Figure (9.4) we show the collision strengths for three forbidden transitions from the ground state in $e^- - Fe^+$ scattering. We compare results from an R-matrix calculation by Pradhan and Berrington[46] with distorted wave calculations by Nussbaumer and Storey[47]. In the R-matrix calculation, 38 states in LS coupling were included in expansion (9.51), corresponding to all $3d^7$, $3d^6 4s$ and $3d^6 4p$ quartet and sextet spin states, giving rise to collision strengths for 703 transitions. The results shown in Figure (9.4), which correspond to the forbidden transitions $a^6D \rightarrow a^4F$, $a^6D \rightarrow a^4D$ and $a^6D \rightarrow a^4P$, illustrate the importance of resonances in low energy electron scattering. These resonances are not represented in the distorted wave calculations which provide only an approximation to the non-resonant background at low energies. It is found that the resonances substantially increase the effective collision strengths for all three transitions at low temperatures. This result is typical of results obtained for many complex ions showing the importance of adopting theoretical methods where the resonance contributions to the cross sections and collision strengths are accurately represented.

4.3 Electron Impact Ionization of C^{3+}

It is possible to use the RMPS, IERM and other close coupling approaches where the target state expansion has been augmented by the inclusion of pseudo-states to account for the target continuum, to calculate an estimate of the cross section for electron impact ionization. This can be obtained by the summing the excitation cross sections to pseudo-states whose energy levels lie above the ionization threshold. However, a more accurate determination of the scattering amplitude can be obtained by determining the overlap of the pseudo-states with the discrete and continuum spectrum of the target atom or ion. As well as the direct ionization process a number of important indirect ionization processes can also occur. These become progressively more important for heavier ions and may even dominate the ionization cross section. The importance of such processes was first demonstrated by Peart and Dolder[48] who observed a three fold increase in the ionization cross section of Ba^+ due to excitation of inner-shell electrons which led to excited resonance states which subsequently autoionized yielding Ba^{2+}.

To illustrate these ionization processes we consider electron impact ionization of C^{3+}. In this case, in addition to direct ionization, which can be written as

$$e^- + C^{3+}(1s^2 2s) \rightarrow C^{4+}(1s^2) + e^- + e^-, \tag{9.87}$$

three important indirect ionization processes can occur which proceed through the formation of autoionizing states. These are excitation autoionization (EA) given by

$$\begin{aligned} e^- + C^{3+}(1s^2 2s) \rightarrow\ & C^{3+*}(1s2sn\ell) + e^- \\ & \downarrow \\ & C^{4+}(1s^2) + e^- + e^-, \end{aligned} \tag{9.88}$$

resonant excitation double autoionization (REDA) given by

$$e^- + C^{3+}(1s^2 2s) \rightarrow C^{2+*}(1s2sn\ell n'\ell')$$

$$\downarrow$$
$$C^{3+*}(1s2sn''\ell'') + e^-$$
$$\downarrow$$
$$C^{4+}(1s^2) + e^- + e^- \tag{9.89}$$

and resonant excitation auto double ionization (READI) given by

$$e^- + C^{3+}(1s^22s) \rightarrow C^{2+*}(1s2sn\ell n'\ell')$$
$$\downarrow$$
$$C^{4+}(1s^2) + e^- + e^-, \tag{9.90}$$

where the stars in these equations refer to autoionizing states. Processes (9.88), (9.89) and (9.90) give rise to resonance structure on top of a smooth direct ionization background due to process (9.87).

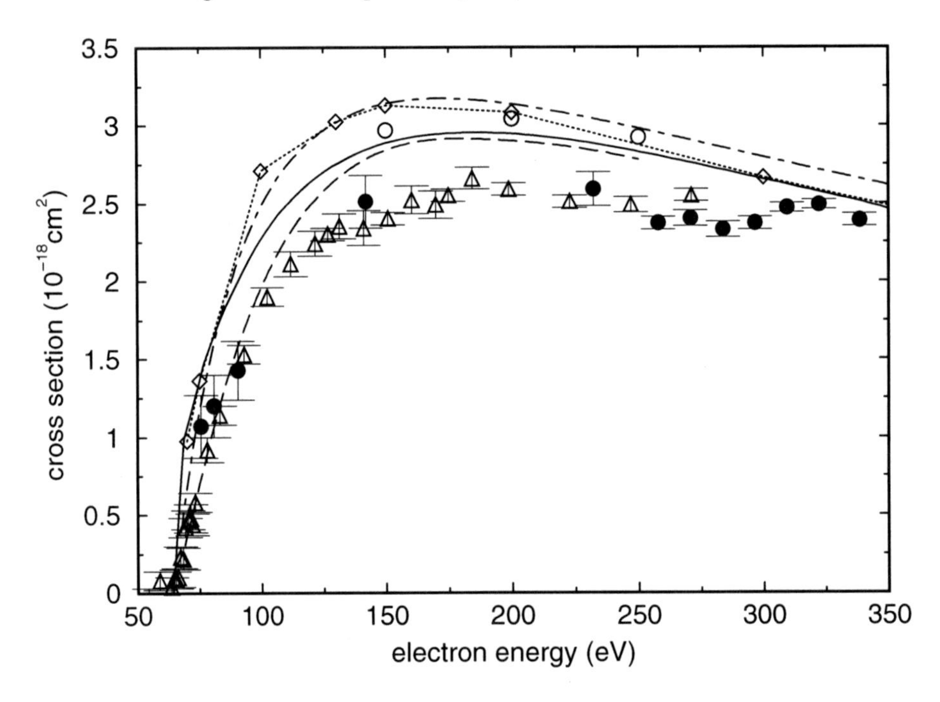

Figure 9.5. Electron impact ionization of C^{3+} (50-350eV). Theoretical calculations : ——, RMPS (Scott *et al*[49]); $\cdots\diamond\cdots$ CCC (Bray[50]); — — —, RMPS calculation of Mitnik *et al*[51] —·—·—, distorted wave calculation of Younger[52]; ∘, TDCC (Mitnik *et al*[51]). Experimental data:•, Crandall *et al*[53, 54]; △, Bannister[55].

In Figures (9.5) and (9.6) we compare the results of a recent RMPS calculation by Scott *et al.*[49], which included the five lowest lying physical states of C^{3+}, 21 pseudo-states representing the continuum and 16 autoionizing states,

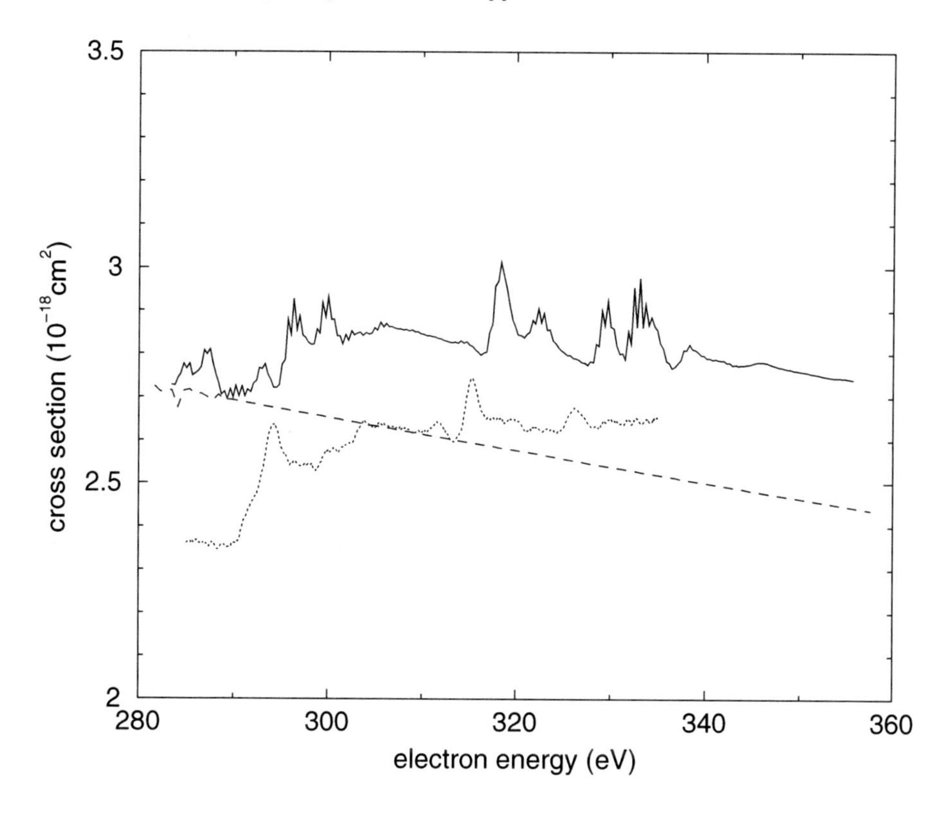

Figure 9.6. Electron impact ionization of C^{3+}. ——, convoluted RMPS results, including direct and indirect ionization processes and interference effects (Scott *et al*[49]); – – –, RMPS results including direct ionization process only (Scott *et al*[49], see Figure 9.5); · · ·, experimental data of Müller *et al*[56].

with experimental measurements and other theoretical calculations. In Figure (9.5) we compare results for the direct ionization process in the energy range between 50eV and 350eV for different theoretical calculations and experimental measurements. The theoretical calculations include the RMPS calculations of Scott *et al.*[49]; the CCC data of Bray[50]; the time-dependent close-coupling (TDCC) results of Mitnik *et al.*[51]; distorted wave results of Younger[52] and a further RMPS calculation by Mitnik *et al.*[51]. The experimental data shown in Figure (9.5) includes the measurements of Crandall *et al.*[53, 54] and, more recently, those of Bannister[55]. We note that for comparison of the direct ionization process the autoionizing states were omitted from the RMPS calculation of Scott *et al.*. Finally, in Figure (9.6) we compare the theoretical results of Scott *et al.*at higher energies with the experimental measurements of Müller *et al.*[56]. In this energy range indirect processes denoted by equations (9.88) - (9.90) can occur as well as direct ionization. The good agreement between the

RMPS calculations of Scott *et al.*[49] and the experimental measurements in both Figures (9.5) and (9.6) illustrates the validity of the pseudo-state approach for both direct and indirect ionization processes.

References

[1] I. C. Percival and M. J. Seaton, Proc. Camb. Phil. Soc. **53**, 654 (1957).

[2] M. E. Rose Elementary Theory of Angular Momentum New York:John Wiley & Sons,Inc. (1957).

[3] P. G. Burke and K. Smith, Rev. Mod. Phys. **24**, 458 (1962).

[4] N. F. Mott and H. S. W. Massey *The Theory of Atomic Collisions 3rd Edition* Oxford:Clarendon Press, (1965).

[5] P. G. Burke and W. B. Eissner, Low-energy electron collisions with complex atoms and ions. In *Atoms in Astrophysics* Eds. P.G. Burke,W.B. Eissner,D.G.Hummer and I.C.Percival, 1–54, New York, London:Plenum Press, (1983).

[6] R. J. W. Henry, Reports on Progress in Physics **56**, 327 (1993).

[7] P. G. Burke and K. A. Berrington *Atomic and Molecular Processes – An R-matrix Approach.* Bristol, Philadelphia:IOP Publishing, (1993).

[8] K. Bartschat, Ed. *Computational Atomic Physics. Electron and Positron Collisions with Atoms and Ions* Berlin, Heidelberg, New York:Springer, (1996).

[9] H. E. Saraph, Comput. Phys. Commun. **3**, 256 (1972).

[10] H. A. Bethe and E. E. Salpeter, *Quantum Mechanics of One- and Two-Electron Atoms* Berlin, Göttingen, Heidelberg:Springer-Verlag, (1957).

[11] K. A. Berrington, W. B. Eissner and P. H. Norrington, Comput. Phys. Commun. **92**, 290 (1995).

[12] P. H. Norrington and I. P. Grant, Comput. Phys. Commun. *to be published* (2002).

[13] H. Feshbach, Ann. Phys. (N.Y.) **5**, 357 (1958).

[14] H. Feshbach, Ann. Phys. (N.Y.) **19**, 287 (1962).

[15] U. Fano, Phys. Rev. **124**, 1866 (1961).

[16] P. G. Burke, Adv. At. Mol. Phys. **4**, 173 (1968).

[17] G. J. Schulz, Rev. Mod. Phys. **45**, 378 (1973).

[18] G. J. Schulz, Rev. Mod. Phys. **45**, 423 (1973).

[19] M. J. Seaton, Mon. Not. Roy. Astron. Soc. **118**, 504 (1958)

[20] M. J. Seaton, Reports on Progress in Physics **46**, 167 (1983).

[21] U. Fano, Phys. Rev. A **2**, 353 (1970).

[22] E. P. Wigner and L. Eisenbud, Phys. Rev. A **72**, 29 (1947).

[23] P. G. Burke and M. J. Seaton, Methods Comput. Phys. **10**, 1 (1971).

[24] P. G. Burke, A. Hibbert and W.D. Robb, J. Phys. B: At. Mol. Phys. **4**, 153 (1971).

[25] C. Bloch Nucl. Phys. **4**, 503 (1957).

[26] P. J. A. Buttle, Phys. Rev. **160**, 719 (1967).

[27] F. W. Byron Jr. and C. J. Joachain, J. Phys. B: At. Mol. Phys. **14**, 2429 (1981).

[28] B. H. Bransden and J. P. Coleman, J. Phys. B: At. Mol. Phys. **5**, 537 (1972).

[29] B. H. Bransden, T. Scott, R. Shingal and J. Raychoudhury, J. Phys. B: At. Mol. Phys. **15**, 4605 (1982).

[30] I. Bray, I. McCarthy, J. Mitroy and A. T. Stelbovics, Phys. Rev. A **32**, 166 (1989).

[31] H. R. J. Walters, Phys. Rep. **116**, 1 (1984).

[32] M. S. Pindzola and D. R. Schultz, Phys. Rev. A **53**, 1525 (1996).

[33] M. S. Pindzola and F. Robichaux, Phys. Rev. A **54**, 2142 (1996).

[34] I. Bray and A. T. Stelbovics, Phys. Rev. A **46**, 6995 (1992).

[35] J. Callaway and J. W. Wooten, Phys. Rev. A **9**, 1924 (1974).

[36] J. Callaway and J. W. Wooten, Phys. Rev. A **11**, 1118 (1975).

[37] P. G. Burke, C. J. Noble and P. Scott, Proc. Roy. Soc. A **410**, 287 (1987).

[38] K. Bartschat, E. T. Hudson, M. P. Scott, P. G. Burke and V. M. Burke, J. Phys. B: At. Mol. Opt. Phys. **29**, 115 (1996).

[39] K. M. Aggarwal, K. A. Berrington, A. E. Kingston and A. Pathak, J. Phys. B: At. Mol. Opt. Phys. **24**, 1757 (1994).

[40] K. M. Dunseath, M. Terao-Dunseath, M. LeDourneuf and J-M Launay, J. Phys. B: At. Mol. Opt. Phys. **30**, L865 (1997).

[41] I. Bray, I. McCarthy, J. Wigley and A. T. Stelbovics, J. Phys. B: At. Mol. Opt. Phys. **26**, L831 (1993).

[42] L. A. Morgan, J. Phys. B: At. Mol. Phys. **12**, L735 (1979).

[43] D. F. Dance, M. F. A. Harrison and A. C. H. Smith, Proc. Roy. Soc. A **290**, 74 (1966).

[44] P. G. Burke and A. J. Taylor, J. Phys. B: At. Mol. Phys. **2**, 44 (1969).

[45] K. T. Dolder and B. Peart, J. Phys. B: At. Mol. Phys. **6**, 2415 (1973).

[46] A. K. Pradhan and K. A. Berrington, J. Phys. B: At. Mol. Opt. Phys. **26**, 157 (1993).

[47] H. Nussbaumer and P. J. Storey, Astron. Astrophys. **89**, 308 (1980).

[48] B. Peart and K. T. Dolder, J. Phys. B: At. Mol. Phys. **1**, 872 (1968).

[49] M. P. Scott, Huaguo Teng and P. G. Burke, J. Phys. B: At. Mol. Opt. Phys. **33**, L63 (2000).

[50] I. Bray, J. Phys. B: At. Mol. Opt. Phys. , **28** L247 (1995).

[51] D. Mitnik, M. S. Pindzola, D. C. Griffin and N. R. Badnell, J. Phys. B: At. Mol. Opt. Phys. **32**, L479 (1999).

[52] S. M. Younger, J. Quant. Spectrosc. Radiat. Transfer **26**, 329 (1981).

[53] D. H. Crandall, R. A. Phaneuf, B. E. Hasselquist and D. C. Gregory, J. Phys. B: At. Mol. Phys. **12**, L249 (1979).

[54] D. H. Crandall, Phys. Scr. **23**, 153 (1981).

[55] M. Bannister, *Private Communication*, (2000).

[56] A. Müller, G. Hofmann, K. Tinschert and E. Salzborn, Phys. Rev. Lett. **61**, 1352 (1988).

Chapter 10

ELECTRON IMPACT IONISATION OF HYDROGEN-LIKE IONS

B. E. O'Rourke, H. Watanabe and F. J. Currell
School of Mathematics and Physics,
The Queen's University of Belfast,
Belfast BT7 1NN, UK.
f.j.currell@qub.ac.uk

Abstract Removal of the final tightly bound electron from an ion by electron impact is discussed. The experimental methods used to obtain absolute or relative cross sections for this process are reviewed. Several analytical formulations which may be used to calculate these cross sections are considered and compared to the limited data available for high-Z systems.

1. Introduction

Electron impact ionisation (EI) is the removal of a bound electron by electron impact. If the incident electron has an energy greater than the ionisation potential of the least tightly bound electron then ionisation may occur. Since EI is the dominant ion-creation process in almost all plasmas, its cross sections are of key importance, particularly for high temperature laboratory and astrophysical plasmas. The simplest system available in which to study EI is a hydrogen-like ion (i.e. having only one bound electron) because the reaction unambiguously involves an incident electron removing the bound electron to leave a bare nucleus. Multi-electron effects, e.g. excitation auto-ionisation, are completely absent and only direct single-electron-impact ionisation contributes to the cross section. The nuclear charge Z of the ion is a parameter of interest, with different physics being revealed in different Z ranges.

Using just the Bohr model, one can begin to get a feel for the physics involved. The threshold for EI is $E \approx RZ^2 = 13.6Z^2$[eV], ranging from 13.6 eV for H to 137 keV for U^{91+} giving us the opportunity to examine the same physical phenomenon over four orders of magnitude of energy. Prior to the interaction, the Bohr model's orbital velocity of the bound electron is $v = Z\alpha c$, where

F.J. Currell (ed.), The Physics of Multiply and Highly Charged Ions, Vol. 1, 333-349.

$\alpha = e^2/\hbar c \approx 1/137$ is the fine structure constant. Its characteristic orbital radius is $r = a_0/Z$ where a_0 is the Bohr radius. It is no surprise that the cross-section decreases with increasing Z, approximately as Z^4, since the collision must impart more energy to remove an increasingly tightly bound electron orbiting closer to the nucleus. For U^{91+}, the classical velocity associated with this electron is about 2/3rds of the speed of light indicating mandatory relativistic treatment for high Z.

Experimental methods used to measure these cross sections are outlined in section 2 with various theories being outlined in section 3. Comparison and discussion follow in section 4, with reference to all the high Z ($Z > 20$) data available for hydrogen-like ions.

2. Experimental Methods

Compared to spectroscopic determination of energy intervals, cross sections are notoriously difficult to determine. Rather than simply determining peak positions in some measured spectra, one must measure several separate factors, with the least precise dictating the final error bar. Alternatively some normalization scheme can be applied to relative cross sections although in this case the problem is transferred to measuring some other cross section in an absolute fashion and errors are bound to propagate due to the normalization procedure. Accordingly it is no surprise that the first measurements of the ionisation cross section of atomic hydrogen [1] occurred five decades later (1958) than the classic measurements of energy intervals of the hydrogen atom [2, 3] so important in the formation of quantum theory.

2.1 Crossed Beams Methods

The first cross section measurements used a 'crossed beams method' in which a beam of atomic hydrogen was intersected at 90° by an electron beam with chopping techniques being used to enhance the signal to noise ratio. The pioneering use of this technique was in the study of the photodetachment of H^- ions by Branscomb and Fite [4]. Dolder *et al* [5] used a crossed beams technique with both the electron and ion beam being modulated to make the first measurements of He^+ EI. Simultaneous measurement of both beams was necessary because the electron beam partially neutralises the space charge of the ion beam and changes its spatial distribution. First results for the next ion in the hydrogen-like sequence, Li^{2+}, were not reported until almost 30 years later [6], obtained using the animated crossed beams method [7, 8] whereby the electron beam is swept across the ion beam at a constant speed with sufficient amplitude to ensure that there is no overlap of the beams in the extreme positions.

Following the analysis of Müller *et al.* [8], the ionisation rate R obtained when collimated ion and electron beams, with total currents I_e and I_i, intersect is given by,

$$R = \sigma I_e I_i M / F \tag{10.1}$$

where M is a kinematic factor which depends on the velocities of the ions and electrons and their angle of intersection and σ is the ionisation cross section. F is a from factor which describes the overlap of the beams.

$$F = \frac{I_e I_i}{\int_{-\infty}^{\infty} i_e(z) i_i(z) \mathrm{d}z} \tag{10.2}$$

For a given slice of the ion beam of height $\mathrm{d}z$ and total ion current $I_i = i_i(z)\mathrm{d}z$. Equation 10.1 can then be written as,

$$\mathrm{d}R(z) = \sigma M i_e(z) I_i \tag{10.3}$$

Assume now that the electron beam is swept across this slice of the ion beam with a constant speed u, the number of ions produced during one sweep is then given by,

$$\begin{aligned} \mathrm{d}N &= \int \mathrm{d}R(z')\mathrm{d}t = \sigma M \tfrac{i_i(z)\mathrm{d}z}{u} \int_{z_1}^{z_2} i_e(z')\mathrm{d}z' \\ &= \sigma M I_e i_i(z)\mathrm{d}z / u \end{aligned} \tag{10.4}$$

When the electron beam is moved across the whole ion beam the total number of ionised ions per sweep is

$$\begin{aligned} N &= \sigma M (I_e/u) \int_{z_1}^{z_2} i_i(z)\mathrm{d}z \\ &= \sigma M I_e I_i / u. \end{aligned} \tag{10.5}$$

The electron beam position and sweep rate are thus correlated with the ionised ion count rate. This is a convenient method of measuring electron impact ionisation cross sections without determining the beam overlap factor F.

The same technique, coupled with advances in ion source technology has given rise to EI cross sections of B^{4+}, C^{5+}, N^{6+} and O^{7+} [9], representing the state of the art for absolute measurements. Figure 10.1 shows the experimental crossed beams apparatus used to measure hydrogen-like electron impact ionisation cross sections for ions up to O^{7+}. Ions are produced in an electron cyclotron resonance (ECR) ion source and momentum analysed by a magnet. The ion beam is then transported to the interaction region and crossed with a high current electron beam. Ionised ions are separated from the parent ion beam by a second magnet and detected by a channel electron multiplier single particle detector.

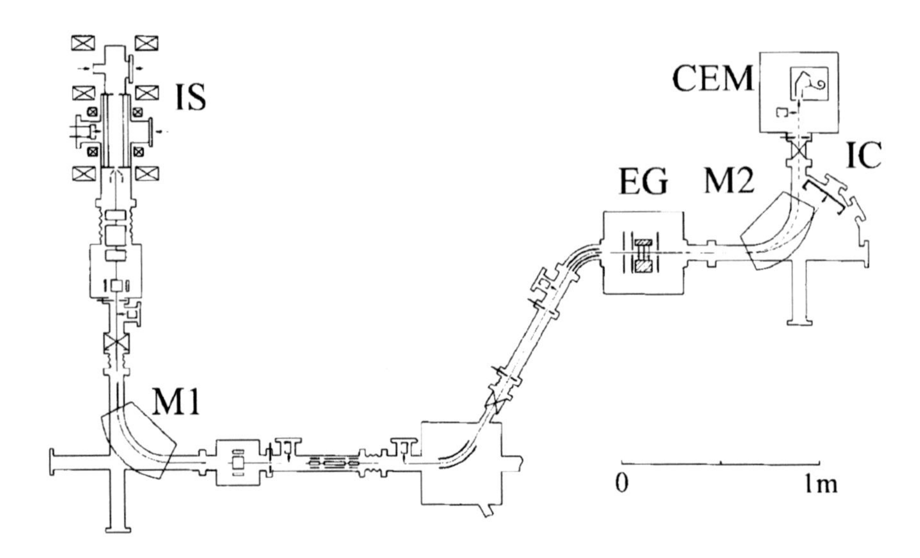

Figure 10.1. Schematic view of the Giessen crossed beams apparatus [6]. The main features include the ECR ion source, IS, and the high power electron gun, EG. M1 and M2 are magnets for momentum analysis and charge state separation respectively. IC is a large Faraday cup for collecting the parent ion beam and CEM a channel-electron-multiplier single particle detector to detect the ionised product ions.

To study more highly charged ions, trap-based techniques can be used and the idea of absolute measurement abandoned or a large dose of experimental modelling applied. Even a cursory glance at the collected literature [10] shows that there is far less measured data available for high Z systems, due to the difficulties associated with the production of highly charged ions.

2.2 Trap Based Measurements

In 1981 Donets and Ovsyannikov [11] reported EI cross section measurements for all C, N, O and Ne positive ions and for Ar^{3+} to Ar^{17+}, obtained by modelling the charge state evolution of ions in an electron beam ion source (EBIS) [12] as measured by time of flight of ions extracted after different electron beam interaction times. To obtain absolute cross sections either modelling or measurement must take account for the size of the electron beam and the ion cloud whilst processes such as double ionisation, charge exchange, radiative recombination and ion escape must be either accounted for or neglected; hence the cross sections measured in this way are labelled 'effective'.

When restricting measurements to hydrogen-like systems, an electron beam ion trap's (EBIT's) [13, 14] innate spectroscopic capabilities can be used in a cleaner manner [15], even extending to U^{91+} [16]. At equilibrium, the rates of creation and destruction of bare ions are equal. Hence, the EI cross section (a factor in the creation rate) can be inferred from the relative abundance of hydrogen-like and bare ions in the trap at equilibrium. Detection rates of characteristic x-rays due to radiative recombination is used to infer this relative abundance.

At equilibrium, the rate of change of the population of bare ions (in a regime where escape is negligible) is written

$$\frac{dN_B}{dt} = 0 = -\left(\frac{J}{e}\right) N_B f_{e,B} \sigma_B^{\sum RR} + \left(\frac{J}{e}\right) N_H f_{e,H} \sigma_H^{EI} - N_0 N_B \left\langle v\sigma_B^{CX} \right\rangle, \tag{10.6}$$

where N_B and N_H are the axial number densities of bare and hydrogen like ions, $f_{e,B}$ and $f_{e,H}$ the overlap factors of the ion clouds with the electron beam and N_0 the number density of neutral atoms. $\sigma^{\sum RR}$, σ^{EI} and $\langle v\sigma^{CX}\rangle$ are the total radiative recombination, electron impact ionisation cross sections and charge exchange rate respectively. Rearranging eq. 10.6 leads to

$$\sigma_H^{EI} = \frac{N_B f_{e,B} \sigma_B^{\sum RR}}{N_H f_H} + \frac{N_0 N_B \left\langle v\sigma_B^{CX} \right\rangle}{N_H f_{e,H} \left(J/e\right)}. \tag{10.7}$$

N_B and N_H can be related to the yield of RR x-ray photons due to capture into n=1 observed at 90° into solid angle Ω, using the relations,

$$N_{h\nu_B} = \frac{I_e}{r_e^2} N_B \frac{\mathrm{d}\sigma_B^{RR=1}}{\mathrm{d}\Omega} \mathrm{d}\Omega \tag{10.8}$$

and

$$N_{h\nu_H} = \frac{I_e}{r_e^2} N_H \frac{\mathrm{d}\sigma_H^{RR=1}}{\mathrm{d}\Omega} \mathrm{d}\Omega \tag{10.9}$$

where N_{hv_B} and N_{hv_H} are the respective photon yields as measured by a solid state detector. The beam radius (r_e) has been shown to be weakly dependent on current [17]. For sufficiently shallow trap depth, the overlap factors $f_{e,B}$ and $f_{e,H}$ are approximately equal with values close to 1. Then, equation 10.7 can then be written

$$\sigma_H^{EI} = \frac{N_{h\upsilon_B}}{N_{h\upsilon_H}} \frac{\sigma_B^{\sum RR} \mathrm{d}\sigma_H^{RR=1}/\mathrm{d}\Omega}{\mathrm{d}\sigma_B^{RR=1}/\mathrm{d}\Omega} \left(1 + \frac{N_0 r_e^2}{I\sigma_B^{\sum RR}} \left\langle v\sigma_B^{CX} \right\rangle \right) \qquad (10.10)$$

To take account of charge exchange, experiments are performed at various neutral gas densities, N_0, and beam currents, I_e. The ratio $N_{h\upsilon_B}/N_{h\upsilon_H}$ is extrapolated to either the limit $N_0 \to 0$ or $1/I_e \to 0$.

In either limit the change exchange contribution to the ratio disappears when the desired electron impact ionisation cross section can be obtained from

$$\sigma_H^{EI} = \frac{\sigma_B^{\sum RR} \mathrm{d}\sigma_H^{RR=1}/\mathrm{d}\Omega}{\mathrm{d}\sigma_B^{RR=1}/\mathrm{d}\Omega} \left(\frac{N_{h\upsilon_B}}{N_{h\upsilon_H}}\right)_{lim \langle v\sigma_B^{CX}\rangle \to 0}. \qquad (10.11)$$

Figure 10.2 shows the RR x-ray signal obtained from hydrogen-like and bare iron ions as measured at the Tokyo EBIT. Measurements are shown at an energy of 15 keV and at three different beam currents, 60 mA, 100 mA and 160 mA. It is clear that as the current is increased the charge balance in the trap is driven more towards bare ions. The ratio of hydrogen-like to bare ions decreases and this can be used to infer the charge exchange contribution.

All relevant radiative recombination cross sections can be calculated theoretically to high accuracy [18]. This is because RR is the inverse of photoionisation, a process which has been extensively studied [19]. Using theoretical RR cross sections, ionisation cross sections can be determined from equation 10.11 based on the equilibrium behaviour of the machine.

Figure 10.3 shows the measured cross sections in hydrogen-like iron ions compared to distorted wave calculations [18]. Molybdenum ions have also been studied in separate experiments [20, 15]. Cross sections are plotted in figure 10.4 and compared to theoretical cross sections as reported by Fontes *et al* [21] (see section 10.3.2). The two separate molybdenum experiments show excellent agreement both with each other and with the theoretical calculation. A summary of these measurements and all other published measurements of the EI cross sections of high Z hydrogen-like ions using EBIT's is given in table 10.1.

It is readily apparent from figures 10.3 and 10.4 that the experimental error bars are biggest for higher electron energy. This can be understood by equation 10.11 without taking the limit $\langle v\sigma_B^{CX}\rangle \to 0$. The error in the measured electron impact ionization cross section can be written

$$\epsilon_H^{EI} = \alpha \times \epsilon^{N_{h\nu_H}/N_{h\nu_B}} \qquad (10.12)$$

$$\alpha = \left(\frac{\sigma_H^{EI^2}}{\sigma_B^{\sum RR}} \frac{\frac{d\sigma_B^{RR=1}}{d\Omega}}{\frac{d\sigma_H^{RR=1}}{d\Omega}}\right) \qquad (10.13)$$

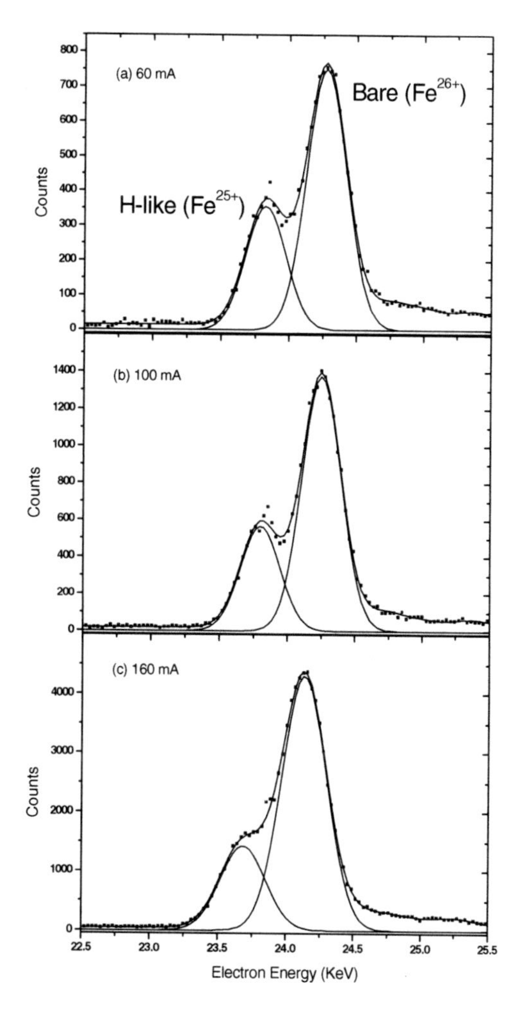

Figure 10.2. X-ray spectra of the RR of H-like and bare iron ions measured at several electron beam currents, (a) 60 mA, (b) 100 mA, (c) 160 mA. All measurements were performed at an energy of 15 keV. The ratio of hydrogen-like to bare ions is seen to decrease as the current is increased.

where ϵ_H^{EI} is the error of the electron impact ionization cross section and $\epsilon^{N_{h\nu_H}/N_{h\nu_B}}$ that of the intensity ratio in the limit. Figure 10.5 shows the electron energy dependence of α for iron and molybdenum, calculated with the values used and obtained in this study. As seen in the figure α grows rapidly as the electron energy increases.

Hence, even if the factor $\epsilon^{N_{h\nu_H}/N_{h\nu_B}}$ can be measured with equally good statistics across the whole energy range, the errors tend to be larger at higher energies. Note this analysis provides a lower bound on the error. As is discussed

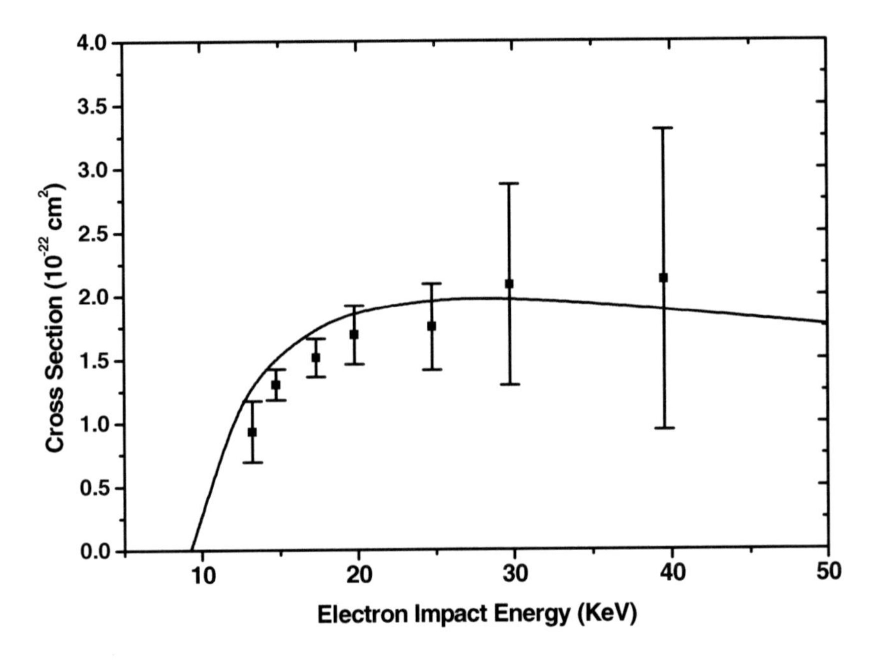

Figure 10.3. Measured electron impact ionisation cross sections for H-like iron (squares) compared to a distorted wave calculation (solid line) [18].

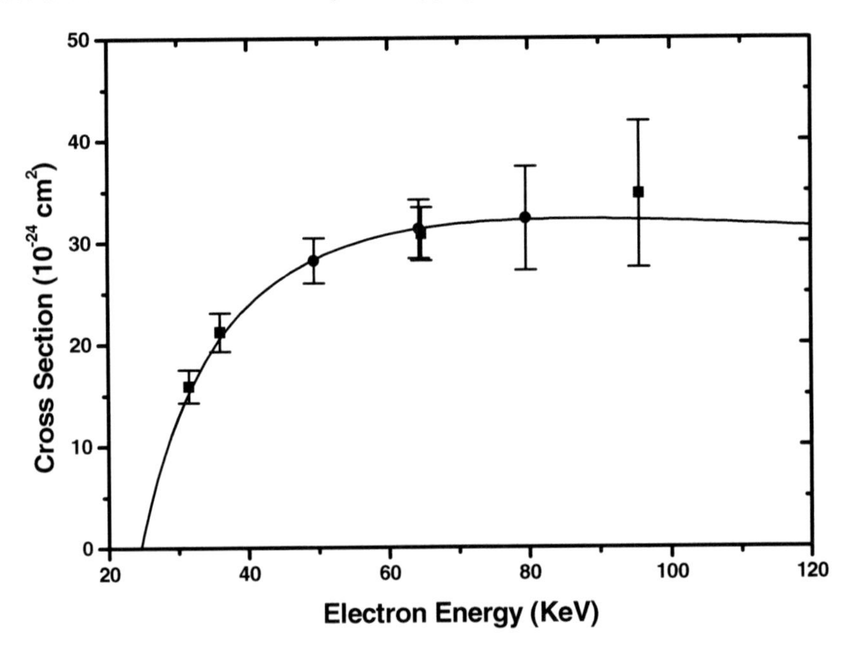

Figure 10.4. Measured electron impact ionisation cross sections for H-like molybdenum measured in two different experiments, Marrs *et al* [15](squares) and Watanabe *et al* [20](circles), compared to the Fontes-Sampson-Zhang formula [21] (solid line).

Table 10.1. Measured cross sections for the electron impact ionisation of high Z hydrogen-like ions, $(Z > 20)$. $\sigma_B^{\sum RR}$ is the total radiative recombination cross section into bare ions and σ_H^{EI} is the H-like electron impact ionisation cross section. Cross section units are 10^{-24} cm^2.

Element (Z) [ref]	*H-like I.P. (keV)*	*Ee (keV)*	*Ee/I.P.*	$\sigma_B^{\sum RR}$	σ_H^{EI}
Fe(26)[18]	9.27	13.3	1.43	77.32	93±24
		14.8	1.60	64.81	130±12
		17.3	1.87	49.73	151±15
		19.8	2.14	39.58	169±23
		24.8	2.68	26.67	175±34
		29.8	3.21	19.18	208±79
		39.6	4.27	11.28	212±118
Mo(42)[15]	24.6	31.5	1.28	90.33	15.9±1.6
		36.1	1.47	72.39	21.2±1.9
		64.8	2.63	26.56	30.8±2.6
		95.6	3.89	13.01	34.7±7.2
Mo(42)[20]		49.4	2.01	42.76	28.2±2.2
		64.4	2.62	26.87	31.3±2.9
		79.6	3.24	18.31	32.3±5.1
Dy(66)[15]	63.1	95.1	1.51	67.20	4.17±0.58
		153.1	2.43	30.20	6.29±0.83
Au(79)[15]	93.3	153.1	1.64	58.60	2.33±0.33
Bi(83)[15]	104.1	191.6	1.84	48.50	2.37±0.19
U(92)[16]	131.8	198.0	1.50	67.50	1.55±0.27

in section 10.3.2, at higher electron energy, relativistic effects become more pronounced so it is important to make good quality measurements in this region. However as the above analysis indicates this is a difficult challenge requiring a long period of beam time.

3. Theory and Analytical Formulae

In this chapter, the main emphasis is placed on theoretical formulations expressed in some closed analytical form. Such formulations have the merits of being compact, directly comparable to future measurements and formalisms, and directly applicable to modeling calculations, either in their raw form or when integrated over some electron energy distribution to give rates.

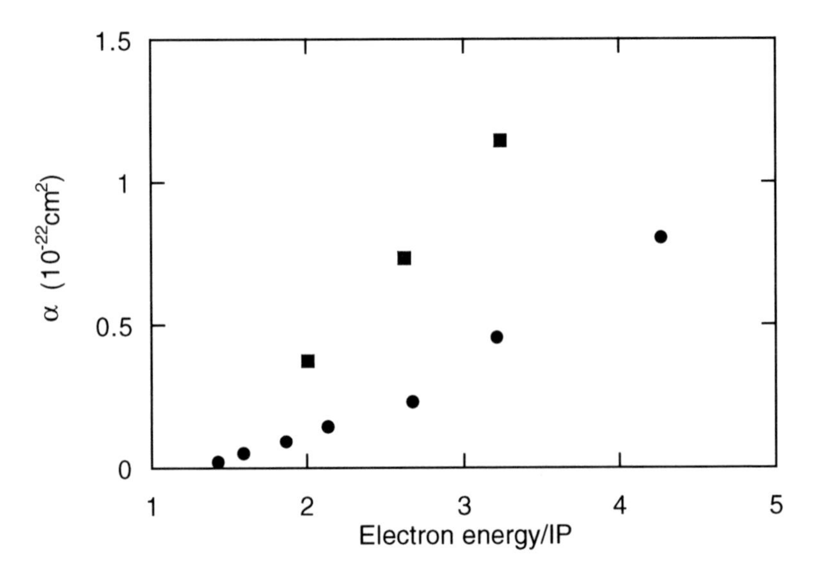

Figure 10.5. The error propagation factor α in equation 10.12 for iron (circles) and molybdenum (squares). The abscissa is the electron energy in ionization threshold units.

3.1 Non-Relativistic Calculations

Thomson adopted the first classical model for describing electron impact ionisation in 1912 [22]. For hydrogen-like ions this classical approach predicts the simple scaling law,

$$\sigma I^2 = f(E/I). \tag{10.14}$$

If the cross section is scaled by the square of the ionisation potential then a universal curve is expected. Aichele *et al* [9] showed that this scaling was in quite good agreement with measured cross section in the low Z regime studied (up to O^{7+}).

By the 1960's the approximations most widely used in the theory of ionising collisions were those suggested by Born and Bethe [23]. For electron ion collisions, allowance should be made for the long-range Coulomb attraction generated by the ion. Both the Born and Bethe approximations are only valid for fast projectile electrons, i.e. when the energy of the incident electron is large compared with the binding energy of the atomic electrons. Lotz [24] proposed a semi-empirical formula based on the high energy Bethe limit, with adjustable parameters being determined with reference to experimental results for He^+ [5]. This fruitful approach has since been repeated a number of times, with analytically based (high or low energy limit) forms being adjusted with reference to specific data determined either theoretically or experimentally [9, 25, 26]. Table 10.2 presents a summary of several non-relativistic formalisms

Table 10.2. Non-relativistic theoretical formulations for the ionisation cross sections of hydrogen-like ions.

Name(s) & Ref.	*Formulae describing the formalism* *E=electron impact energy, I=ionisation energy, Z =nuclear charge,* $u = E/I$	*many-* e^-	χ^2
Lotz [24]	$\sigma^{Lotz} = 4.5 \times 10^{-14} \frac{\ln[E/I]}{EI} = 4.5 \times 10^{-14} \frac{\ln[u]}{uI^2}$	yes	33.3
Younger [30] [31]	$\sigma^{Younger} = \frac{1}{uI^2} \left\{ \begin{array}{c} A\left(1-\frac{1}{u}\right) + B\left(1-\frac{1}{u}\right)^2 + \\ C\ln u + \frac{D}{u}\ln u \end{array} \right\}$ see [30] for calculation of parameters A, B, C and D	yes	32.7
Deutsch [26]	$\sigma^{Deutsch} Z^{4.155} = 5.08 \times 10^{-16} f(u)$ $f(u) = u^{-1}[(u-1)/(u+1)]^a \times$ $\{b + c[1-(2u)^{-1}] \ln[2.7 + (u-1)^{0.5}]\}$ $a = 1.30, b = 0.75, c = 0.68$	yes	339
Rost & Pattard [9] [25]	$\sigma^{Rost-Pattard}(x) = \sigma_M[\cosh(\beta \ln x)]^{-1/\beta}$ $x = (E-I)/E_M, \beta = 0.4$ $E_M(Z) = I_H\left(3.31 + 5.07(Z-1) + 1.65(Z-1)^2\right)$ $\sigma_M(Z) = \pi a_0^2 \left(\frac{1.0278}{1.0975+Z-1}\right)^4, I_H = 13.616\text{eV}$	no	42

which predict EI cross sections for hydrogen-like systems while table 10.3 presents the same information for several relativistic formalisms. Some of these formalisms are also capable of predicting EI cross sections for multi-electron systems although sometimes not in the simplified form presented here. This extensibility is indicated in the table and the reader is referred to the original papers for extension to multi-electron cases. The last column of tables 10.2 and 10.3 gives a chi-squared value based on comparing the measured data for $Z > 20$ with the formulation concerned.

The Lotz formula was found to agree well with the Coulomb-Born-exchange (CBE) calculations of Rudge and Schwartz [27] for high Z hydrogen-like ions. Scaled cross sections for ionisation of hydrogen-like ions in the limit $Z \to \infty$ were also calculated in the CBE approximation by Golden and Sampson [28]. Rudge and Schwartz [27] calculated similar cross sections for a fictitious ion with $Z = 128$. Together these results were fitted to an analytical expression that gave the correct Bethe approximation results of Omidvar [29]

Table 10.3. Relativistic theoretical formulations for the ionisation cross sections of hydrogen-like ions.

Name(s) & Ref.	*Formulae describing the formalism* E=*electron impact energy*, I=*ionisation energy*, Z =*nuclear charge*, $u = E/I$	*many-* e^-	χ^2
Relativistically corrected Lotz [34]	$\sigma^{RelLotz} = \sigma^{Lotz}\left(\frac{\tau+2}{\varepsilon+2}\right)\left(\frac{\varepsilon+1}{\tau+1}\right)^2 \times \left(\frac{(\tau+\varepsilon)(\varepsilon+2)(\tau+1)^2}{\varepsilon(\varepsilon+2)(\tau+1)^2+\tau(\tau+2)}\right)^{3/2}$ $\varepsilon = E/(m_0c^2)$, $\tau = I/(m_0c^2)$, m_0c^2=rest mass of electron $\approx$511keV	yes	20.7
Relativistic Deutsch [26]	$\sigma^{RelDeutsch} = \sigma^{Deutsch} R(u)\left[1 + (2u)^{0.25}/J^2\right]$ $R(u) = \left[\frac{1+2J}{u+2J}\right]\left[\frac{u+J}{1+J}\right]^2\left[\frac{(1+u)(u+2J)(1+J)^2}{J^2(1+2J)+u(u+2J)(1+J)^2}\right]^{1.5}$ $J = (m_ec^2)/I$	yes	221
Fontes, Sampson and Zhang [21]	$\sigma^{Fontes} = \frac{\pi a_0^2}{I(Ry)^2} Q'[u,Z]F[Z]$ $F[Z] = \left[140 + (Z/20)^{3.2}\right]/141$ $Q'[u,Z] = \frac{1}{u}\left\{\begin{array}{l} 1.13\ln u + 3.7059\left(1-\frac{1}{u}\right)^2 + \\ C[Z]u\left(1-\frac{1}{u}\right)^4 + \\ \left(\frac{1.9527}{u^2} - \frac{0.28394}{u}\right)\left(1-\frac{1}{u}\right) \end{array}\right\}$ $C[Z] = \{(Z-20)/55\}^{0.92} + 0.20594$ $I\,(Ry) = Z^2\{1 + 3.2\times10^{-3}Z + 1.5\times10^{-6}Z^2 + 1.4\times10^{-7}Z^3\}$	yes	11.1

at high energies [28]. Younger [30, 31] used several variants of the Coulomb-Born and distorted wave approximations to calculate ionisation cross sections of hydrogen-like ions.

Absolute and scaled electron impact ionisation cross sections for hydrogen-like ions were calculated by Deutsch *et al* [26] using the semiclassical Deutsch-Märk [33] approach. This formalism, which is based on a combination of the classical binary encounter approximation and the quantum mechanical Born-Bethe approximation, is also given in table 10.2. It predicts a scaling law in close agreement with the widely used Z^4 scaling law for hydrogen-like targets. Rost and Pattard [9, 25] derived a universal shape function onto which all non-relativistic hydrogen-like EI cross sections can be mapped through parameterisation of the maximum value $\sigma_M(Z)$ of the cross section and the excess energy above threshold $E_M(Z)$ where this maximum occurs. In common with many of the other formalisms, this universal shape function goes to the Bethe limit at high energies. Distinctively, this approach also has well defined low energy behaviour, following the Wannier law [32]. This scaling was seen to be in good agreement with all hydrogen-like ions studied up to O^{7+} [9].

3.2 Relativistic Calculations

As indicated in the introduction, for more highly charged ions, relativistic effects must be included in the calculations. A first attempt to include relativistic effects was made by introducing a relativistic correction factor to the Lotz formula [34]. In the 1980's relativistic distorted wave methods were applied to electron impact ionisation. Kao *et al* [35] calculated relativistic ionisation cross section for a number of hydrogen-like species (H, He^{+}, C^{5+}, Ne^{9+}, Fe^{25+} and Ag^{46+}), using a distorted wave approximation. Results for higher charge species deviate from the universal trend at high energies.

At these high energies QED effects become important. Accurate expressions for the energy of two electrons interacting with a quantized electro-magnetic field should include the Breit interaction, a term which arises due to the transfer of virtual photons between the electrons. When the energy difference between two levels becomes very high, as in a very highly charged ion, the energy of a virtual photon is large and its oscillation must be included in the calculation. This is called the generalized Breit interaction (GBI). Alternatively the Møller interaction can be included which considers magnetic interactions and retardation. Inclusion of the Møller interaction is essentially equivalent to use of the GBI.

Moores and Pindzola [36] considered hydrogen-like ions with nuclear charge between 26 and 92, demonstrating the increased importance of relativistic effects with Z and incident electron energy. The effect of including the Møller interaction was investigated in a no-exchange scattering approximation. They found that this effect was only significant when the incident electron energy reaches about 250 keV. When exchange was included the effect was found to be important over a wide range of energies [37].

The range of nuclear charge over which inclusion of the generalized Breit interaction is important was investigated by Fontes *et al* [21]. Cross sections were calculated for hydrogen-like ions with Z values of 10, 20, 30, 40, 50, 66, 79 and 92 and a range of impact energies u. These calculated data were then interpolated with fit functions given in table 10.3.

The energy dependence $I\,(Ry) = Z^2(1 + 3.2 \times 10^{-3}Z + 1.5 \times 10^{-6}Z^2 + 1.4 \times 10^{-7}Z^3)$ used in conjuction with the scaling law of Fontes, Sampson and Zhang [21] was extracted from table 10.3 of that paper by polynomial fitting. Indeed if a simpler expression is used such as $I(Ry) = Z^2$ is used then the approximation gives a considerably higher chi-squared when compared to the experimental data.

4. Discussion

For low Z, the formalism of Rost and Pattard [25] is capable of accounting for all the measurements available [9]. Clearly the combination of low and high energy limits and reference to a significant body of high quality data has given rise to a compact, meaningful formalism capable of describing the EI process for hydrogenic systems. When one looks at the data for $Z > 20$ (where relativistic effects can be expected to play a significant role), there is much less data available (see table 10.1). The chi-squared based on comparing this data ensemble with each of the formalisms is given in tables 10.2 and 10.3. As might be expected, increased theoretical sophistication leads to better capability to account for the available data. The relativistic Lotz formalism does surprisingly well considering it has only one adjustable parameter, which was determined by reference to He^+ measurements made several decades ago. The lowest chi-squared however is achieved by the formalism of Fontes *et al* [21] and is a testimony to the quality of the distorted wave calculations and the careful account taken of the generalised Breit interaction. It is interesting to note that the ionisation potentials used in applying this formalism cannot simply be of the form 13.616 Z^2[eV]. If this formula is used in predicting the cross-sections then the formalism fairs no better than some of the non-relativistic ones.

Figure 10.6 shows a comparison of the the four best formalisms with the data for $Z > 20$. The agreement between theory and experiment extends over a wide range of Z values and impact energies. With so few data points available to guide theory, and with the error bars being typically 10%, it is clear that for highly charged ions, more measurements are required. There is however a hint of an interesting systematic deviation between the 'best' (in terms of the chi-squared value) theory, of Fontes *et al* [21] and the measured data in the energy region between 1.2 and 2.7 ionisation potentials. As shown by eq. 10.12 and fig. 10.5 it should be relatively straight forward to make more measurements in this region. The other theories however have no such suggestion of a systematic deviation. In contrast, the lower charge states have now been comprehensively studied.

Several other frontiers exist where new measurements could stimulate theoretical developments. The goal of genuine absolute cross section determination awaits ingenious experimentalists as does the development of a technique which gives rise to smaller error bars at higher energy. Recent instrumental developments such as Thompson scattering off an EBIT's electron beam [17], imaging of the ion cloud [38] coupled with the measurements of onsets [39] may soon make absolute cross section determination possible. There are almost no measurements of open shell EI cross sections of highly charged ions [40] and ionisation from excited initial states is another area awaiting development.

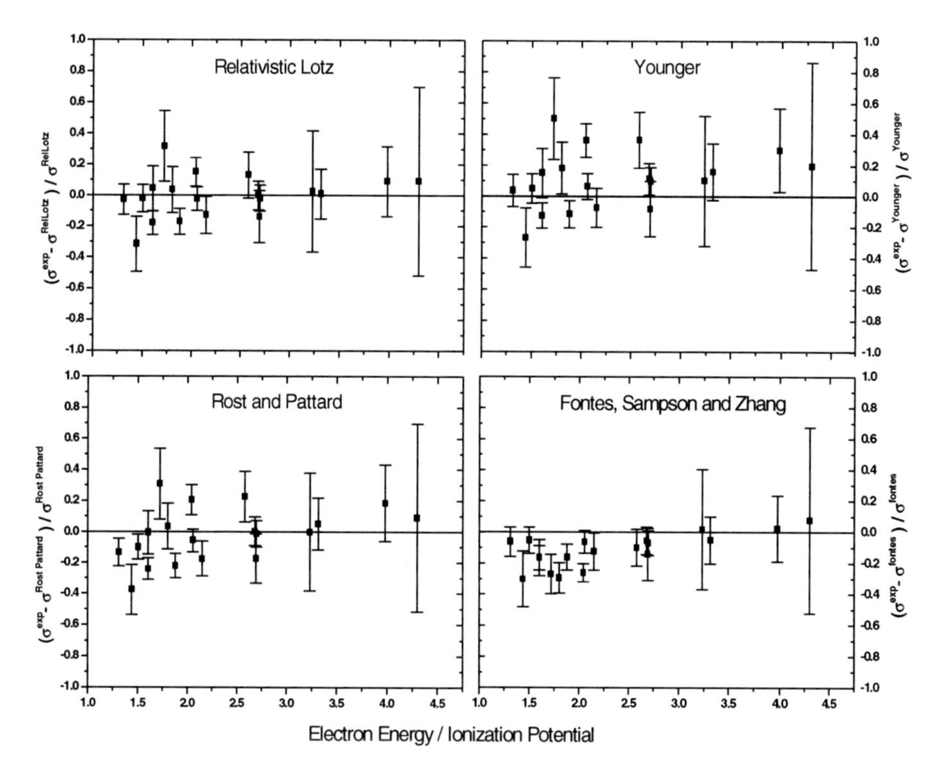

Figure 10.6. A comparison of the four most successful formulations with the measured high Z H-like cross sections. The fractional difference between theory and experiment is plotted as a function of the reduced energy, the electron energy divided by the ionisation potential. The labels refer to the names given to them in tables 10.2 and 10.3

5. Conclusions

Very few measurements have been made on the electron impact ionisation of hydrogen-like ions. Good quality absolute values have been determined up to O^{7+} by crossed beams methods. Further progress with this method has been hampered due the difficulty of producing a sufficient current of hydrogen-like ions from the ion source and intersecting the ion beam with a high current, high energy electron beam.

For highly charged ions, several measurements have been made using EBITs. These overcome the problems associated with crossed beams by trapping the ions, creating a quasi-stationary ion target allowing hydrogen-like and bare ions to be produced. By measuring the population balance of these ions the cross section can be determined. Even these measurements result in a large error bar. More and higher quality measurements are needed to further stimulate and guide theoretical developments.

Acknowledgments

The authors are indebted to Professor Shunsuke Ohtani and collaborators in the EBIT group at the University of Electro-communications, Tokyo and the Japan Science and Technology (JST), Cold Trapped Ions project for helpful discussions. We wish to express our gratitude to Professor Alfred Müller for providing the technical drawing from which figure 10.1 was created.

References

[1] W. L. Fite and R. T. Brackmann, Phys. Rev. **112**, 1141 (1958).

[2] T. Lyman, Astrophys. J. **23**, 181 (1906).

[3] F. Paschen, Ann. Phys. **27**, 537 (1908).

[4] L. M. Branscomb and W. L. Fite, Phys. Rev. **93**, 651 (1954).

[5] K. T. Dolder, M. F. A. Harrison, and P. C. Thonemann, Proc. R. Soc. A **264**, 367 (1961).

[6] K. Tinschert *et al.*, J. Phys. B **22**, 531 (1989).

[7] P. Defrance *et al.*, J. Phys. B **14**, 103 (1981).

[8] A. Müller *et al.*, J. Phys. B **18**, 2993 (1985).

[9] K. Aichele *et al.*, J. Phys. B **31**, 2369 (1998).

[10] H. Tawara and M. Kato, NIFS-DATA **51** (1999).

[11] E. D. Donets and V. P. Ovsyannikov, Zh. Eksp. Teor. Fiz. **80**, 916 (1981), (Engl. transl. *Sov. Phys.-JETP* **53** 466 (1981).

[12] E. D. Donets, in *The Physics and Technology of Ion sources*, edited by I. G. Brown, p. 245, Wiley, New York, 1989.

[13] R. E. Marrs *et al.*, Phys. Rev. Lett. **60**, 1715 (1988).

[14] F. J. Currell, The physics of electron beam ion traps, in *Trapping Highly Charged Ions; Fundamentals and Applications*, edited by J. Gillaspy, Nova science publishers, New York, 2000.

[15] R. E. Marrs, S. R. Elliott, and J. H. Scofield, Phys. Rev. A **56**, 1338 (1997).

[16] R. E. Marrs, S. R. Elliott, and D. A. Knapp, Phys. Rev. Lett **72**, 4082 (1994).

[17] H. Kuramoto *et al.*, Rev. Sci. Instrum. **73**, 42 (2002).

[18] B. O'Rourke *et al.*, J. Phys. B **34**, 4003 (2001).

[19] E. B. Saloman, J. H. Hubbell, and J. H. Scofield, At. Data Nucl. Data Tables **38**, 1 (1988).

[20] H. Watanabe *et al.*, Electron impact ionization of hydrogen-like molybdenum ions, submitted to J. Phys. B.

[21] C. J. Fontes, D. H. Sampson and H. L. Zhang, Phys. Rev. A **59**, 1329 (1999).

[22] J. J. Thomson, Phil. Mag. **23**, 449 (1912).

[23] E. W. McDaniel, *Atomic Collisions: electron and photon projectiles* (Wiley, New York, 1989).

[24] W. Lotz, Z. Phys **216**, 241 (1968).

[25] J. M. Rost and T. Pattard, Phys. Rev. A **55**, R5 (1997).

[26] H. Deutsch, K. Becker, and T. D. Märk, Int. J. Mass Spectrom. Ion Proc. **151**, 207 (1995).

[27] M. R. H. Rudge and S. B. Schwartz, Proc. Phys. Soc. **88**, 563 (1966).

[28] L. B. Golden and D. H. Sampson, J. Phys. B **10**, 2229 (1977).

[29] K. Omidvar, Phys. Rev. **177**, 212 (1969).

[30] S. M. Younger, Phys. Rev. A **22**, 111 (1980).

[31] S. M. Younger, J. Quant. Spectrosc. Radiat. Transfer **26**, 329 (1981).

[32] G. H. Wannier Phys. Rev. **90** 817 (1953).

[33] H. Deutsch and T. D. Märk, Int. J. Mass. Spectrom. Ion Proc. **79** (1987).

[34] M. Gryzinski, Phys. Rev. **138**, A336 (1965).

[35] H.-C. Kao *et al.*, Phys. Rev. A **45**, 4646 (1992).

[36] D. L. Moores and M. S. Pindzola, Phys. Rev. A **41**, 3603 (1990).

[37] D. L. Moores and K. J. Reed, Phys. Rev. A **51** (1995).

[38] J. V. Porto, I. Kink, and J. D. Gillaspy, Rev. Sci. Instrum. **71**, 3050 (2000).

[39] E. Sokell, F. J. Currell, H. Shimizu, and S. Ohtani, Phys. Scr. **T80** (1999).

[40] T. Stöhlker *et al.*, Phys. Rev. A **56**, 2819 (1997).

Chapter 11

TEST OF STRONG-FIELD QED

Th. Stöhlker, T. Beier, H.F. Beyer, Th. Kühl and W. Quint
GSI, Planckstr. 1,
D-64291 Darmstadt, Germany
t.stoehlker@gsi.de

Abstract This review reports on experiments addressing quantum electrodynamical (QED) effects for the ground state in hydrogen- and helium-like ions at high nuclear charges. Such experiments probe QED in the domain of the strongest possible electromagnetic fields. In a short outline, we emphasize the particularities of bound–state QED in the strong field. Experimentally we concentrate on research conducted at the ESR storage ring at GSI Darmstadt where highly charged and even bare ions at high-Z can be studied under unique experimental conditions. As a representative example, the 1s binding energy in H-like uranium, determined with an accuracy of 13 eV, is discussed. At present, this constitutes the most precise 1s Lamb shift measurement available for the high-Z regime. These 1s Lamb Shift studies are discussed in detail as well as the experiments aiming at a precise determination of the ground–state ionization potentials of high-Z helium like ions where the two-electron QED corrections in the strong field limit can be studied. In addition QED effects associated with the strong magnetic fields as present at high Z are reviewed. In this context, the studies of the hyperfine splitting of the ground state in H-like bismuth are presented, as measured precisely by collinear laser spectroscopy. Similar to the hyperfine splitting, highly accurate measurements of the bound state g factor in hydrogen-like ions provide a sensitive tests of bound-state QED. The measurement of the g factor of bound electrons is a new experimental approach to address QED in the strong–field domain which is accomplished by storing a single highly charged ion in an ion trap. This technique is discussed in detail and its potential for future investigations is illustrated by the presentation of a first result already obtained for the case of hydrogen-like carbon.

Keywords: quantum electrodynamics, Lamb shift, hyperfine structure, g factor of electrons, heavy ions, storage rings

F.J. Currell (ed.), The Physics of Multiply and Highly Charged Ions, Vol. 1, 351-386.

1. Introduction

Our present understanding of bound electronic systems is formulated through the theory of quantum electrodynamics (QED). The latter can be regarded as one of the most precisely tested theories of physics. One impressive example is the g factor of the free electron which presently is known as

$$g_{\text{free}} = 2 + 2 \times 1\,159\,652\,188.4(4.3) \times 10^{-12} \quad \text{(Experiment)},$$
$$g_{\text{free}} = 2 + 2 \times 1\,159\,652\,216.0(1.2)(67.8) \times 10^{-12} \quad \text{(Theory)}.$$

The second uncertainty indicated for the theory results from the uncertainty in the fine-structure constant $\alpha \simeq 1/137$. In turn, the g factor experiment allows for the most precise determination of α by comparing experimental result and theoretical value.

Similar accuracies are nowadays obtained in systems like positronium or the Lamb shift in hydrogen where insufficiently known nuclear parameters limit the accuracy of theoretical predictions. The electric field involved in all these systems is rather low, however, compared to the strongest electromagnetic fields accessible to experimental investigation today. In atomic systems, these strong fields are provided by the Coulomb field of the lead or uranium nucleus. The single electron is bound in a system similar to hydrogen, and the whole system is therefore called a hydrogen-like system. The expectation value of the electric field strength experienced by the innermost electron in such a system is depicted in fig. 11.1. The field strength at the nuclear surface is even higher. For example, at the surface of a uranium nucleus, $|\mathbf{E}| \cong 2 \cdot 10^{19}$ V/cm. This is only a factor of 2 less than the field strength in superheavy systems with $Z \geq 170$ where spontaneous pair production is predicted to take place if the total charge can be confined in a sufficiently small volume for a sufficiently long time. It seems evident that in such strong fields "normal" atomic physics – valid for a hydrogen atom where the field probed by the electron is six orders of magnitudes smaller – may be questioned. An accurate knowledge about the validity of QED in strong external fields is also necessary for the detection of new physics beyond QED. Thus it is a primary goal to explore the behaviour of electrons in some of the strongest electromagnetic fields accessible to experimental investigation. The results are also important for the energy levels and related x-ray spectra of heavy neutral or singly ionized heavy elements. The correlation effects among the many contributing particles, however, render the dedicated study of the fundamental QED effects impossible.

The advent of powerful ion sources like the superEBIT at Livermore and of heavy-ion storage rings like the ESR at GSI Darmstadt has precipitated both the experimental and theoretical study of very heavy ions with only one or few bound electrons where correlation effects are either absent or under control for

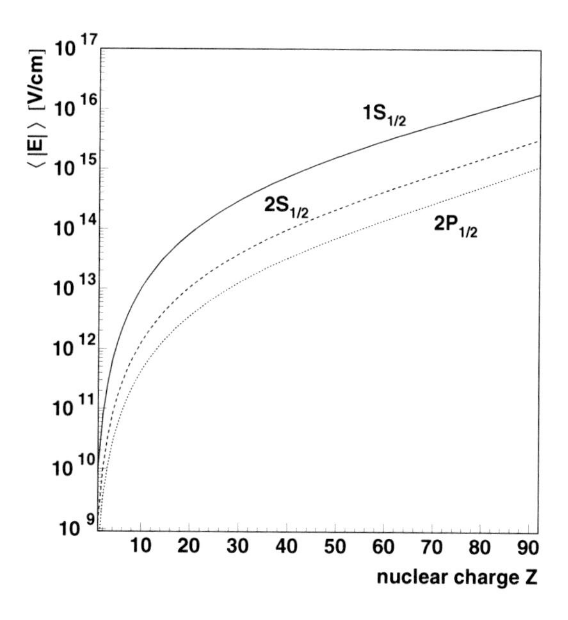

Figure 11.1. Expectation value of the electric field strength for the lowest-lying states of a hydrogen-like atom in the range $Z = 1 - 92$.

specific studies[1]. This will enable tests of QED in the new domain of strong electromagnetic fields.

In this chapter we sketch the specific steps necessary for a meaningful comparison between experiment and theory and summarize how far both have progressed.

Section 11.2 is dedicated to the theoretical foundations and peculiarities at strong fields. The QED effects manifest in the Lamb shift of the innermost energy levels of heavy one and two-electron ions as will be discussed in sections 11.3 and 11.4. For hydrogen-like U^{91+}, for instance, the Lamb shift of the $1s_{1/2}$ ground state amounts to approximately 470 eV as compared to $\simeq 3.4 \times 10^{-5}$ eV for hydrogen. Other quantities used for tests of QED are the hyperfine splitting of the $1s_{1/2}$ ground state accessible through Laser spectroscopy which can now be performed with high accuracy and which will be covered in section 11.5. Also the Zeeman effect and the corresponding g factor of the bound electron are governed by QED effects and can be measured with electromagnetic ion traps as will be discussed in section 11.6.

2. Theory of QED

Quantum electrodynamics is based on the quantization of the photon field and takes the simplest form of a field theory for Dirac particles with spin 1/2, e.g., electrons. The mathematical foundations of this theory will not be laid out here but can be found in numerous excellent textbooks on the subject, e.g., [2]. For the purpose of this chapter it is sufficient to realize that all interactions of charged particles can be reduced to an exchange of photons between them, and that in many cases the consideration of a few photon exchanges is sufficient to describe the whole process with sufficient accuracy. For illustrative purposes, we introduce Feynman diagrams which allow to depict these fundamental processes in a simple picture-like scheme. Two of the simplest processes

Figure 11.2. Feynman diagram for the one-photon exchange between two free electrons (left) and the one-photon exchange of a free electron with an external field (right). The free electrons are represented by solid lines, the photon by a wavy line. The cross denotes an external field such as the binding Coulomb field of a nucleus, and the dots denote vertices, where a photon is absorbed or emitted by an electron.

are depicted in fig. 11.2. Each vertex or external-field interaction contributes with a quantity proportional to the total charge involved there, i.e. e or Ze for an electron or a nucleus of nuclear charge number Z. As $e^2/(4\pi) = \alpha$ in relativistic units (with $\hbar = c = m_e = 1$), the contributions of diagrams with more than one photon involved decrease rapidly except for the case of highly charged ions where $Z\alpha$ (the coupling constant for an interaction with the external field) is no longer small but amounts to $Z\alpha \approx 0.6$ for uranium. Therefore the binding cannot be treated as a perturbation to the free case, contrary to light systems like hydrogen where only charges e or Ze, Z small, are involved. Bound-state QED provides the relativistic description of the electrons in highly charged atomic systems. Contrary to the pure perturbative approach in light systems, where the free electron can be represented by plane waves, and analytical (but difficult) expressions can be obtained for every diagram under consideration, the permanent consideration of binding requires some of the most elaborate numerical techniques available to date. To test the predictions of this theory with utmost precision is a major challenge in today's atomic physics.

The important 'new' feature of *quantum* electrodynamics is the interaction of the electron with its own photon field, to lowest order in α described by the two diagrams given in fig. 11.3. For free particles, the mathematical expressions

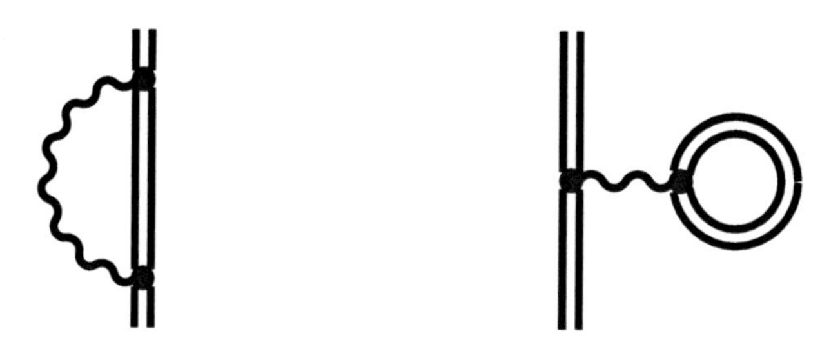

Figure 11.3. Feynman diagrams for the self energy (left) and the vacuum polarization (right) of order α for a bound electron. To denote the binding state, the electron lines in the Feynman diagrams are doubled. On the left, an electron emits and reabsorbs a photon. The loop on the right represents a virtual electron-positron pair (also in the field of the nucleus) that is created and reannihilated.

for these diagrams lead to the infinities for which QED is notorious. These are overcome by redefinition of mass and charge of the electron, respectively. For bound particles, there is an additional finite contribution to the binding energy that depends on the state of the electron. In particular, it is different for s- and p-states and gives rise to the famous Lamb shift between the $2s_{1/2}$ and the $2p_{1/2}$ state in hydrogen. Nowadays, all deviations from the Dirac binding energy are subsumed under the term "Lamb shift" even for one single state, and it is a permanent race between experiment and theory to verify/predict more and more accurate numbers. In heavy hydrogen-like ions the ratio of these contributions to the total binding energy is much pronounced, compared to hydrogen, as depicted by fig. 11.4. Note that apart from the radiative contributions self energy and vacuum polarization, also the nuclear size gives a considerable contribution in heavy systems. The Dirac energy eigenvalue is based on the Coulomb field of a point nucleus. If this charge distribution becomes smeared out, the Coulomb field at the origin changes and therefore so also do the energy eigenvalues. For the $1s_{1/2}$ state of hydrogen-like uranium, insufficient knowledge of the nuclear charge distribution causes an uncertainty of 0.1 eV in the Lamb shift predictions, which is as much as the total uncertainty of all ten diagrams of order α^2 for the radiative contributions. Furthermore, the finite nuclear mass affects the energy value since a finitely heavy nucleus also moves when surrounded by an electron and this effect has to be described in the proper QED language. In addition, virtual excitations of the nucleus also contribute like 'radiative' contributions ('nuclear polarization') but are much more difficult to predict since the nuclear parameters are not sufficiently well known. These features limit the QED tests

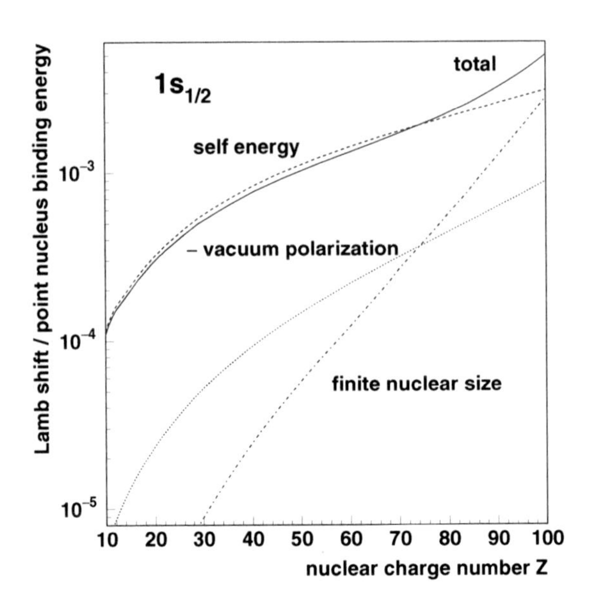

Figure 11.4. Ratio of Lamb shift to total binding energy for the $1s_{1/2}$ state and the dominant contributions.

in heavy one-electron atoms. In fig. 11.5 we depict the level scheme of the $n = 1$ and $n = 2$ states of hydrogen-like uranium, according to Bohr's and Dirac's theories together with those obtained from QED.

One way to overcome the difficulty resulting from nuclear parameters at least partially, is to compare one- and two- or one- and three-electron atoms of the same nucleus and consider only the *differences* between them. At first glance, this seems to be a difficult undertaking. If more than one electron is present in a system, the wave functions are not known analytically and the desired accuracy seems out of reach, since we have to employ many-body perturbation theory with lots of approximations and limitations. But this is not so for very heavy ions. As the binding to the nucleus scales with $Z\alpha$ but a one-photon exchange between two electrons is always of the order α, the wave function of each electron in a heavy system is very close to a hydrogen-like one, and the influence of the other electrons can be considered as a perturbation series in α again (one-photon exchange, two-photon exchange, and so on). For lithium-like uranium, the complete set of all diagrams of order α^2 is at present sufficient to explain all of the experimental results. Of course this becomes worse for lighter ions. A very clean investigation of the two-electron effects is possible in the ground state of helium-like atoms, where all nuclear effects are the same for both electrons and the complete difference between the binding

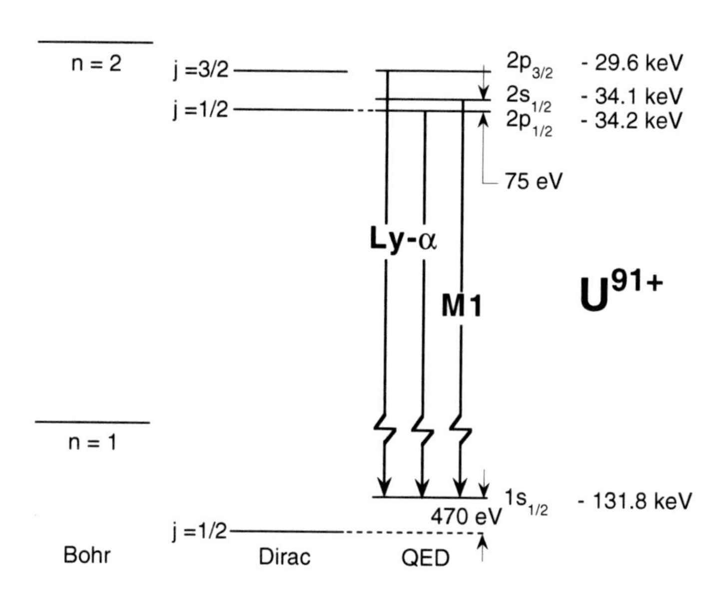

Figure 11.5. Level scheme and principal decay modes of U^{91+}. The numbers on the right indicate the electronic binding energies.

energy of the first and second electron is due to two-electron QED effects, i.e., either pure photon exchange as in fig. 11.2 with one or more photons, or 'screened' radiative corrections like those depicted in fig. 11.6. A very accurate measurement of the binding energies in both hydrogen- and helium-like heavy

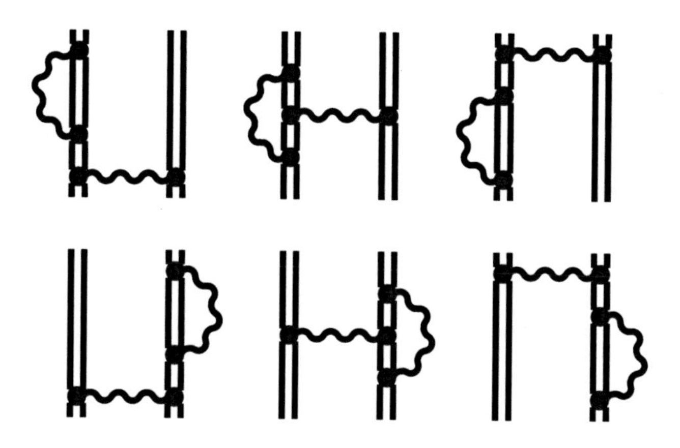

Figure 11.6. Screened-self energy diagrams of order α^2 in a more-than-one electron atom. A similar set exists for the vacuum polarization contributions.

ions would exactly yield the contribution of these diagrams unaffected from the nuclear structure, since those vanish for the difference .

In the present chapter, we do not focus on the actual mathematical formulations of QED in strong fields. During the last years, a number of comprehensive overview articles were published to which we refer. Fundamentals and one-electron QED is thoroughly discussed in [3], with the newest developments given in [4] and [5] and references therein. In addition, [6] provides a good overview about a new approach including the discussion of two- and three-electron atoms. For readers interested in just a basic overview, we also recommend [7].

In addition to the shift of energy levels discussed so far, the investigation of effects due to the presence of an additional *magnetic* field offers some new possibilities. The magnetic field can either be generated by the nucleus itself which leads to the hyperfine structure splitting of levels where QED can also be investigated, or an external homogeneous magnetic field can be applied in which case the Zeeman splitting of levels is the subject of consideration. By this latter method g factors of highly charged ions are studied. Although the magnetic field B at the surface of a bismuth nucleus is of the order $10^9 - 10^{10}$ T, it still can be considered as a perturbation compared to the Coulomb field of the nucleus which mediates the binding. The QED corrections of order α which have to be considered are given in fig. 11.7 where the magnetic field is either nearly point-dipole like (HFS) or homogeneous (Zeeman effect, g factor investigations). If the splitting of an energy level due to a hyperfine

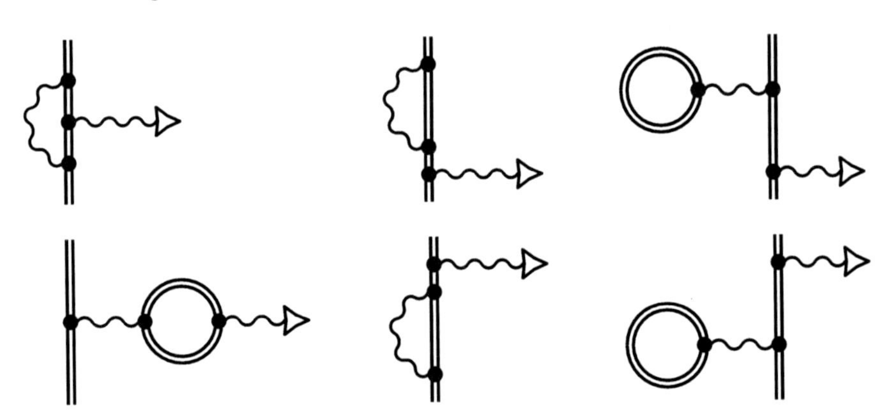

Figure 11.7. QED contributions of order α to the interaction of a magnetic field with a bound electron. The external magnetic field is indicated by a triangle.

interaction is considered, the size of these contributions has to be added in addition to the relativistic prediction for the splitting energy. In particular, the value for the nuclear magnetic moment linearly enters into this quantity, and many nuclear moments are known with an accuracy of only $\simeq 10^{-5}$. There are two other nuclear effects which also enter the numerical value: the finite size of the nucleus that affects the wave functions similar to the Lamb shift as discussed above and the deviation of the nuclear magnetic field from a point dipole, also known as "Bohr-Weisskopf effect". The first of these corrections is relatively easy to handle if the size of the nucleus is known with sufficient accuracy. It is the second effect that at present hinders the further investigation of QED by hyperfine structure studies in heavy ions since the size of the effect has to be obtained from a nuclear model. As an example, in Table 11.1 we present for ^{209}Bi values derived from both the single-particle model and the Dynamical-correlation model.

Table 11.1. Contributions to the hyperfine structure splitting of the ground state in hydrogen-like and lithiumlike ^{209}Bi according to different models. The prediction for the Bohr-Weisskopf effect in lithiumlike bismuth is based on the comparison of the theoretical prediction by the single-particle model and the experimental value obtained for hydrogen-like bismuth (see below). The uncertainties for the Bohr-Weisskopf effect are 'in-model' uncertainties and do not account for the possibility of a failure of the model. Details are given in [8]. Values are given in electron volts [eV].

	$^{209}\mathrm{Bi}^{82+}$	$^{209}\mathrm{Bi}^{82+}$	$^{209}\mathrm{Bi}^{80+}$
ΔE_{HFS}	5.8393(3)	5.8393(3)	0.95849
nuclear size	–0.6482(7)	–0.6482(7)	–0.1138(2)^{a}
Bohr-Weisskopf effect	–0.061(27)^{b}	–0.109(8)^{c}	–0.0134(2)^{a,d}
One-electron QED (order α)	–0.0298	–0.0298	–0.0051(2)^{a}
Two-and-more electron QED			–0.02906
Total Theory	5.100(27)^{b}	5.052(11)^{c}	0.7971(2)

^{a}Uncertainty margins not independent.
^{b}Single-particle model.
^{c}Dynamical-correlation model.
^{d}Value by comparing theory and experiment for ^{b} and scaling.

From the difference of the numerical values introduced via the Bohr-Weisskopf effect the difficulties in investigating QED in these systems become clear. However, at present these investigations also form the most precise access to the nuclear magnetization distribution. Theoretical investigations indicate that it would be possible to eliminate this problem by the measurement of the 2s-hyperfine structure in the corresponding lithium-like system. The transition wavelength is predicted with high accuracy using the result from the 1s-hyperfine splitting in the hydrogen-like case as input for an elimination of the Bohr-Weisskopf-effect [9].

To study the QED contributions in the presence of a magnetic field separately, investigations on the Zeeman splitting in an external magnetic field form a complement to the HFS experiments. The Zeeman-splitting experiments focus on the determination of the g factor of the system, which has already been mentioned, and which for a charge q with mass m_q and angular momentum $\mathbf{J}$ is defined via the relation

$$\boldsymbol{\mu} = g\frac{q}{2m_q}\mathbf{J} \tag{11.1}$$

where $\boldsymbol{\mu}$ is the corresponding magnetic moment. For the spin of the free electron $g = 2$ from Dirac theory, and the deviations presented at the beginning of the chapters are due to QED. For an electron bound in the $1s_{1/2}$ state of a hydrogen-like ion, the Dirac value of 2 changes to

$$g = (2/3)\left(1 + 2\sqrt{1-(Z\alpha)^2}\right) \quad . \tag{11.2}$$

In order to completely predict the value for g further corrections have to be performed in addition to the radiative corrections of order α depicted in fig. 11.7. This is similar to the case of the Lamb shift. Eq. (11.2) is valid only for point nuclei and a correction for the nuclear size is necessary. Like for the Lamb shift, also the finite nuclear mass affects the value considerably. In addition, QED contributions of order α^2 and higher should be considered if compared to an accurate measurements. In Table 11.2 we present the most recent numerical values for the g factor of the ground state of hydrogen-like carbon and uranium. Compared to all other fields of investigation, experiments on the g factor in

Table 11.2. Contributions to the electronic g factor in $^{12}C^{5+}$ and $^{238}U^{91+}$. QED contributions of order α^2 and higher are not yet completely evaluated. The error margin given there represents an order-of-magnitude estimate. Part of the numerical values were communicated by V. M. Shabaev and V. A. Yerokhin and are not published at the time of writing this section.

	$^{12}C^{5+}$	$^{238}U^{91+}$
Dirac value	1.998 721 354 4	1.654 846 173(3)^a
nuclear size	0.000 000 000 4	0.001 275 0(25)^b
finite nuclear mass	0.000 000 087 6	0.000 002 491
QED (order α)	0.002 323 663 7(9)^c	0.003 088 93(3)^c
QED (order α^2 and higher)	–0.000 003 516 2(2)^d	–0.000 003 8(9)^d
Total Theory	2.001 041 589 9(9)	1.659 208 9(27)

^{a}Uncertainty caused by current uncertainty of α, $\alpha = 1/137.035\,999\,76(50)$
^{b}Includes size and shape uncertainty.
^{c}Numerical error.
^{d}Error estimates due to yet incomplete calculations.

hydrogen-like atoms allow study of quantum electrodynamics to at least four digits even in uranium. The enormous theoretical precision for carbon even allows redetermination of the mass of the electron, as will be discussed below.

The study of magnetic interactions together with the QED of strong fields is a rather new subject of investigation. A good overview is presented by [8], in addition the latest developments are given in [10] and [11] and in the references therein.

3. The $1s_{1/2}$ Lamb Shift

The influence of quantum electrodynamics on electrons bound to a highly charged nucleus can be studied through a measurement of binding and transition energies via x-ray spectroscopy[12]. The accuracy to which the x-ray energy can be experimentally determined is largely related to the way the ions are prepared and to the measurement strategies which are intended to defeat inherent limitations mainly set by the large Doppler corrections.

3.1 Experimental Methods

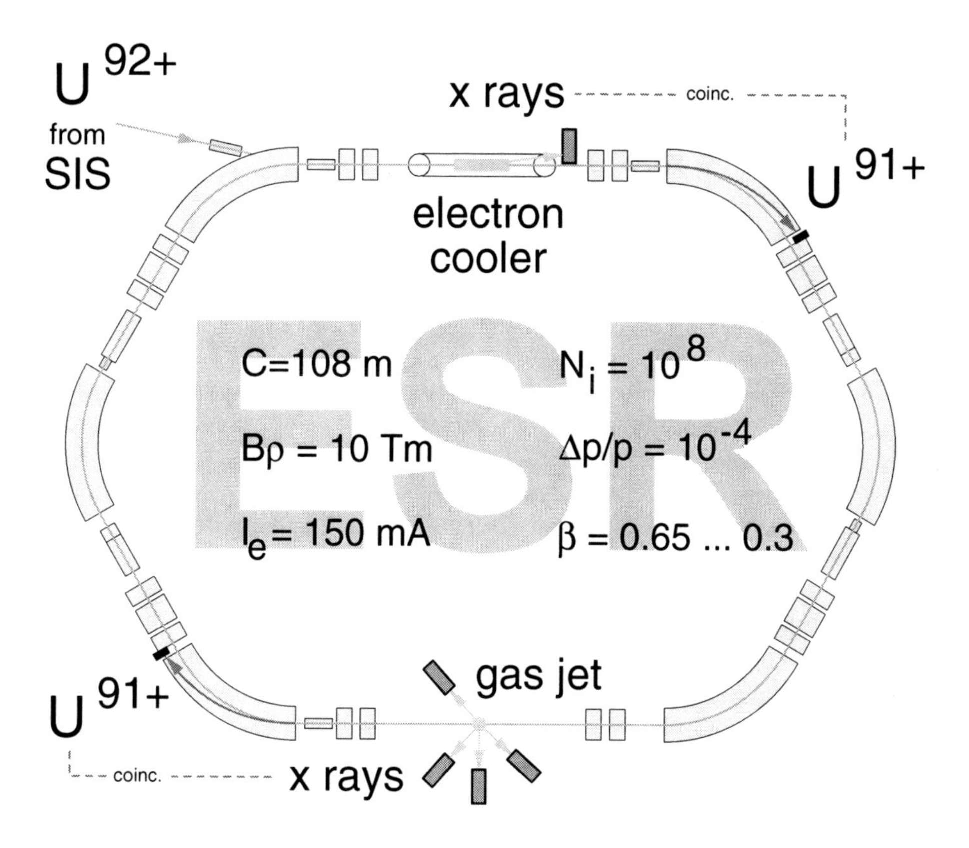

Figure 11.8. Layout of the ESR storage ring with typical parameters applying for spectroscopy experiments. As x-ray sources serve the electron-cooling section and the internal gas target. The x rays have been measured in coincidence with those ions that have lost one unit of charge by recombination or capture processes.

3.1.1 Preparation of Highly Charged Ions. A large fraction of highly charged and even bare ions can be obtained in an ion beam accelerated to a high velocity penetrating a stripper target. At GSI completely stripped very heavy ions up to uranium can be injected into the Experimental Storage Ring (ESR) a sketch of which is shown in fig. 11.8. This storage ring is used for further manipulation of the injected ions such as ion cooling to low momentum spread and low angular divergence plus eventually the deceleration to low velocity. The main operational parameters are listed in the figure.

It has proven very useful to start with a bare ion to which a single electron is added in a recombination process in the electron cooler or in a single electron capture event in the gas-jet target. The subsequently emitted x rays are measured in coincidence with those ions that have lost one unit of charge during the interaction with the target and which are deflected out of the main ion stream in the following dipole magnet of the storage ring.

3.1.2 X-ray Spectroscopy. It turns out that a large fraction, exceeding 0.5, of the stored inventory of bare ions in the storage ring can be converted into x rays useful for spectroscopy.

The characteristic K x rays emitted by the heavy ions are in the 100-keV energy range. In the laboratory they are observed Doppler shifted according to the Lorentz transformation

$$E_{\mathrm{o}} = E_{\mathrm{x}}\,\gamma(1 - \beta\cos\vartheta), \tag{11.3}$$

where E_{o} and E_{x} denote the x-ray energy in the emitter and laboratory frames of reference respectively and β and γ are the usual relativistic parameters whereas ϑ denotes the observation angle relative to the ion-beam direction. At an ion-beam velocity of 50% of the speed of light there is a factor of three variation in the laboratory x-ray energy going from forward ($\vartheta = 0°$) to backward ($\vartheta = 180°$) observation. The x ray spectra are recorded with high efficiency employing intrinsic germanium, Ge(i), detectors. They are arranged near the electron cooler and the gas target as indicated in fig. 11.8.

3.1.3 Systematic and Practical Limits. The experimental accuracy to which the Lamb shift can be determined is governed by several factors. The measurement quantity for the $1s_{1/2}$ level shift is burdened by a huge offset between the $1s_{1/2}$ and the $2p_{1/2}$ levels (cf. the level scheme of fig. 11.5). This means that the quantity of interest is not directly measured but manifests as a fraction of about 0.5 keV out of 100 keV when dealing with the Lyman-α transitions in hydrogen-like uranium. The systematic limit represented by the natural line width Γ, amounting to roughly 30 eV for uranium, is relatively unimportant for the Lyman transitions as opposed to the classical $2s_{1/2}$–$2p_{1/2}$ Lamb-shift interval where the same natural width is imposed on an approximately eight times smaller measurement effect.

The Doppler effect remains as one of the most serious difficulties. As implied from equation 11.3 an accurate measurement of the photon energy in the emitter frame of reference requires, besides the energy measurement in the laboratory, an accurate knowledge of both the ion velocity and the observation angle. A first order estimate of these uncertainties are obtained, by differentiation of equation 11.3, as

$$\left[\frac{\Delta E_{\rm o}}{E_{\rm o}}\right]^2 = \left[\frac{\beta \sin\vartheta}{1-\beta\cos\vartheta}\Delta\vartheta\right]^2 + \left[\left(\beta\gamma^2 - \frac{\cos\vartheta}{1-\beta\cos\vartheta}\right)\Delta\beta\right]^2 + \left[\frac{\Delta E_{\rm x}}{E_{\rm x}}\right]^2 . \quad (11.4)$$

The terms containing $\Delta\vartheta$ and $\Delta\beta$, respectively, are strongly dependent on the observation angle and are much reduced with decelerated ions. The uncertainties introduced by angular uncertainties vanish at zero and 180 degrees observation. At the same time the velocity uncertainties become most important. Conversely, for side-on observation near $\cos\vartheta = \beta$ velocity uncertainties become unimportant but angular uncertainties are critical. In practice a velocity-sensitive measurement near zero degree was realized at the electron cooler and an angular-sensitive geometry was realized at the gas-jet target. This way absolute observation angles are either not critical or they are spectroscopically determined by using several detectors viewing the same x-ray source simultaneously.

3.2 Results from the Electron Cooler

The only way a free electron and a bare ion can recombine is via radiative recombination, the time reverse process of the photo effect. For a storage ring employing electron cooling it is a loss mechanism. However, each ion that looses a unit of charge that way emits one or more photons useful for precision spectroscopy. The situation is explained by the term scheme sketched in fig. 11.9 including also a measured spectrum emerging from stored Au^{79+} ions. Because of the friction forces between ions and electrons both beams have identical mean velocities in the laboratory. As a consequence the electrons are energetically located just at the boarder to continuum. The strongest transition in the recombination process is the one into the $1s_{1/2}$ ground state where the photon energy is just the $1s_{1/2}$ ground-state binding energy. The intensity of the other recombination transitions drops off with the principal quantum number n as approximately n^{-1}. As a result there is a strong cascade feeding the Lyman-α lines. With the experimental line width near 900 eV the $2s_{1/2} \rightarrow 1s_{1/2}$ M1 transition is not spectroscopically resolved from the $2p_{1/2} \rightarrow 1s_{1/2}$ E1 Lyman-α_2 transition. However, the M1 transition is nearly absent because at the lower end of the cascade chain only the $2p_{1/2}$ states are noticeably populated. Experimentally this is reflected in the nearly statistical intensity ratio of 2:1 for the Lyman-α doublet.

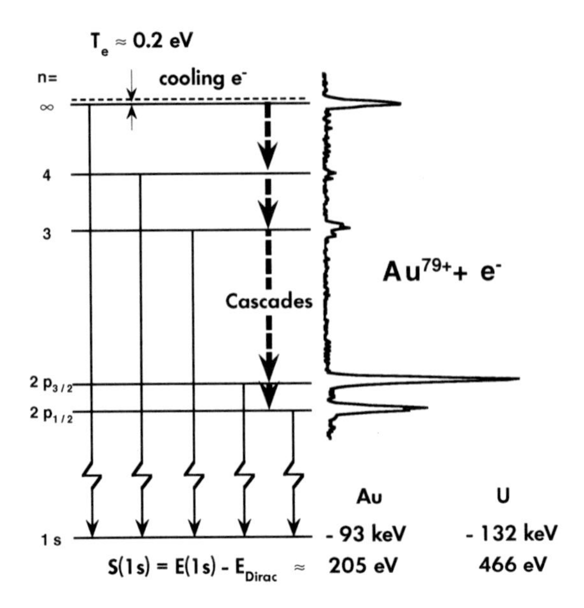

Figure 11.9. Level scheme and x-ray spectrum of hydrogen-like Au^{78+} measured at the electron cooler when a beam of bare Au^{79+} ions was stored. The line with the highest photon energy corresponds to a direct radiative transition of a cooling electron into the $1s$ ground state. The other lines, notably the Lyman-α doublet, originate from recombination into higher $n\ell$ states plus subsequent cascade transitions [13].

The $1s_{1/2}$ binding energy was experimentally determined from the three strongest lines in the spectrum, i.e. the Lyman-α doublet and the $1s_{1/2}$ recombination line. The three results obtained independently were averaged and the uncertainty was obtained by linearly adding the individual uncertainties. It turned out that the limited counting statistics and the independent measurement of the ion-beam velocity introduced approximately equal energy uncertainties. Three different methods of velocity measurement were considered: (i) spectroscopically the velocity emerges from a comparison of the measured and theoretical interval between the $1s_{1/2}$ recombination and the Lyman-α_1 line, (ii) collinear laser spectroscopy with two counterpropagating laser beams can be used to calibrate the ion velocity, (iii) the terminal voltage at the electron cooler is a direct measure of the electron and hence the ion average velocity. For the particular experiments the method using the cooler voltage gave the most accurate results.

Two different ion species were measured[13, 14]. For Au^{78+} the final uncertainty amounted to ± 7.9 eV and for U^{91+} it was ± 16 eV. The measurements were carried out at a beam velocity corresponding to $\beta \approx 0.65$.

3.3 Results from the Gas Target

Four different germaniun detectors were used simultaneously for a measurement of the uranium $1s_{1/2}$ Lamb shift at the gas target. Their arrangement is sketched in fig. 11.8. Two detectors were placed symmetrically left and right of the ion beam near $48°$. A back–forward symmetry was obtained by another detector at $132°$ in addition to the one at $90°$. Two of the detectors, one at $48°$ and the other at $90°$, were segmented into seven vertical stripes. Spectroscopic measurements were carried out after the uranium beam was injected and cooled and decelerated to 68 and 49 MeV/u, respectively. A sample spectrum is shown in figure 11.10. At these velocities there is a high probability for electron capture into excited states of the projectile producing the Lyman spectrum directly or via cascades.

The $1s_{1/2}$ binding energy was obtained from this highly redundant arrangement using the Lyman-α_1 line measured with all detectors where the stripes

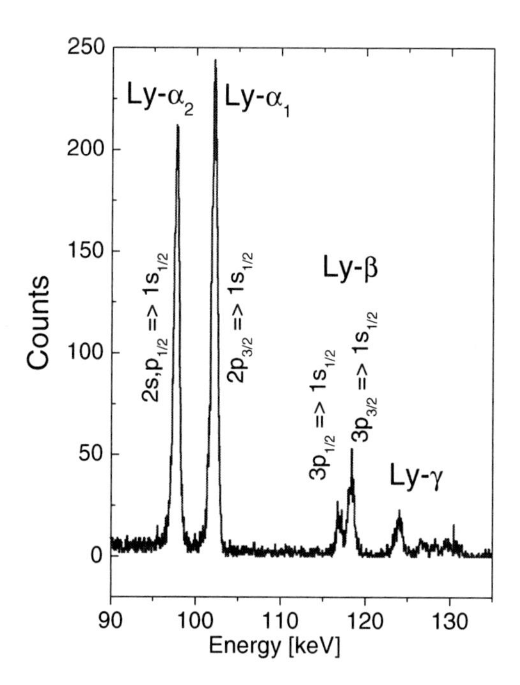

Figure 11.10. Lyman lines of hydrogen-like U^{91+} observed at the gas target of the ESR at a specific beam energy of 68 MeV/u. The spectrum is from the Ge(i) detector mounted at 132° and the energy scale has already been transformed into the emitter frame[15].

of the segmented detectors were treated as independent detectors. The relative angles between the detectors have been carefully assessed by computer-assisted leveling and trigonometry to an uncertainty of close to $0.01°$. The relative angles together with the well known angular separations of the stripes were used in the final analysis. The absolute direction of the ion beam and the overlap region between ion beam and gas jet, i.e. the location of the x-ray source, were treated as free parameters. With these constraints a consistent fit of the spectral data was performed resulting in one Lyman-α_1 energy in the emitter frame for each beam velocity. The combined final uncertainty of the $1s_{1/2}$ Lamb shift amounts to ± 13 eV. Dominating sources of uncertainty were the limited knowledge of the geometry and the combined uncertainties from calibration and the fit procedure.

3.4 Comparison with Theory

Table 11.3. Experimental and theoretical $1s$ Lamb shift in hydrogen-like gold and uranium.

	Z=79	Z=92
Experiment		
Beyer et al. 1994 [13, 14]	202.3±7.9	
Briand et al. 1990 [16]		325±130
Stöhlker et al. 1993 [17]		429±63
Lupton et al. 1994 [18]		508±98
Beyer et al. 1995 [14]		470±16
Stöhlker et al. 2000 [15]		469±13
Theory		
Beier et al. 1997 [19]	205.7	
Yerokhin et al. 2001 [5]		463.95±0.50

In Table 11.3 the experimental results obtained at the ESR are compared with theoretical calculations. There is only one measurement for gold from the electron cooler. It agrees well with the theoretical value within the experimental uncertainty of ± 7.9 eV. In case of uranium a steady decrease in the uncertainty bars can be observed starting from the first measurement at the BEVALAC by Briand *et al.* [16] and ending with the two measurements at the ESR [14, 15]. The latter yielded nearly identical results. If combined they reach a level of uncertainty near $\pm$ 10 eV. Within this uncertainty excellent agreement with theory is observed.

An overview of the present status of experiment in comparison with theory is given in figure 11.11 for all atomic numbers. The values are scaled by the

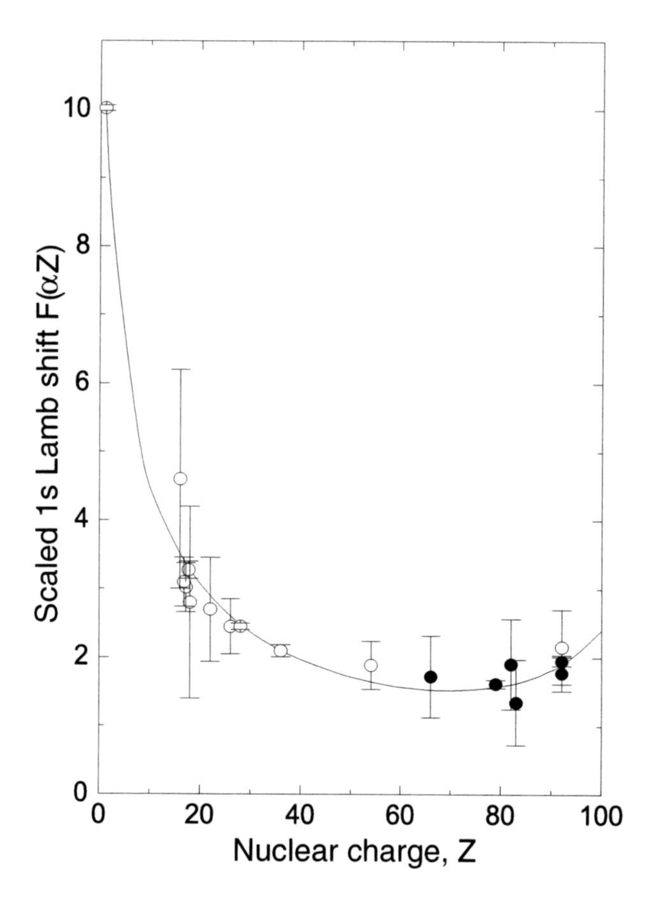

Figure 11.11. The $1s_{1/2}$ Lamb shift of hydrogen and hydrogen-like ions, scaled according to relation (11.5), as a function of the nuclear charge Z. The experimental results are represented by the data points and the theoretical calculations by the continuous curve. Solid circles: Data measured at the ESR storage ring.

function

$$S = \frac{\alpha}{\pi} \frac{(Z\alpha)^4}{n^3} F(\alpha Z)\, m_e c_o^2\,, \tag{11.5}$$

that contains the major $Z\alpha$ dependence separately and the function $F(\alpha Z)$ varies only slowly with Z. The most accurate measurement has been performed on atomic hydrogen with a relative uncertainty of ± 3 ppm. For heavy ions the most accurate results were obtained for slow argon recoil ions and for decelerated nickel ions of the UNILAC yielding uncertainties below the 2% level. The newer results obtained at the ESR for $Z = 79$ and $Z = 92$ are now approaching the uncertainties of experiments with ions of intermediate atomic number. A further reduction of the uncertainties is anticipated when higher spectral resolution will be applied using bolometric low-temperature detectors and crystal spectrometers. Then the experiments start to become sensitive also to the second-order radiative corrections which are now going to be completed from the theoretical side.

4. Two-Electron QED

Besides the one-electron systems, the two-electron ions are of particular interest as they represent the simplest multi-electron systems. Investigations of these ions along the isoelectronic sequence uniquely probe our understanding of correlation, relativistic, and quantum electrodynamical effects. The theoretical and experimental investigations of these fundamental systems achieved a considerable accuracy. For the ground state even the two-electron QED effects have been calculated completely to second order [20–22].

The calculations include the so called non-radiative QED part of the *electron-electron* interaction as well as the two-electron Lamb shift, i.e. the two-electron self energy and the two-electron vacuum polarization (see the Feynman diagrams displayed in the introduction in fig. 11.6). Note, that compared to the QED calculations for high-Z H-like systems, where some higher-order QED effects are still uncalculated, the claimed theoretical uncertainty for the two-electron QED contributions is very small and, for the particular case of He-like uranium, estimated to be of the order of only 0.1 eV. Most importantly, as has been shown in detail by Persson *et al.* [20], the two-electron QED effects are almost completely unaffected by the uncertainties of the nuclear charge radius, one of the most serious limitations for the QED tests in high-Z one-electron systems (see section 11.2).

Experimentally, the progress achieved manifests itself by a novel approach where the two-electron contributions to the binding energies in He-like ions can be experimentally isolated [24, 25]. This technique exploits the x-ray transitions from the continuum into the vacant K-shell of H- and He-like high-Z ions in order to measure the ionization potentials of He-like species with respect to that of the H-like ions. Since the two-electron contributions are experimentally isolated, all one-electron contributions such as the effects of the finite nuclear size cancel out almost completely in this type of experiment [20, 25].

4.1 Studies at SuperEBIT

At superEBIT bare and H-like ions of almost any element can be produced and trapped in an electron beam of arbitrary energy up to 200 keV and currents up to 200 mA [23, 25]. At such collision conditions, the fast moving free electrons may undergo a direct radiative recombination transition into the vacant K-shell of the bare or H-like species. Since RR is the time reversal of the photoelectric effect, the energy carried away by the photon is just given by $\hbar\omega = E_{kin} + V$ where E_{kin} denotes the kinetic energy of the electron captured and V is the ionization potential of the ionic system after undergoing radiative capture. Since both the bare and H-like ions are simultaneously trapped, i.e. both ion species are interacting with the same electron beam, the difference in the photon energies between radiative transitions into the bare and H-like ions is independent of the electron beam energy. It corresponds just to the difference in the ionization potentials between the H- and He-like species formed by the RR process which gives exactly the two-electron contribution to the ground state binding energy in He-like ions. A schematic presentation of this experimental situation at the EBIT is shown in fig. 11.12.

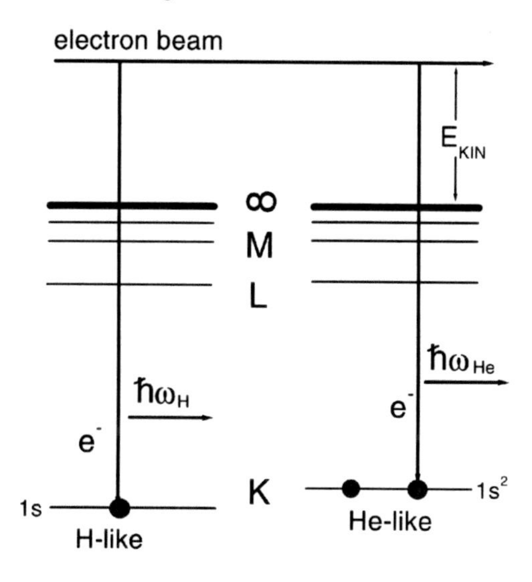

Figure 11.12. Schematic presentation of the RR process of free electrons into the initally bare and H-like ions. The energy difference $\hbar\omega_H - \hbar\omega_{He}$ gives exactly the two-electron contribution to the ionization potential in He-like ions.

By applying the experimental method described above, data were obtained for Ge (Z=32), Xe (Z=54), Dy (Z=66), W (Z=74), Os (Z=76) and Bi (Z=83). For the experiments, x-ray spectra were taken with a solid state Ge(i) detector viewing the electron-beam/ion interaction [24]. In fig. 11.13, sample spectra for the x-ray regime of RR into the vacant K-shell of germanium, dysprosium and bismuth are given separately. In all cases, the RR line splitting between RR into the bare and H-like ions, respectively, appears well resolved. As observed in the spectra, the relative intensity of the peak of the bare ions changes drastically between the two elements as the K-shell ionization cross section decreases rapidly as $1/Z^4$. Up to now, this has prevented the extension of the experiments to elements beyond Z=83. This limitation is not present at the ESR.

The experimental results obtained for the two-electron contributions to the binding energy of some He-like ions are given in table 11.13 [24] in comparison

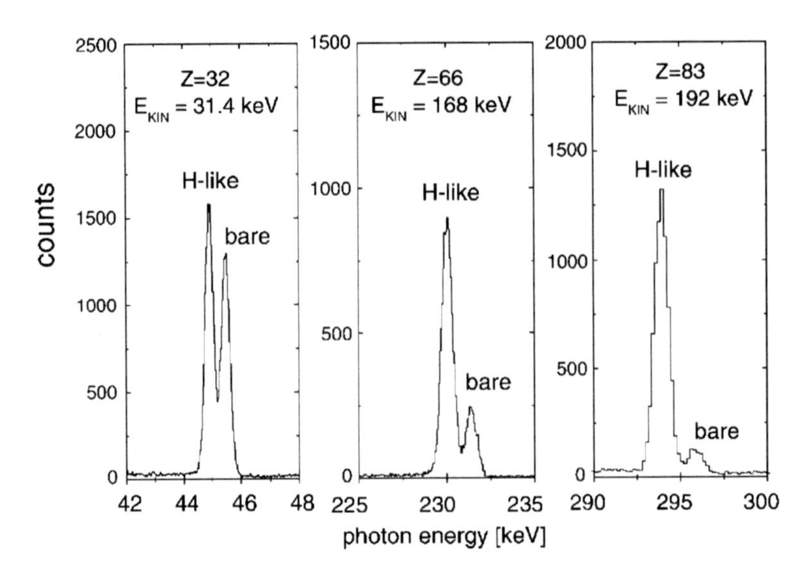

Figure 11.13. Sample K-shell RR line recorded for germanium (Z=32), dysprosium (Z=66), and bismuth (Z=83) at the superEBIT [24].

with the theoretical calculations of Persson *et al.* [20] performed recently. The experimental uncertainty quoted in the table is entirely determined by counting statistics. The predictions are based on relativistic many-body perturbation calculations [5]. In particular, all two-electron QED contributions are considered for the first time complete to second order in α. As can be deduced from the experimental and theoretical results presented in table 11.13 the experimental data provide already a meaningful test of the many-body non-QED part of the *electron-electron* interaction. Moreover, the data are already at the threshold of a sensitive test of the two-electron QED contributions. For this purpose only a moderate increase in accuracy of a factor of three is required.

Table 11.4. Experimental (SEBIT) [24] and theoretical [20] two-electron contribution to the binding energy of some He-like ions (in eV). In addition, the predicted total two-electron QED contribution is given.

	Z=32	Z=54	Z=66	Z=74	Z=83
SEBIT (eV)	562 (1.6)	1027.2 (3.5)	1341.6 (4.3)	1568.9 (15)	1876 (14)
Theory (eV)	562.0	1028.2	1336.6	1573.9	1881.5
2eQED (eV)	– 0.4	– 1.4	– 2.3	– 3.1	–4.3

4.2 Studies at the ESR Storage Ring

The experimental technique established at the superEBIT can also be applied for the same kind of experiments utilizing the ESR storage ring at GSI Darmstadt. For this purpose, the experimental setup at the electron cooler section seems to be most appropriate, a setup which has already been used in former 1s Lamb shift experiments [13, 14]. The main difference between the superEBIT experiment and our investigation at the ESR storage ring is that the RR transition into the initially bare and H-like ions have to be measured in alternate order. Consequently, the requirement of the experiment is that both the bare and H-like species must travel with identical energies. Moreover, within the cooler section, the trajectory for both ion species must be the same. Since the beam energy at the ESR is determined by the cooler voltage, identical beam energies for both ion species are guaranteed. Also, the trajectories of the ion beams inside the cooler section are well controlled. Even, a slight misalignment between the beams of bare and H-like ions (e.g. 1 mm) does not seem to effect the final accuracy of the experiment. Here, we profit from the 0° geometry of our x-ray setup which is rather insensitive to an uncertainty in the observation angle [13, 14]. Up to now, the accuracy in the Doppler shift corrections of $\Delta E/E \approx 5 \cdot 10^{-5}$ at the ESR is determined by the uncertainty in the determination of the absolute beam velocity. This constitutes the most serious limitation for precision x-ray experiments involving ground state transitions in H-like ions. For the particular case of uranium, however, where the splitting between RR transitions into bare and the H-like species amounts to 2.2 keV (emitter frame), this velocity uncertainty would introduce an uncertainty of less than $\pm$ 0.1 eV and can therefore be neglected.

In particular the experimental study at the ESR benefits from the deceleration technique [26] established. At low energies, all uncertainties associated with Doppler corrections are strongly reduced compared to high-energy beams. Also, for decelerated ions the bremsstrahlung intensity caused by the cooler electrons is strongly reduced (due to the comparably small cooler voltage of 27 kV and current of 100 mA). Consequently, very clean conditions for x-ray spectroscopy are present at the cooler section. This is depicted in fig. 11.14 where a preliminary calibrated, coincident x-ray spectrum is plotted as observed for initially bare uranium ions at an energy of 43 MeV/u. Note, the low bremsstrahlung intensity allowed us for the very first time to observe even RR transition into the L-shell, located at the low-energy part of the spectrum. In the spectrum, the x-ray transitions for RR into the ground state show up in the high energy part of the spectrum which are of particular interest for the current study. Most important, the tails of the Lyman-α transition lines, caused by cascade feeding of the L-shell levels [27], are consequently not present in the case of the RR photon emission. In the inset of fig. 11.14 the energy regime relevant for K-RR transitions is displayed. There the line intensities as measured for initially bare- and H-like uranium are compared.

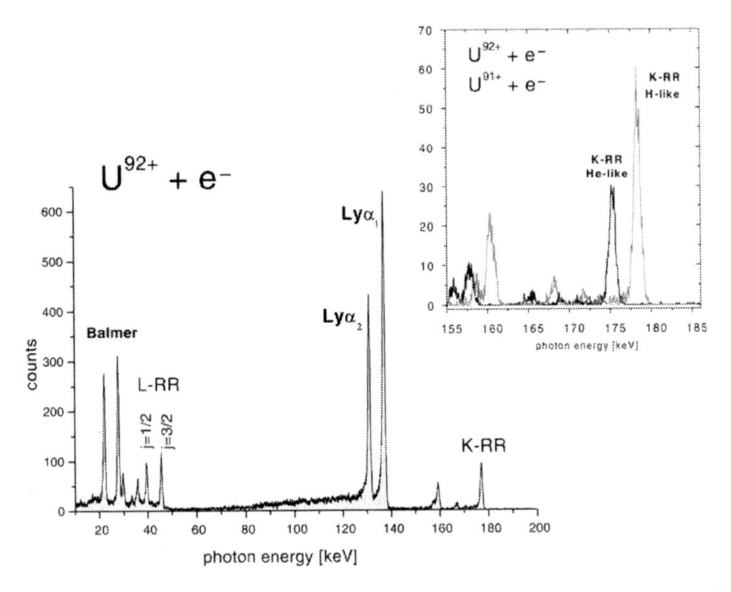

Figure 11.14. x-ray spectrum for H-like uranium as observed for decelerated ions at the electron cooler of the ESR storage ring. In the inset, the x-ray spectra for recombination into bare and H-like uranium are compared whereby the energy regime relevant for K-RR transitions is given.

5. Ground-State Hyperfine Structure of Heavy Hydrogen-Like Ions

In neutral and low-charge atoms, the predictions of QED are in perfect agreement with experiment. Radiative QED effects in highly charged hydrogenic ions increase by a scaling factor of Z^4, and the size of these higher-order contributions makes these systems exceptionally interesting for testing possible limits of the validity of QED. This opportunity has given rise to renewed experimental effort to investigate the Lamb shift in heavy ions. Completely complementary is the possibility to test QED in the combination of strong electric and magnetic fields by precision laser spectroscopy of the ground–state hyperfine splitting in high-Z hydrogenic systems.

The magnetic interaction of the 1s electron in the nucleus causes a hyperfine splitting in the ground state of hydrogen and heavier isotopes. In particular, the 1.4 GHz splitting frequency in hydrogen, corresponding to a transition wavelength of 21 cm, has been extensively studied by precision microwave spectroscopy, and in hydrogen masers the splitting frequency has been determined to an accuracy of 7×10^{-13} [28].

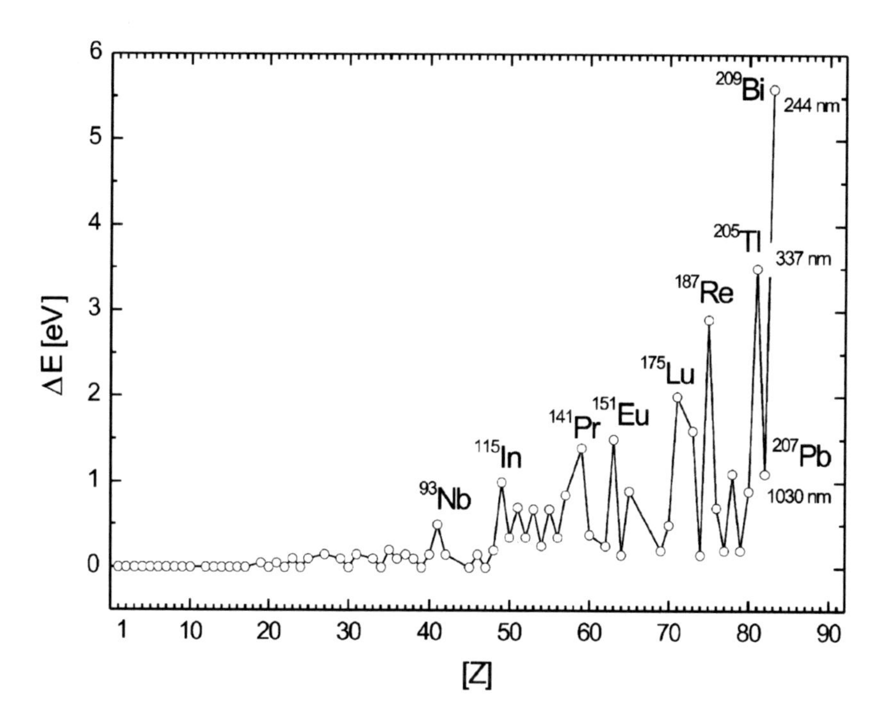

Figure 11.15. Possible candidates for laser spectroscopy.

Transition energies for highly-charged ions are shown in Tab. 11.1 For the highest-Z atoms the increase of the splitting with Z^3 shifts the HFS transition into the optical region of the spectrum, and makes it accessible to precision laser spectroscopy [29].

5.1 Storage Ring Laser Spectroscopy

Unique access for laser spectroscopy of such highly charged ions has been made possible by the advent of heavy-ion cooler rings. In the past, poor beam quality and short interaction time had restricted the application of lasers at accelerators to exotic single events, and laser spectroscopy appeared as a more or less parasitic use of these installations. The key experimental features of heavy-ion cooler rings in comparison to standard accelerators are the increase in interaction time and the dramatic improvement of the beam quality achieved by electron cooling. Electron cooling was demonstrated for the first time in 1969 at the proton storage ring in Novosibirsk [30] and was applied to heavy-ion storage rings starting with the TSR at the Max-Planck Institut für Kernphysik in Heidelberg.

There are a number of unique possibilities accessible for laser-based experiments in heavy-ion storage rings using the exotic properties of the stored ions to address fundamental problems of physics. Even laser-saturation spectroscopy is possible. This allows for resolution close to the natural line width of the transition, and was used to test Special Relativity at the TSR. The much higher velocity, compared to conventional experiments, makes this test particularly sensitive to higher–order contributions. The full spectrum of possibilities for storage-ring laser spectroscopy is reached at the very high energies available at the ESR, where intense beams of highly-charged ions are available for precision laser spectroscopy, and give novel opportunities for the study of quantum electro-dynamics in high magnetic and electric fields. The stripping of uranium ions in a fixed metal foil requires a beam energy of 300 MeV per nucleon to reach an efficiency of 10 percent. So far only the SIS (magnetic rigidity 18 Tm) – ESR (magnetic rigidity 9 Tm) facility, at GSI-Darmstadt, can meet this requirement. The alternative production scheme by electron-bombardment in an electron-beam ion trap (EBIT) presently yields around 100 to 1000 ions per filling cycle for hydrogen-like uranium, compared to 10^7 particles per pulse at SIS. For heavy ions like bismuth and uranium, beam intensities of 10^8 stored ions are reached.

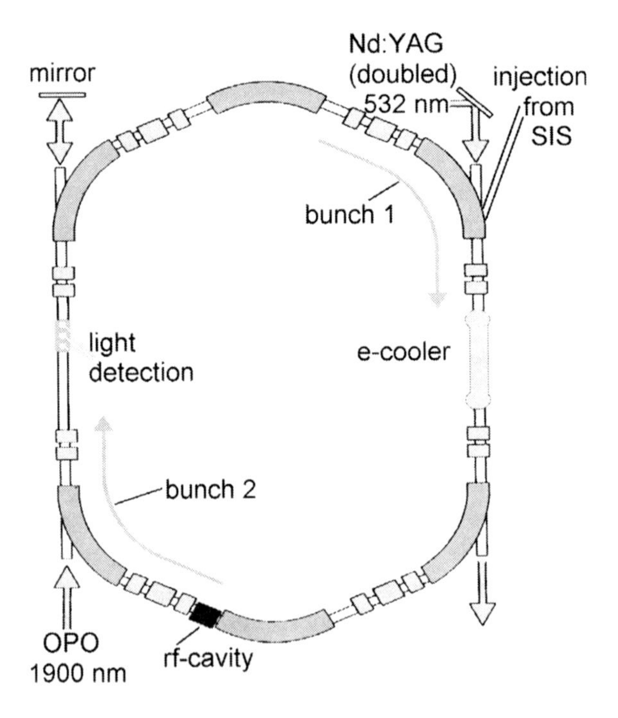

Figure 11.16. Laser spectroscopy set-up at the ESR facility at GSI.

Heavy ions are pre-accelerated in the UNILAC linear accelerator to energies up to 20 MeV per nucleon. Before injection into the SIS heavy-ion synchrotron, the charge-state of these ions can be increased by passing them through a stripping foil. The final energy is limited by the bending fields of 18 Tm to, for example, 1 GeV per nucleon for uranium. At this energy naked ions can be prepared by additional stripping. Laser light is introduced either parallel or anti-parallel to the direction of the ions.

An additional benefit for future experiments at the SIS – ESR facility is the availability of intense beams of unstable nuclei produced by fragmentation of the primary ion beam. At GSI, these secondary ions are separated from the primary beam in flight in the FRS (fragment separator), and the resulting beam of unstable nuclei can be stored in the ESR.

5.2 Experimental Procedure

For laser light impinging onto an absorber moving at relativistic velocity, the resonance frequency is shifted relative to the transition frequency ν_0 in the rest frame of the ions according to the relativistic Doppler formula

$$\nu = \nu_0 \left[\gamma (1 - \beta \cos \theta) \right]^{-1} \tag{11.6}$$

where θ denotes the angle between the ions and the laser beam ($\theta = 0$ for the parallel case). This relation also causes the light emitted from the moving ion to be observed in the laboratory frame at a strongly shifted frequency. The large Doppler shift due to the relativistic ion velocity can be used to advantage to excite transitions in the ultraviolet and infrared spectral regimes, even when no suitable lasers would be available at the transition wavelength in the rest frame. The detection of fluorescence after laser excitation of an infrared transition is facilitated by the large Doppler blue-shift for observation at forward angles. This shifts and compresses the wavelength of the emitted light into the region detectable by single-photon sensitive photomultipliers. As a disadvantage, the velocity spread $\delta\beta/\beta$ of the ion beam leads to a Doppler width of

$$\delta\nu_{Dopp} \approx \nu \cdot \gamma^2 \delta\beta \,. \tag{11.7}$$

This Doppler width can be avoided by typical sub-Doppler laser spectroscopy techniques. Laser saturation spectroscopy with a resolution close to the natural line width was used for a test of Special Relativity at the TSR. For such sub-Doppler resolution one must also take into account the small additional broadening and shift arising from the angle θ between laser beam and ion beam in the Doppler formula. At an interaction length of 10 meters and more, angles are easily controlled to be better than 1 mrad. This limits a possible shift, which enters as

$$\Delta\nu_\theta = -\nu\beta^2\theta^2 , \qquad (11.8)$$

to be $2 \cdot 10^{-7}\ \nu$ or smaller.

In the same way the transverse beam temperature leads to a broadening via the betatron movement of the particles. For the straight sections this effect can be estimated to be of minor importance for cooled beams.

In order to increase the signal-to-noise ratio, experiments can be performed on a bunched beam. In this way a pulsed laser can cover a much larger percentage of the ions stored in the ring, compared to the coasting beam condition. An intrinsic background correction can be provided by using two bunches in the ESR, where only one bunch was excited by the laser. As an example a result from hydrogen-like lead is given in fig. 11.17. The resonance was found at λ_0 = 1019.7(2) nm. The transition was excited by a frequency-doubled Nd:YAG laser in collinear geometry.

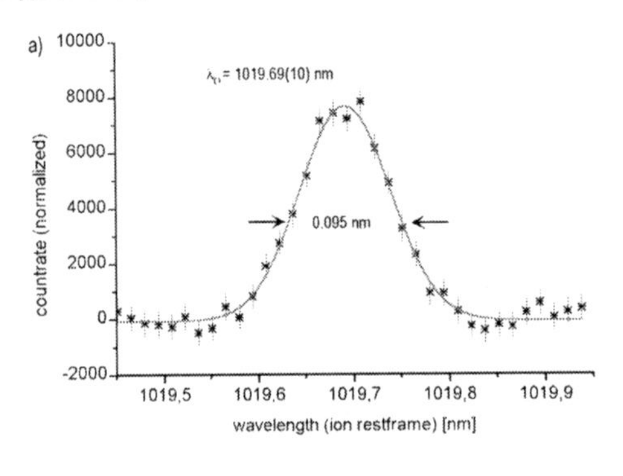

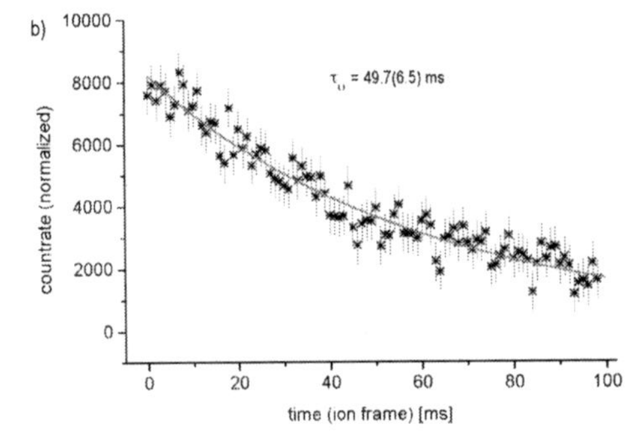

Figure 11.17. Upper trace: Background corrected and normalized fluorescence signal. Lower trace: Time dependence of fluorescence count rate after excitation by the laser pulse.

5.3 Results

So far data on the energy difference of the hyperfine splitting and of the transition lifetime were measured by storage ring laser spectroscopy for hydrogen-like bismuth [31] and lead [32]. In addition data were measured by conventional spectroscopy at the superEBIT in Livermore for the transition wavelength in holmium, rhenium, and thallium[33]. Excellent agreement was found for the transition lifetime in hydrogen-like bismuth. The decay probability of an excited 1s hyperfine structure level can be given in terms of the g factor of the electron bound in the electric field of the nucleus g_e. First measurements of g were performed only in rather light systems (H, D, He and C^{5+}). The earlier lifetime measurements of the excited hyperfine state in hydrogen-like bismuth and lead performed by laser spectroscopy in the ESR suffered from a lack in accuracy. A remeasurement of the transition lifetime resulted in an accuracy sufficient to test the relativistic correction. In addition there is good agreement with the theoretical value including QED but some disagreement, if QED effects are not included in theory, which would lead to g_e (no QED) = 1.7281 for hydrogen-like bismuth. Unfortunately, the agreement of the transition wavelengths with theoretical values is not very convincing [34]. This is probably due to the fact that the accuracy of the nuclear magnetic moment data and also of the theoretically deduced Bohr-Weisskopf values is hard to assess. Especially in the case of lead, two different values for the magnetic moment are found in the literature. Independent experimental verification is difficult to achieve. A major improvement of this situation would arise from an accurate determination of the nuclear magnetic moments with an independent method, which would also support or disproof the Bohr-Weisskopf calculations. As indicated in section 11.1, a way around this problem is given by a combination of the data for the hydrogen-like and lithium-like ion, which allows for evaluation of the nuclear parameters. However, experimental access is difficult due to the lower transition energy, where the transition wavelength is no longer well accessible with typical photomultipliers. The large Doppler-shift still allows for reasonable conditions in the case of lithium-like bismuth. In an experiment performed at the ESR so far no resonance line was found within the limits given by the present theoretical estimates, but the experimental feasibility seems well established. The exciting prospect of such an elimination of nuclear parameters lies in the fact that ultimately the full resolution possible by laser spectroscopy comes into play. Accepting more complicated laser systems, the problem caused by the uncertainty of the ion velocity can be eliminated to a large degree by combining parallel and anti-parallel excitation. This should allow a test of QED at a percent level[9].

6. The g Factor of the Bound Electron in Hydrogen-Like Ions

High-accuracy measurements of the g factor of the bound electron in hydrogen-like ions are sensitive tests of bound-state QED. We have developed a Penning trap apparatus for investigating g factors of highly charged ions. In our Penning trap a single hydrogen-like ion (e. g. $^{12}C^{5+}$) can be stored for up to a few months at a temperature of 4 Kelvin. The electronic spin state of the ion is monitored via the continuous Stern-Gerlach effect in a quantum non-demolition measurement [35]. Quantum jumps between the two spin states (spin *up* and spin *down*) are induced by a microwave field at the spin precession frequency of the bound electron. The g factor of the bound electron is obtained by varying the microwave frequency and counting the number of spin flips for a fixed time interval. We determined the g factor of the bound electron in hydrogen-like carbon $^{12}C^{5+}$ with an accuracy on the level of 10^{-9}.

6.1 Penning Trap Apparatus

In a Penning trap a charged particle is stored in a combination of a homogeneous magnetic field B_0 and an electrostatic quadrupole potential [36, 37]. The magnetic field confines the particle in the plane perpendicular to the magnetic field lines, and the electrostatic potential in the direction parallel to the magnetic field lines. The three eigenmotions (fig. 11.18) that result are the trap-modified cyclotron motion (frequency ω_+), the magnetron motion (frequency ω_-), which is a circular E $\times$ B drift motion perpendicular to the magnetic field lines, and the axial motion (parallel to the magnetic field lines, frequency ω_z). The free-space cyclotron frequency $\omega_c = (Q/M)B_0$ of an ion with charge Q and mass M in a magnetic field B_0 can be determined from a combination of the trapped ion's three eigenfrequencies ω_+, ω_z and ω_- with the formula

$$\omega_c^2 = \omega_+^2 + \omega_z^2 + \omega_-^2. \tag{11.9}$$

In our experiment, the magnetic field ($B_0 = 3.8$ Tesla) is provided by a superconducting solenoid. The electrostatic quadrupole potential is produced by a stack of five cylindrical electrodes with the same inner diameter (7 mm): the so-called ring electrode at the center and the compensation electrodes and endcaps placed on either side of the ring electrode. With a positive voltage U_0 applied between the two endcaps and the ring electrode, a potential well $V_{el}(z) = QU_0z^2/d^2$ is created along the magnetic field lines for a positively charged ion, where d is a characteristic size of the trap electrodes. The voltage at the compensation electrodes is tuned to minimize anharmonic terms of the electric potential. The axial motion of the trapped ion is a harmonic oscillation with frequency

$$\omega_z = \sqrt{\frac{Q}{M}\frac{U_0}{d^2}}. \tag{11.10}$$

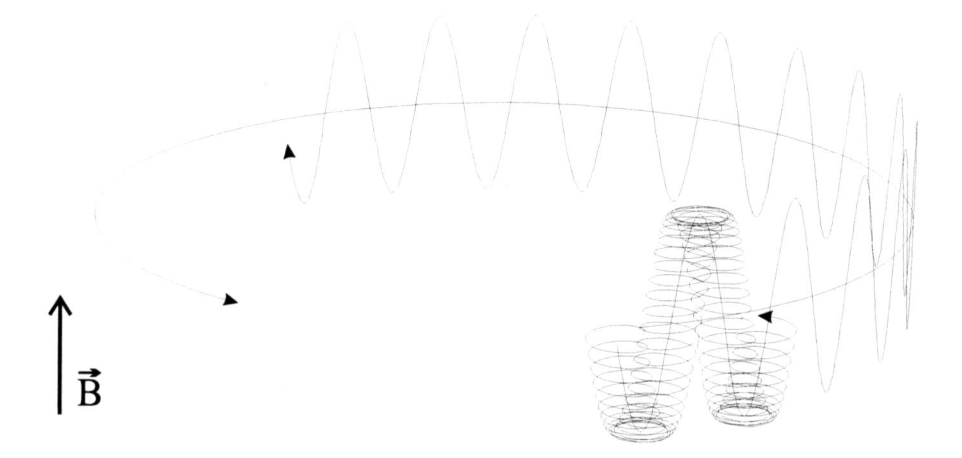

Figure 11.18. Three eigenmotions of a particle in a Penning trap: axial oscillation, cyclotron motion (small circles), and magnetron motion.

6.2 Continuous Stern-Gerlach Effect

The principle of the continuous Stern-Gerlach effect is based on a coupling of the magnetic moment μ of the particle to its axial oscillation frequency ω_z in the Penning trap [38]. This coupling is achieved by a quadratic magnetic field component ("magnetic bottle") superimposed on the homogeneous magnetic field B_0 of the Penning trap.

$$B(z) = B_0 + \beta_2 z^2 \tag{11.11}$$

Due to the interaction of the z-component μ_z of the magnetic moment with the "magnetic bottle" term the trapped ion possesses a position-dependent potential energy $V_m = -\mu_z(B_0 + \beta_2 z^2)$, which adds to the potential energy V_{el} of the ion in the electrostatic well. Therefore, the effective trapping force is modified by the magnetic interaction, and the axial frequency of the trapped ion is shifted upwards or downwards, depending on the sign of the z-component μ_z of the magnetic moment. This axial frequency shift is given by

$$\delta\omega_z = \frac{\beta_2 \mu_z}{M\omega_z} \tag{11.12}$$

where ω_z is the unshifted axial frequency. The determination of the spin orientation via the continuous Stern-Gerlach effect is a quantum non-demolition measurement because the spin state is measured non-destructively.

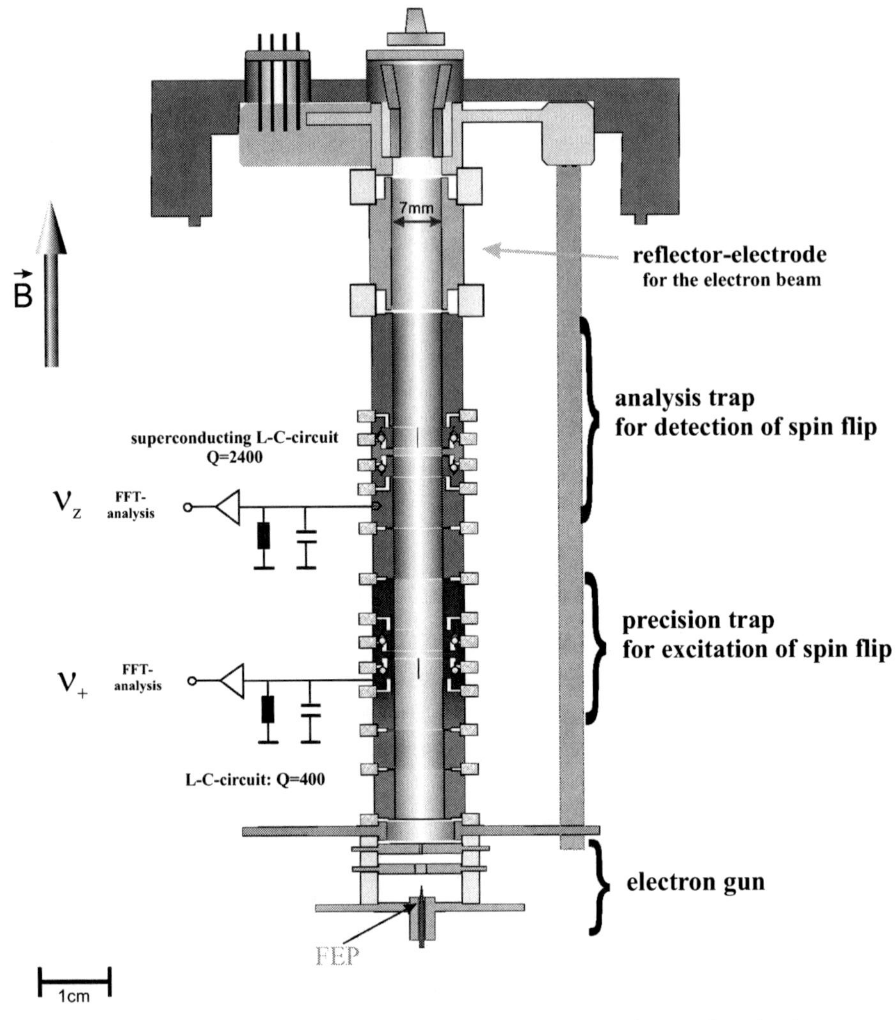

Figure 11.19. Penning trap set-up with the cylindrical electrodes for trapping, the electron gun for production of hydrogen-like ions by electron ionization, and L-C-circuits for detection of trapped ions.

6.3 Detection of a Single Hydrogen-Like Ion

In our Penning trap apparatus there are two positions in the stack of cylindrical electrodes where ions can be trapped (fig. 11.19). In the precision trap ions are created in different charge states by electron impact ionization of neutral atoms, and unwanted ion species are removed by selectively heating their axial motion. Consecutively, the number of $^{12}C^{5+}$ ions is reduced to one particle by lowering the electrostatic potential well. In the precision trap the spin-flip transition of the bound electron is excited by applying a microwave field. The single

$^{12}C^{5+}$ ion is then transported to the analysis trap. The ring electrode of this trap is made out of ferromagnetic material (nickel) to produce the quadratic component of the magnetic field ($\beta_2 = 0.01$ Tesla/mm^2) which is necessary to observe the continuous Stern-Gerlach effect. The trap electrodes are housed in a sealed ultra-high vacuum chamber which is kept at a temperature of 4 Kelvin. With the effect of cryopumping a vacuum pressure better than 10^{-16} mbar is reached.

The axial oscillation frequency ω_z of the single $^{12}C^{5+}$ ion in the analysis trap is measured non-destructively with an electronic detection method through the image currents which are induced in the trap electrodes by the particle motion. An electronic resonance circuit at $\omega_z = 2\pi \times 364$ kHz is attached to one of the trap electrodes to optimise the detection sensitivity. The ion's axial frequency is determined in a frequency analysis of the signal across the resonance circuit.

6.4 *g* factor Measurement

The quantum state of the $^{12}C^{5+}$ ion, i.e. the magnetic quantum number m_s of the bound electron in the $1s_{1/2}$ ground state, can be monitored non-destructively in the analysis trap in repeated measurements of the axial frequency of the trapped ion. Transitions between the two spin states $m_s = \pm 1/2$ are induced by a microwave field (at 104 GHz) resonant with the Larmor precession frequency ω_L of the bound electron.

$$\hbar\omega_L = g\frac{e\hbar}{2m_e}B = g\mu_B B \tag{11.13}$$

Here, g is the g factor of the bound electron and $\mu_B = e\hbar/2m_e$ is the Bohr magneton. The spin-flip transitions are observed as discrete changes of the ion's axial frequency. Fig. 11.20 shows a clear demonstration of such quantum jumps observed via the continuous Stern-Gerlach effect. The measured axial frequency shift for a transition between the two quantum levels is $\omega_z(\uparrow) - \omega_z(\downarrow) = 2\pi \times 0.7$ Hz, in excellent agreement with the expected value calculated from Equ. 11.12.

The observation of the continuous Stern-Gerlach effect on the hydrogen-like carbon ion $^{12}C^{5+}$ makes it possible to measure its electronic g factor to high accuracy. Using the cyclotron frequency $\omega_c = (Q/M)B$ of the ion for the calibration of the magnetic field B, the g factor of the bound electron can be expressed as the ratio of the Larmor precession frequency ω_L of the electron and the cyclotron frequency ω_c of the $^{12}C^{5+}$ ion.

$$g = 2 \cdot \frac{\omega_L}{\omega_c} \cdot \frac{Q/M}{e/m_e} \tag{11.14}$$

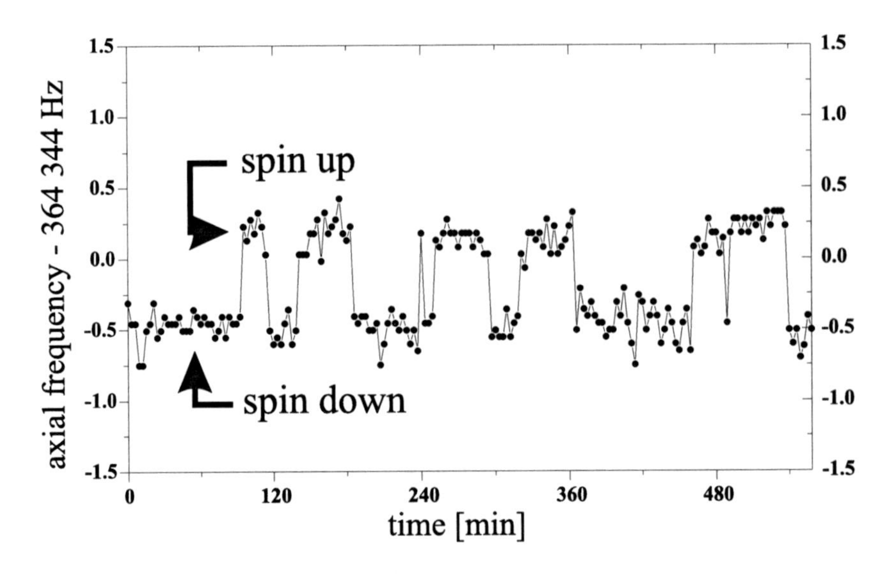

Figure 11.20. Quantum non-demolition measurement of the spin state: the spin-flip transitions are observed as small discrete changes of the axial frequency of the stored $^{12}C^{5+}$ ion.

The ratio of charge-to-mass ratios of the carbon ion (Q/M) and of the electron (e/m_e) was measured in a Penning trap to an accuracy of $2{\cdot}10^{-9}$ [39].

A resonance spectrum of the Larmor precession frequency ω_L of the bound electron is obtained in the following way. In the precision trap a microwave field at ω_L is applied to excite a spin-flip transition. Then the $^{12}C^{5+}$ ion is transferred to the analysis trap where the spin state is analysed via the continuous Stern-Gerlach effect. The ion is moved back to the precision trap, and the measurement cycle starts again. The number of spin-flip transitions which occurred in the precision trap is counted. Then the microwave frequency is varied and the measurement is repeated at different excitation frequencies. Finally, the plot of the quantum jump rate versus excitation frequency yields the resonance spectrum. A g factor resonance is obtained by dividing the Larmor frequency by the cyclotron frequency of the stored $^{12}C^{5+}$ ion which is determined simultaneously in a fast Fourier transform of the image currents induced in the trap electrodes by the ion motion (fig. 11.21). The experimental data points are fitted to a theoretical lineshape (fig. 11.22).

6.5 Experimental Results

In our measurements we determined the g factor of the electron bound in hydrogen-like carbon with a fractional accuracy of $2 \cdot 10^{-9}$ [40]. The experimental error in the determination of the g factor of the bound electron is mainly given by the uncertainty of the atomic mass of the electron. Our experimental

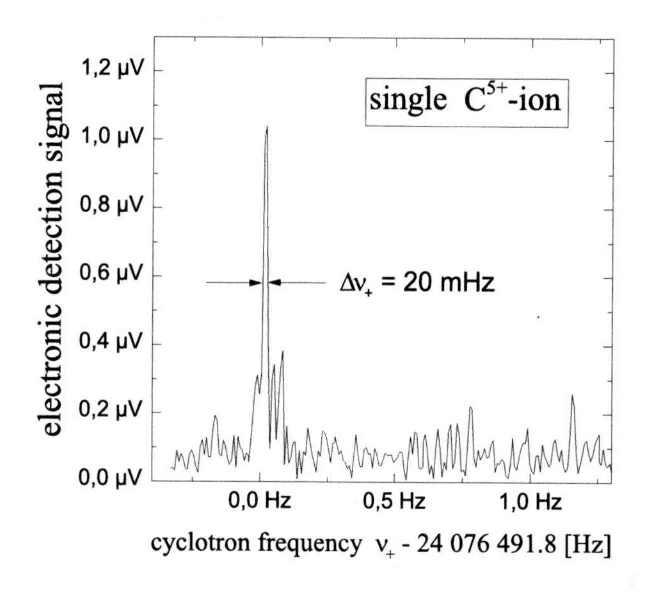

Figure 11.21. Measurement of the trap-modified cyclotron frequency ω_+ of a $^{12}C^{5+}$ ion. The fractional line width is smaller than 10^{-9}.

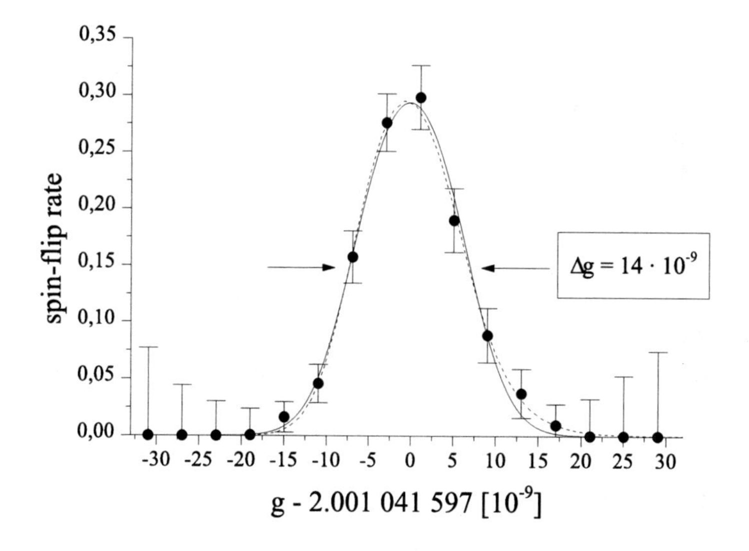

Figure 11.22. *g* factor resonance measured in the precision trap.

value for the g factor of the bound electron in $^{12}C^{5+}$ is in very good agreement with the theoretical value $g_{th} = 2.001\,041\,589\,9(9)$. The theoretical uncertainty is given by numerical errors and uncalculated Feynman graphs on the two-loop level. Together experiment and theory test the bound-state QED contributions to the g factor of the electron bound in hydrogen-like carbon with high accuracy. An interesting application of this g factor measurement is a redetermination of the electron's mass in atomic units u. The electron's mass is obtained by inserting the measured frequency ratio ω_L/ω_c and the theoretical g factor into Equ. 11.14. Our value for the electron's mass, $m_e = 0.000\,548\,579\,909\,2(4)\,u$ is more accurate by a factor of three compared to previous measurements [41]. Future measurements of the g factor of the bound electron in heavier ions up to hydrogen-like uranium (U^{91+}) will provide more stringent tests of bound-state Quantum Electrodynamics in extremely strong electromagnetic fields.

References

[1] H.F. Beyer, V.P. Shevelko (eds.), *Atomic Physics with Heavy Ions* (Springer, Berlin Heidelberg 1999).

[2] C. Itzykson and J. Bernard Zuber, *Quantum Field Theory*, McGraw-Hill, New York, 1980.

[3] P. J. Mohr, G. Plunien and G. Soff, Physics Reports **293**, 227 (1998).

[4] V. A. Yerokhin et al., Phys. Rev. A **64**, 032109 (2001).

[5] V. A. Yerokhin and V. M. Shabaev, Phys. Rev. A **64**, 062507 (2001).

[6] V. M. Shabaev, Physics Reports **356**, 119 (2002).

[7] G. Soff et al., Hyperfine Interactions **132**, 75 (2001).

[8] T. Beier, Physics Reports **339**, 79 (2000).

[9] V.M. Shabaev, M.B. Shabaeva, I.I. Tupitsyn, Phys. Rev.A **52** 3686 (1995)

[10] V. A. Yerokhin and V. M. Shabaev, Phys. Rev. A **64**, 012506 (2001).

[11] V. M. Shabaev, Phys. Rev. A **64**, 052104 (2001).

[12] H.F. Beyer, H.-J. Kluge and V.P. Shevelko, *X-Ray Radiation of Highly Charged Ions* Springer Ser. Atoms Plasmas, Vol. 19 (Springer, Berlin, Heidelberg 1997).

[13] H.F. Beyer, D. Liesen, F. Bosch, K.D. Finlayson, M. Jung, O. Klepper, R. Moshammer, K. Beckert, H. Eickhoff, B. Franzke, F. Nolden, P. Spädtke, M. Steck, G. Menzel and R.D. Deslattes, Phys. Lett. A **184**, 435 (1994)

[14] H.F. Beyer, G. Menzel, D. Liesen, A. Gallus, F. Bosch, R.D. Deslattes, P. Indelicato, T. Stöhlker, O. Klepper, R. Moshammer, F. Nolden, H. Eickhoff, B. Franzke and M. Steck, Z. Phys. D **35**, 169 (1995)

[15] T. Stöhlker, P.H. Mokler, F. Bosch, R.W. Dunford, B. Franzke, O. Klepper, C. Kozhuharov, T. Ludziejewski, F. Nolden, H. Reich, P. Rymuza, Z.

Stachura, M. Steck, P. Swiat and A. Warczak, Phys. Rev. Lett. **85**, 3109 (2000)

[16] J.P. Briand, P. Chevallier, P. Indelicato, D. Dietrich and K. Ziock, Phys. Rev. Lett. **65**, 2761 (1990)

[17] T. Stöhlker, P.H. Mokler, K. Beckert, F. Bosch, H. Eickhoff, B. Franzke, M. Jung, T. Kandler, O. Kleppner, C. Kozhuharov, R. Moshammer, F. Nolden, H. Reich, P. Rymuza, P. Spädtke and M. Steck, Phys. Rev. Lett. **71**, 2184 (1993)

[18] J.H. Lupton, D.D. Dietrich, C.J. Hailey, R.E. Stewart and K.P. Ziock, Phys. Rev. A **50**, 2150 (1994)

[19] T. Beier, P.J. Mohr, H. Persson, G. Plunien, M. Greiner and G. Soff, Phys. Lett. A **236**, 329 (1997)

[20] H. Persson, S. Salomonson, P. Sunnergren and I. Lindgren, Phys. Rev. Lett. **54**, 2805 (1996)

[21] V.A Yerokhin and V.M. Shabaev, Phys. Lett. A **207**; 274 (1995), A **210**, 437 (1996)

[22] V.A. Yerokhin, A.N. Artemyev and V.M. Shabaev, Phys. Lett. A **234**, 361 (1997)

[23] R.E. Marrs, S.R. Elliott and D.A. Knapp, Phys. Rev. Lett. **72**, 4082 (1994)

[24] R.E. Marrs, S.R. Elliott and Th. Stöhlker, Phys. Rev. **A52**, 3577 (1995)

[25] R.E. Marrs, P. Beiersdorfer, S.R. Elliott, D.A. Knapp and Th. Stöhlker, Physica Scripta **59**, 183 (1995)

[26] Th. Stöhlker, T. Ludziejewski, H. Reich, F. Bosch, R.W. Dunford, J. Eichler, B. Franzke, C. Kozhuharov, G. Menzel, P.H. Mokler, F. Nolden, P. Rymuza, Z. Stachura, M. Steck, P. Swiat and A. Warczak, Phys. Rev. A **58**, 2043 (1998)

[27] H.F. Beyer, D. Liesen, F.Bosch, K.D. Finlayson, M. Jung O. Klepper, R. Moshammer, K. Beckert, F. Nolden, H. Eickhoff, B. Franzke, and M. Steck, Phys. Lett. A **184**, 435 (1994).

[28] L. Essen, R.W. Donaldson, M.J. Bangham and E.G. Hope, Nature **229** (1971)

[29] V. M. Shabaev, J. Phys.B **27**, 5825 (1994)

[30] G.I. Budker and N. Skrinsky, Us. Fiz. Nauk **124**, 561 (1978); Sov. Phys. **21**, 277 (1978)

[31] I. Klaft *et al.*, Phys. Rev. Lett. **73**, 2425 (1994)

[32] P. Seelig, *et al.*, Phys. Rev. Lett. **81**, 4824 (1998)

[33] J.R. Crespo López-Urrutia *et al.*, Phys. Rev. Lett. **77**, 826 (1996)
J.R. Crespo López-Urrutia *et al.*, Phys. Rev.A **57**, 879 (1998)
Beiersdorfer *et al.*, Phys. Rev.A **64**, 32506 (2001)

[34] V. Shabaev *et al.*, Lecture Notes in Physics **Vol. 570**, 714 (2001)

[35] N. Hermanspahn, H. Häffner, H.-J. Kluge, S. Stahl, W. Quint, J. Verdú and G. Werth, Phys. Rev. Lett. **84**, 427 (2000)

[36] H. Dehmelt, Rev. Mod. Phys. **62**, 525 (1990)

[37] L.S. Brown and G. Gabrielse, Rev. Mod. Phys. **58**, 233 (1986)

[38] H. Dehmelt, Proc. Natl. Acad. Sci. U.S.A. **83**, 2291 (1986)

[39] R.S. Van Dyck Jr., D.L. Farnham and P.B. Schwinberg, Physica Scripta **T59**, 134 (1995)

[40] H. Häffner, T. Beier, N. Hermanspahn, H.-J. Kluge, S. Stahl, W. Quint, J. Verdú and G. Werth, Phys. Rev. Lett. **85**, 5308 (2000)

[41] T. Beier, H. Häffner, N. Hermanspahn, S. G. Karshenboim, H.-J. Kluge, W. Quint, S. Stahl, J. Verdú, and G. Werth, Phys. Rev. Lett. **88**, 011603 (2002)

Index

Breinigsville, PA USA
23 February 2011
256173BV00007B/1/P

9 781402 015656